Lecture Notes in Mathematics

Edited by A. Dold and B. Eckmann

1078

A. J. E. M. Janssen
P. van der Steen

Integration Theory

Springer-Verlag
Berlin Heidelberg New York Tokyo 1984

Authors

A. J. E. M. Janssen
Philips' Research Laboratories
P.O.Box 80.000, 5600 JA Eindhoven, The Netherlands

P. van der Steen
Department of Mathematics, University of Technology
P.O. Box 513, 5600 MB Eindhoven, The Netherlands

AMS Subject Classification (1980): 28-01, 26-01

ISBN 3-540-13386-0 Springer-Verlag Berlin Heidelberg New York Tokyo
ISBN 0-387-13386-0 Springer-Verlag New York Heidelberg Berlin Tokyo

Printing and binding: Beltz Offsetdruck, Hemsbach / Bergstr.
2146/3140-543210

PREFACE

Presenting yet another book on integration theory requires some justification. When writing the present material, we had in mind to explain what Lebesgue integration is and how it can be developed. An important point was to reconcile the various methods to introduce the integral. Many of the ideas used occur already in papers by Stone in 1948-1950. But the form we present them in, and much else as well, springs from a series of lectures by N.G. de Bruijn, around 1964. The General Introduction extensively explains our intentions.

Thanks are due to N.G. de Bruijn: without him the book would never have been written; to J.W. Nienhuys, who critically read portions of the manuscript; and to David Klarner, who read the whole manuscript and suggested many improvements. Parts of the manuscript were typed at the Mathematical Department of the Technological University Eindhoven. The final typing was done by Mrs. Elsina Baselmans-Weijers, who did a superb job, as usual.

A.J.E.M. Janssen
P. van der Steen

CONTENTS

GENERAL INTRODUCTION

The main purpose of this book is to present and compare various ways to introduce Lebesgue integration. The underlying observation is that the usual methods (catchwords: Carathéodory, Bourbaki, Daniell) from a certain point on follow similar courses. We shall show that these methods can in fact go together a considerable part of the way.

This common part can be roughly described as follows. In each of the methods one obtains somehow a space B of basic functions on a set S, together with a positive linear functional I on B that has a certain continuity property. Then one enlarges B to a complete space L and extends I to a positive linear functional (again denoted by I) on L. In the situation thus obtained, denoted symbolically by $\mathcal{L}(L,I)$ or $\mathcal{L}$, one has Lebesgue's dominated convergence theorem, which is really at the heart of any theory of integration.

How one gets at the class B depends on the point of departure chosen. In the text the following three possibilities are discussed. First there is Carathéodory's method, the starting point of which is denoted by $\mathcal{R}$. In this case one has a ring (or a semiring) Γ of subsets of S and a measure μ defined on Γ. From the pair (Γ,μ) the space B and the appropriate functional I are readily constructed. Then one has Bourbaki's point of view, here denoted by $\mathcal{T}$. In this case B consists of the real-valued continuous functions of compact support on a locally compact Hausdorff space S, and I is assumed to be given from the outset as a positive linear functional on B (without further requirements). Finally, in Daniell's method as extended by Stone both B and I are assumed to be given a priori. As noted before, I must satisfy some continuity condition. There are two feasible possibilities for this condition; they are denoted by $\mathcal{D}$ or $\mathcal{D}'$, where $\mathcal{D}'$ is the stronger of the two. It turns out that $\mathcal{R}$ leads to $\mathcal{D}$, and $\mathcal{T}$ to $\mathcal{D}'$, so in a way the first two methods are concrete, and the last one is an abstraction of the first two.

Suppose we have a class B of basic functions on S with a positive linear functional I and an appropriate continuity condition $\mathcal{D}$ or $\mathcal{D}'$, how do we get at $\mathcal{L}$? As said before, B must be completed and I suitably extended. For this completion we use a norm defined on the class Φ of the $\mathbb{R}^*$-valued functions on the underlying set S. One of the requirements for such a norm is that it coincides with I for the non-negative members of B, and another is that it satisfies a strong triangle inequality. A situation where we have B, I and a suitable norm $\|\cdot\|$ is called a station $\mathcal{N}(B,I,\|\cdot\|)$, or $\mathcal{N}$. (There will appear many such stations; they always consist of function classes, functionals, and certain relationships between these objects. The reason for using the term "station" will be clarified later.)

Now the norm can be defined in two ways, depending on whether the starting point was $\mathcal{D}$ or $\mathcal{D}'$, but both are instances of a more general construction, indicated in the diagram below.

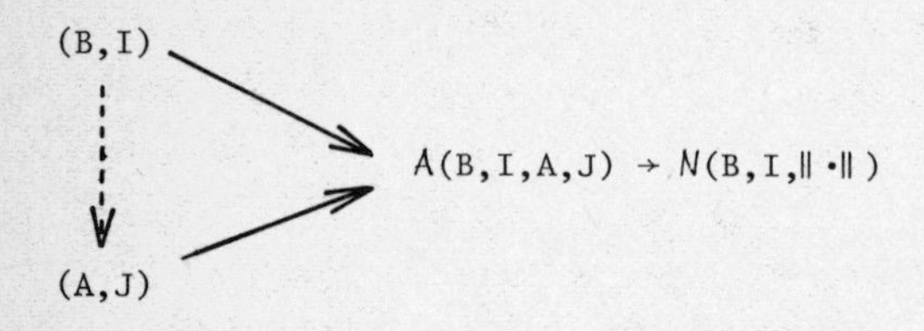

In this construction we assume that besides the class B of basic functions and the functional I, we are given a class A of auxiliary functions (certain non-negative $\mathbb{R}^*$-valued functions on S) together with a positive functional J on A. Such a situation is denoted by $\mathcal{A}$, or $\mathcal{A}$(B,I,A,J), and it is easy to get from $\mathcal{A}$ to $\mathcal{N}$. Of course, A and J must be connected in some way with B and I, and cannot be chosen arbitrarily. One of the restrictions is that the values of the norm and the functional I should coincide for non-negative basic functions. The dotted arrow in the diagram indicates that in some situations the pair (A,J) may be derived from the pair (B,I). In the $\mathcal{D}$- or $\mathcal{D}'$-case this can in fact be done. Due to the restriction mentioned above there is in the $\mathcal{D}$-case virtually only one possible choice for A and J: here A must consist of the non-negative members of B, and J must be a restriction of I. This is equivalent to Daniell's original approach. Since $\mathcal{D}'$ is stronger than $\mathcal{D}$, so that any station $\mathcal{D}'$(B,I) is also a station $\mathcal{D}$(B,I), the same procedure may be applied in the $\mathcal{D}'$-case. In this case, however, there is a second possibility (an abstraction of part of the Bourbaki version of integration theory), which quite often furnishes a bigger class of auxiliary functions and a richer theory of integration.

Departing from $\mathcal{N}$, the class of integrable functions is defined: it consists of all real-valued functions on S that can be approximated arbitrarily closely (with respect to the norm) by basic functions. The integral itself is then derived from I on B by arguments of continuity, and the result is a station $\mathcal{L}$(L,I) where Lebesgue's dominated convergence theorem holds, which was the ultimate aim.

The diagram below depicts the stations described up to now and their connections.

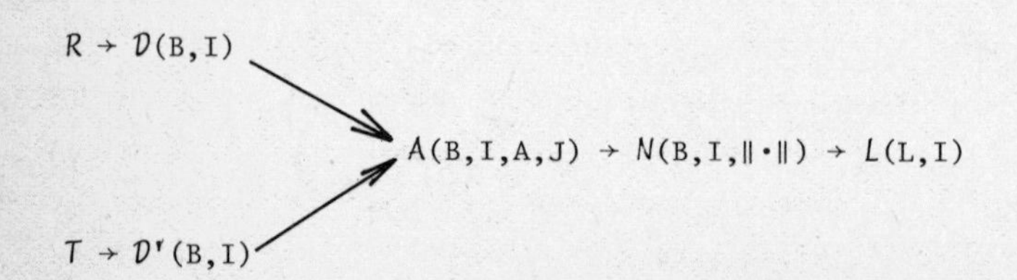

Eventually all paths lead to L, from which point the theory may be developed further. This is not the whole truth, however, for it will turn out that every station L is also a $\mathcal{D}$, that is, an arrow may be added from L to $\mathcal{D}$. The new diagram then contains the following substructure:

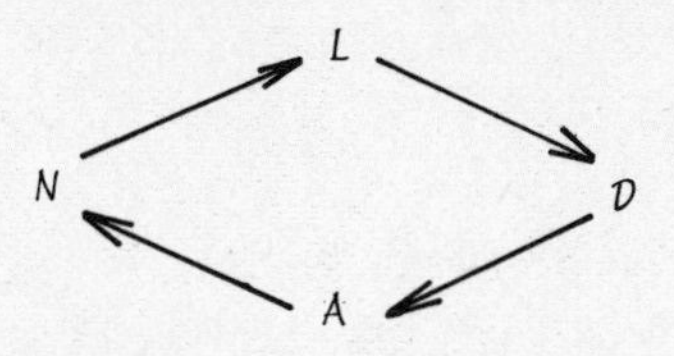

The way it is drawn suggests the name "circle line" for this structure and the term "station" for $\mathcal{D}$, A, N and L.

The name "circle line" itself suggests several questions. For instance, one can start from $\mathcal{D}$ and develop the theory by means of $\mathcal{D} \to A \to N \to L$. What happens if one makes the transition $L \to \mathcal{D}$, and repeats the process? The answer is simple: nothing changes. In particular, if one starts from T or $\mathcal{R}$ and makes the transitions indicated by the arrows, then the class of integrable functions and the integral will be definitively fixed at the first confrontation with L.

Now we give a short description of the contents of each of the seven chapters. Chapter 0 contains some preliminary material: the fundamental notion of a Riesz function space, enough topology to read the chapter on integration on locally compact Hausdorff spaces, something about Riemann-Stieltjes integration, and unordered summation. Most of this will be familiar to most readers. In Chapter 1 the stations of the circle line are developed, the connections between stations are described, and some questions are studied that arise from the possibility of travelling more than once in the circle line. In Chapter 2 the theory is further developed; here L or N is the starting point; one meets measurability, L^p-spaces and local null functions. The next chapter is devoted to measure theory. It contains the description of $\mathcal{R}$ and the connection between $\mathcal{R}$ and $\mathcal{D}$. Since $\mathcal{R}$ has more structure than $\mathcal{D}$ (or A, N, or L), there are more refined results connected with approximation of integrable or measurable functions and sets. Since there is also a simple connection $L \to \mathcal{R}$, the circle line is present in this chapter, too. Chapter 4 is about station T and the connections $T \to \mathcal{D}$, $T \to \mathcal{D}'$. The results of the two ways to derive a norm are compared, and the theory is developed further for the Bourbaki method. As in the $\mathcal{R}$-case, there are approximation results for integrable or measurable functions. Moreover, the important Riesz representation theorem is discussed, which establishes a connection between T and $\mathcal{R}$.

The final two chapters are not related to the circle line. Chapter 5 is about signed measures. The main result is the Radon-Nikodym theorem; an important application of this theorem is the identification of continuous linear functionals on L^p $(1 \le p < \infty)$. Classical questions about the relationship between differentiation and integration on $\mathbb{R}$ are treated here, and something is said about change of variables in integrals. Product spaces are the subject of Chapter 6. First appears Stone's version of the Fubini and Tonelli theorems, which is then specialized for the $\mathcal{R}$-case and the $\mathcal{T}$-case. There is a section on Fourier theory in $L^2(\mathbb{R})$ to show how useful the Fubini theorems are. The final section contains an application of the Radon-Nikodym theorem together with product measures to a question about stochastic processes. (The reader need not know anything about stochastic processes, though.)

Much of the contents of this book is standard, and can be found in many other textbooks as well (except, perhaps, the section about the relation between differentiation and integration, and Section 6.5). It is the presentation of the material, which allows us to describe and compare the various approaches to Lebesgue integration, that distinguishes this work from many other books. The central idea of relating the various approaches to integration in the circle line is due to N.G. de Bruijn. Lecture notes by him were the starting point for this book.

NOTATION

$\mathbb{N}$, $\mathbb{Z}$, $\mathbb{Q}$, $\mathbb{R}$ and $\mathbb{C}$ have their usual meaning, denoting the sets of natural numbers, integers, rational numbers, real numbers and complex numbers, respectively. The symbol $\mathbb{R}^*$ denotes the set of extended real numbers, that is $\mathbb{R}$ together with ∞ and $-\infty$, while $\mathbb{R}^+$ and $\mathbb{R}^*_+$ denote the non-negative members of $\mathbb{R}$ and $\mathbb{R}^*$, respectively.

":=" means "is defined by". For instance, $p := x^2$ defines p as the square of x.

Let S and T be non-empty sets, and let $f : S \to T$ be a function. For any subset B of T the inverse image of B under f is written as $f^{\leftarrow}(B) := \{s \in S \mid f(s) \in B\}$. If f is an injection (that is, if f is one-to-one), then the inverse function of f, which is defined on the range of f, is also denoted by $f^{\leftarrow}$.

Let S and T be sets. If for each $s \in S$ there is in some way given an $f(s) \in T$, then this defines a function $f : S \to T$. We write $f := \curlyvee_{s \in S} f(s)$, which is read as: f is defined to be that function that takes at S the value f(s). (Note that the range space T need not be mentioned explicitly.) For instance, we may write $s := \curlyvee_{x \in \mathbb{R}^2} x^2$, which defines s on $\mathbb{R}$ as the squaring function, or $\cos := \curlyvee_{x \in \mathbb{R}} \sum_{n=0}^{\infty} (-1)^n x^{2n}/(2n)!$, which defines the cosine function on $\mathbb{R}$.

Composition of functions is denoted by means of $\circ$. Thus $f \circ g := \curlyvee_{s \in S} f(g(s))$, if S is the domain of g and g maps S into the domain of f.

CHAPTER ZERO
PRELIMINARIES

This chapter contains some preliminary material, most of which will be familiar. Not everything presented is necessary for each of the following chapters. We indicate briefly what the reader should minimally do.

The first section tells what a vector space is, and in particular what a Riesz function space is. Every reader should know the contents of this section. The second section is about topology. Except for the notion of completeness, its contents are not needed before Chapter 3, and the product topology is used only in Section 6.3. Then there is a short section giving the definition of a normed space and of an inner product space; the terminology suffices and specific results are not needed. The next section deals with summation. In a certain sense it exemplifies what happens in Chapter 1, but only the facts about change of order of summation in series are really necessary; these may as well be taken on faith. The final section, about Riemann and Riemann-Stieltjes integration, may be read through quickly. In order to follow our development of integration the knowledge of calculus is almost sufficient. What one further needs is knowledge of the real numbers, and pencil and paper.

0.1. Algebraic preliminaries

Most readers will meet only one new concept in this section, namely that of a Riesz function space in 0.1.6. Since this concept is fundamental in our approach to integration, one should at least read its definition.

0.1.1. In integration theory it is convenient to extend the set $\mathbb{R}$ of real numbers with the symbols ∞ and $-\infty$; the resulting set is denoted by $\mathbb{R}^*$ and called the *extended real number system*. The algebraic operations are partially extended to $\mathbb{R}^*$ in an obvious way, roughly by thinking of ∞ as a very large positive number. For instance, if $x \in \mathbb{R}$, then $x + \infty = \infty + x = \infty$, and if in addition $x > 0$, then $x \cdot \infty = \infty \cdot x = \infty$, while if $x < 0$, then $x \cdot \infty = \infty \cdot x = -\infty$. Moreover, we use the convention $0 \cdot \infty = \infty \cdot 0 = 0 \cdot (-\infty) = (-\infty) \cdot 0 = 0$, and this will turn up quite often in the following. Expressions like $\infty - \infty$ remain undefined, because there seems to be no way to handle them consistently. The notions of order, and of supremum and infimum are similarly extended. For instance, if S is a subset of $\mathbb{R}$ which is not bounded above (so there is no $a \in \mathbb{R}$ such that $s \leq a$ for all $s \in S$), then we write $\sup S = \infty$.

0.1.2. A *linear space* (or a *vector space*) V *over* $\mathbb{R}$ is a non-empty set V supplied with two operations, called *addition* and *scalar multiplication*. The members of $\mathbb{R}$ are called *scalars*, and $\mathbb{R}$ is the *scalar field* of V. Addition is an operation that takes an element $(x,y) \in V \times V$ into the *sum* $x + y \in V$, while scalar multiplication takes an element $(\alpha,x) \in \mathbb{R} \times V$ into the *scalar multiple* $\alpha x \in V$. These operations are assumed to satisfy the following conditions. If $x,y,z \in V$ and $\alpha,\beta \in \mathbb{R}$, then

$$x + y = y + x \,, \qquad (\alpha + \beta)x = \alpha x + \beta x \,,$$

$$(x + y) + z = x + (y + z) \,, \qquad \alpha(x + y) = \alpha x + \alpha y \,,$$

$$\alpha(\beta x) = (\alpha\beta)x \,, \qquad 1 \cdot x = x \,,$$

while there is a unique element $0 \in V$ such that $u + 0 = u$ for all $u \in V$. This element 0, which behaves neutrally with respect to addition, is called the zero element of V. It follows that $-x$, which is an abbreviation for $-1 \cdot x$, satisfies $x + (-x) = 0$, and that $0 \cdot x = 0$, whenever $x \in V$. All this looks a little bit formal, but it is just the natural generalization of the way we operate with vectors in vector calculus.

What we have defined is commonly called a real vector space, because the scalars are taken from $\mathbb{R}$. In one or two places in the following we shall have occasion to use complex vector spaces, where $\mathbb{C}$ acts as the scalar field, that is, where multiplication by complex numbers is allowed. The formal definition of a complex vector space is the same as that for a real vector space, except for the larger scalar field.

0.1.3. Let V be a vector space (real or complex). A *linear subspace* W of V is a non-empty subset W of V such that $x + y \in W$ and $\alpha x \in W$ for all $x,y \in W$ and all scalars α. Obviously, W is a vector space in its own right with the operations inherited from V.

0.1.4. The nicest functions on a linear space are the linear functions. Let V and W be vector spaces (both real or both complex). A mapping $f : V \to W$ is called *linear* if $f(\alpha x + \beta y) = \alpha f(x) + \beta f(y)$ for all $x,y \in V$ and all scalars α,β. Since the field of scalars is itself a vector space, it may be taken in the role of W, and in this particular case f is called a *linear functional*.

0.1.5. Most vector spaces that we shall consider are of a special kind, which we now describe. Let S be a non-empty set, and let W be a vector space (real or complex). Denote the scalar field of W by K. Let f and g be two functions from S into W. The *sum function* $f + g$ is defined by

$$f + g := \Upsilon_{s \in S}(f(s) + g(s)) \,,$$

and if $\alpha \in K$, then αf is defined by

$$\alpha f := \curlyvee_{s \in S} \alpha f(s) .$$

If V is a non-empty set of functions from S into W such that $f + g \in V$, $\alpha f \in V$ whenever $f,g \in V$ and $\alpha \in K$, then V is called a *function space*. It is a little bit clumsy, but not difficult, to check that with these operations V is a vector space over K. The zero element of V is the function that is identically zero on S.

0.1.6. The most important function spaces in the sequel consist of real-valued functions, and have additional structure. Once again, let S be a non-empty set. If f and g are real-valued functions on S, then the functions $\sup(f,g)$, $\inf(f,g)$, $|f|$ are defined by

$$\sup(f,g) := \curlyvee_{s \in S} \sup(f(s),g(s)) ,$$

$$\inf(f,g) := \curlyvee_{s \in S} \inf(f(s),g(s)) ,$$

$$|f| := \curlyvee_{s \in S} |f(s)| .$$

A function space V over $\mathbb{R}$ is called a *Riesz function space* if $\sup(f,g) \in V$ and $\inf(f,g) \in V$ for every $f,g \in V$.

If V is a function space over $\mathbb{R}$ and $f \in V$, then $|f| \in V$, since $|f| = \sup(f,-f)$. Conversely, if V is a function space over $\mathbb{R}$, and $|f| \in V$ for every $f \in V$, then V is a Riesz function space, as one sees by noting that

$$\sup(a,b) = \tfrac{1}{2}(a + b) + \tfrac{1}{2}|a - b| ,$$

$$\inf(a,b) = \tfrac{1}{2}(a + b) - \tfrac{1}{2}|a - b|$$

for all $a,b \in \mathbb{R}$.

If f and g are elements of a Riesz function space V, then we write $f \geq g$ (also $g \leq f$) if $\sup(f,g) = f$. For any $W \subset V$ we put $W^+ := \{f \in W \mid f \geq 0\}$. In particular, $\mathbb{R}^+ := \{x \in \mathbb{R} \mid x \geq 0\}$. A linear functional I defined on V is called *positive* if $I(f) \geq 0$ for all $f \in V^+$. A subset W of the Riesz function space V is said to be *directed* if for every $f,g \in W$ there is an $h \in W$ such that $h \geq \sup(f,g)$. Obviously, if $\{f_1,f_2,\ldots,f_n\}$ is a finite subset of the directed set W, then there exists $h \in W$ such that $h \geq \sup(f_1,f_2,\ldots,f_n)$.

Exercises Section 0.1

1. Let B be a Riesz function space consisting of functions on the set S. Let I be a linear functional on B such that

$$\sup\{|I(\psi)| \mid 0 \le \psi \le \varphi\} < \infty$$

for $\varphi \in B^+$. Show that I can be decomposed as $I = I_+ - I_-$, where I_+ and I_- are positive linear functionals on B. Use the following steps.

(i) For $\varphi \in B^+$ put

$$I_+(\varphi) := \sup\{I(\psi) \mid 0 \le \psi \le \varphi\} ,$$

$$I_-(\varphi) := \sup\{-I(\psi) \mid 0 \le \psi \le \varphi\} .$$

Show that I_+ is a non-negative function on B^+ satisfying $I_+(\alpha\varphi_1 + \beta\varphi_2) = \alpha I_+(\varphi_1) + \beta I_+(\varphi_2)$ for $\alpha,\beta \ge 0, \varphi_1,\varphi_2 \in B^+$, and that $I_+(\varphi_1) \ge I_+(\varphi_2)$ if $\varphi_1 \ge \varphi_2$ and $\varphi_1,\varphi_2 \in B^+$. Ditto for I_-.

(ii) Show that $I(\varphi) = I_+(\varphi) - I_-(\varphi)$ for $\varphi \in B^+$. (Hint. Use that $I(\varphi) + I_-(\varphi) = \sup\{I(\varphi - \psi) \mid 0 \le \psi \le \varphi\}$.)

(iii) For $\varphi \in B$ put $\varphi_+ := \sup(\varphi,0)$, $\varphi_- := \sup(-\varphi,0)$, $I_+(\varphi) := I_+(\varphi_+) - I_+(\varphi_-)$, $I_-(\varphi) := I_-(\varphi_+) - I_-(\varphi_-)$. Show that both I_+ and I_- are positive linear functionals on B and that $I(\varphi) = I_+(\varphi) - I_-(\varphi)$ for $\varphi \in B$. (Hint. Show first that $I_+(\varphi_1) - I_+(\varphi_2) = I_+(\varphi_3) - I_+(\varphi_4)$ if $\varphi_1,\varphi_2,\varphi_3,\varphi_4 \in B^+$ and $\varphi_1 - \varphi_2 = \varphi_3 - \varphi_4$.)

2. Let B, I, I_+ and I_- be as in the preceding exercise. Show that the decomposition $I = I_+ - I_-$ has the following extremal property. If $I = I'_+ - I'_-$ for positive linear functionals I'_+ and I'_-, then $I_+(\varphi) \le I'_+(\varphi)$, $I_-(\varphi) \le I'_-(\varphi)$ for all $\varphi \in B^+$.

0.2. Topological preliminaries

This section contains what we need from general topology. The simpler proofs have been left out, but the three results of fundamental importance (Dini's theorem on uniform convergence in 0.2.12, Urysohn's separation lemma in 0.2.16, and the theorem on the existence of partitions of unity in 0.2.17) are proved in full.

0.2.1. A *metric space* (S,d) is a non-empty set S and a function $d : S \times S \to \mathbb{R}^+$ such that $d(x,y) = d(y,x)$, $d(x,y) \le d(x,z) + d(z,y)$ for all $x,y,z \in S$, and such that $d(x,y) = 0$ if and only if $x = y$. The function d is called a *metric* on S. Sometimes we speak of the metric space S, with metric d. In a metric space (S,d) one has a notion of convergence of sequences. Let $(x_n)_{n\in\mathbb{N}}$ be a sequence in S, let $x \in S$ and assume that $\lim_{n\to\infty} d(x_n,x) = 0$. Then the sequence $(x_n)_{n\in\mathbb{N}}$ is said to be *convergent to* x (it *tends to* x or *converges to* x), and that x is *limit* of the sequence. Notation: $x = \lim_{n\to\infty} x_n$, or $x_n \to x$ $(n \to \infty)$, or just $x_n \to x$. It is easy to see that in a metric space a sequence has at most one limit, so we can speak about *the* limit of a convergent sequence.

If $x \in S$ and $\delta > 0$, then the *open ball* with *center* x and *radius* δ is the set $B_{x,\delta} := \{y \in S \mid d(x,y) < \delta\}$. It is easy to prove that $x = \lim_{n\to\infty} x_n$ if and only if for every $\delta > 0$ one has $x_n \in B_{x,\delta}$ for all but finitely many $n \in \mathbb{N}$. A subset of S is called *bounded* if it is contained in some open ball.

A sequence $(x_n)_{n\in\mathbb{N}}$ is called a *fundamental sequence* or a *Cauchy sequence* if $d(x_n,x_m) \to 0$ $(n \to \infty, m \to \infty)$; that is, if for every $\varepsilon > 0$ there exists $N \in \mathbb{N}$ such that $d(x_n,x_m) < \varepsilon$ whenever $n > N$, $m > N$. A convergent sequence is a fundamental sequence, but a fundamental sequence need not converge. If in a metric space every fundamental sequence is convergent, then it is called *complete*. Important examples of complete metric spaces are the spaces $\mathbb{R}^n$ (where $n \in \mathbb{N}$) with the usual metric d given by $d(x,y) := (\sum_{i=1}^n |x_i - y_i|^2)^{\frac{1}{2}}$ for $x = (x_1,\dots,x_n) \in \mathbb{R}^n$, $y = (y_1,\dots,y_n) \in \mathbb{R}^n$.

0.2.2. A *topological space* (S,T) is a non-empty set S with a collection T of subsets of S satisfying the following conditions:

(i) $\emptyset \in T$, $S \in T$.

(ii) If $U,V \in T$, then $U \cap V \in T$.

(iii) If $\{U_\alpha \mid \alpha \in A\}$ is a collection of members of T, then $\bigcup_{\alpha\in A} U_\alpha \in T$.

The members of T are called *open* sets, and T is called a *topology* for S. The complements in S of open sets are called *closed* sets. The conditions for a topology imply that the intersection of any collection of closed sets is closed, and that the union of finitely many closed sets is closed.

If A is a subset of the topological space S, then its *closure* $\bar{A}$ is the smallest closed set containing A; that is, $\bar{A}$ is the intersection of the closed sets that contain A. The *interior* A° of A is the largest open set contained in A, or also, A° is the union of the open sets contained in A. A *neighborhood* U of a point $x \in S$ is a subset of S such that $x \in U^\circ$. A subset A of S is called *dense in* S if $\bar{A} = S$. A topological space is called *separable* if it has a countable dense subset.

In a topological space (S,T) convergence of sequences is defined as follows. Let $(x_n)_{n\in\mathbb{N}}$ be a sequence in S and x a point of S. We say that the sequence $(x_n)_{n\in\mathbb{N}}$ is *convergent* (*converges* or *tends*) *to* x (with respect to the topology T) if for every neighborhood U of x we have $x_n \in U$ for all but finitely many $n \in \mathbb{N}$. Again, x is called *limit* of the sequence. (It may occur that a sequence has more than one limit, see Exercise 0.2.8(ii).)

0.2.3. Any metric space is also a topological space. For let (S,d) be a metric space. Call a subset O of S open if for every $x \in O$ there exists an open ball $B_{x,\delta} \subset O$ (the δ may of course depend on x). It is now not difficult to see that the collection of open sets thus defined is a topology for S; it is called the *metric topology*. The open balls are open sets for the topology.

Different metrics can give rise to the same topology. For example, the usual metric for $\mathbb{R}^n$ generates what we call the usual topology for $\mathbb{R}^n$, but this topology is also generated by the metric $d'(x,y) := \max\{|x_k - y_k| \mid 1 \le k \le n\}$, or by $d''(x,y) := \sum_{k=1}^{n} |x_k - y_k|$.

In a metric space convergence of a sequence in the metric sense is equivalent to convergence in the topological sense.

0.2.4. Let S be a set, and let $\mathcal{B}$ be a class of subsets of S. If T is the class of all unions of elements of $\mathcal{B}$, we say that $\mathcal{B}$ is a *base* for T, and that $\mathcal{B}$ *generates* T. It is not difficult to show that $\mathcal{B}$ generates a topology of S if and only if the following conditions are satisfied:

(i) If $A \in \mathcal{B}$, $B \in \mathcal{B}$, and $x \in A \cap B$, then there is a $C \in \mathcal{B}$ with $x \in C \subset A \cap B$.

(ii) $S = \bigcup_{A \in \mathcal{B}} A$.

In a metric space the collection of open balls is a base for the metric topology.

A topological space is said to satisfy the *second axiom of countability* if its topology has a countable base.

0.2.5. If (S,T) is a topological space, and A is a non-empty subset of S, there is a natural topology on A: just take $\{A \cap U \mid U \in T\}$ as the collection of open sets. This topology is called the *relative topology* induced on A; it is denoted by $T|A$.

0.2.6. A topological space (S,T) is called a *Hausdorff space* if for any two distinct points $x \in S$, $y \in S$, there exist neighborhoods U of x and V of y such that $U \cap V = \emptyset$. In a Hausdorff space a sequence has at most one limit. A metric space is a Hausdorff space.

0.2.7. Let A be a subset of the topological space (S,T). An *open covering* of A is a collection of open sets whose union contains A. We call A *compact* if every open covering of A contains a finite *subcovering* (a finite subclass of the covering whose union still contains A).

In $\mathbb{R}^n$ with the usual topology we have the important theorem of Heine-Borel: a subset of $\mathbb{R}^n$ is compact if and only if it is closed and bounded (see Exercise 0.2-10).

0.2.8. Proposition. Let (S,T) be a topological space, C a compact subset of S, and F a closed subset of S. Then $C \cap F$ is compact.

Proof. Exercise 0.2-9. □

0.2.9. Compact subsets of a topological space behave more or less like finite sets. An example of this phenomenon is the following result, which will be needed in the proof of Urysohn's lemma.

Proposition. Let (S,T) be a Hausdorff space, C and D disjoint compact subsets of S. Then there exist disjoint open sets U and V in S with $C \subset U$, $D \subset V$.

Proof. First assume that D consists of one point only, x say. For every $y \in C$ there exist disjoint open sets U_y and V_y such that $y \in U_y$, $x \in V_y$. Now $Q := \{U_y \mid y \in C\}$ is an open covering of the compact set C. Hence Q contains a finite subcovering $\{U_y \mid y \in E\}$, where E is a finite subset of C. Now $U := \bigcup_{y\in E} U_y$ and $V := \bigcap_{y\in E} V_y$ are open sets that are clearly disjoint, and they cover C and $\{x\}$, respectively.

To handle the general case, apply the result just proved as follows. For every $x \in D$ there exist disjoint open sets U_x and V_x with $C \subset U_x$, $x \in V_x$. Now $\{V_x \mid x \in D\}$ is an open covering of the compact set D, which therefore contains a finite subcovering $\{V_x \mid x \in F\}$ where F is a finite subset of D. The sets $U := \bigcap_{x\in F} U_x$ and $V := \bigcup_{x\in F} V_x$ satisfy the conditions. □

0.2.10. It is worth noting that the first part of the preceding proof shows that in a Hausdorff space compact sets are closed.

0.2.11. Let (S,T) and (S',T') be topological spaces. A mapping $\varphi : S \to S'$ is called *continuous* if $\overset{\leftarrow}{\varphi}(U) \in T$ for every $U \in T'$. If (S'',T'') is a third topological space, and $\varphi : S \to S'$ and $\psi : S' \to S''$ are both continuous, then the composite function $\psi \circ \varphi$ is continuous.

If the topology T on S is generated by a metric d, then continuity of $\varphi : S \to S'$, where S' is a topological space, is equivalent to the following condition: for every $x \in S$ and every sequence $(x_n)_{n\in\mathbb{N}}$ in S with $x_n \to x$, one has $\varphi(x_n) \to \varphi(x)$.

Let (S,T) be a topological space, and let $\varphi : S \to \mathbb{R}$. The *support* of φ is the closure of $\{s \in S \mid \varphi(s) \neq 0\}$; we denote it by supp φ. We say that φ has *compact support* if supp φ is compact.

0.2.12. The next result is known as Dini's theorem.

Theorem (Dini's theorem). Let (S,T) be a topological space. Let $(\varphi_n)_{n\in\mathbb{N}}$ be a sequence of continuous real-valued functions on S that decreases to zero pointwise, that is, for every $x \in S$ one has

(i) $\lim_{n\to\infty} \varphi_n(x) = 0$,

(ii) $\varphi_{n+1}(x) \leq \varphi_n(x) \qquad (n \in \mathbb{N})$.

Then the convergence is uniform on every compact subset of S.

Proof. Let $C \subset S$ be compact. For each $x \in C$ there exists $n(x) \in \mathbb{N}$ such that $\varphi_{n(x)}(x) < \varepsilon/2$, and since $\varphi_{n(x)}$ is continuous there is an open neighborhood U_x of x such that $\varphi_{n(x)}(y) < \varepsilon$ for $y \in U_x$. Now $\{U_x \mid x \in C\}$ is an open covering of the compact set C, which therefore contains a finite subcovering, $\{U_x \mid x \in E\}$ say, where $E \subset C$ is finite. Let $N := \max\{n(x) \mid x \in E\}$. If $n > N$, $y \in C$, then $y \in U_x$ for some $x \in E$, and then $\varphi_n(y) \leq \varphi_N(y) \leq \varphi_{n(x)}(y) < \varepsilon$. □

0.2.13. A topological space (S,T) is called *compact* if S is compact as defined in 0.2.7; it is called *locally compact* if every point of S has a neighborhood with compact closure. Every compact space is locally compact, and so are the spaces $\mathbb{R}^n$ with the usual topology.

0.2.14. The above definition says that in a locally compact space every point has at least one compact neighborhood. In a locally compact Hausdorff space we can say more: here every point has "small" compact neighborhoods.

Proposition. Let (S,T) be a locally compact Hausdorff space. Let $x \in S$, and let U be a neighborhood of x. Then U contains a compact neighborhood V of x.

Proof. We may assume that U is open. Let C be a closed and compact neighborhood of x (see 0.2.10), and put $D := C\setminus U$. By Proposition 0.2.8, D is compact. Hence, by Proposition 0.2.9, there exist disjoint open sets P and Q such that $D \subset P$, $x \in Q$. Put $0 := Q \cap U \cap C$. Then 0 is a neighborhood of x, contained in U. Let $V := \overline{0}$. Obviously, 0 is a neighborhood of x. We assert that it is still contained in U, and that it is compact. Since $0 \subset S\setminus P$, and $S\setminus P$ is closed, we have $V = \overline{0} \subset S\setminus P$, and a fortiori $V \subset S\setminus D = (S\setminus C) \cup U$. On the other hand, since C is closed and $0 \subset C$, we have $V \subset C$. It follows that $V \subset U$. Since V is a closed subset of the compact set U, V is compact (Proposition 0.2.8). □

0.2.15. The one-point set of 0.2.14 can be replaced by a compact set, which once more shows that compact sets tend to behave like finite sets.

Proposition. Let (S,T) be a locally compact Hausdorff space. Let C be a compact subset of S and let U be an open set with $C \subset U$. Then there exists an open set V with compact closure such that $C \subset V \subset \overline{V} \subset U$.

Proof. For every $x \in C$ there exists a compact neighborhood $V_x \subset U$ (Proposition 0.2.14). Now the collection $\{V_x^\circ \mid x \in C\}$ is an open covering of the compact set C, which contains a finite subcovering, $\{V_x^\circ \mid x \in E\}$ say, where $E \subset C$ is finite. The set $V := \bigcup_{x \in E} V_x^\circ$ satisfies our conditions. □

0.2.16. Traditionally, the next result is called Urysohn's lemma. It is of fundamental importance in the theory of integration on locally compact Hausdorff spaces, so we call it a theorem.

Theorem (Urysohn's lemma). Let (S,T) be a locally compact Hausdorff space. Let C and O be subsets of S, C compact, O open, and $C \subset O$. Then there exists a continuous function $\varphi : S \to [0,1]$ with compact support such that $\varphi(x) = 1$ $(x \in C)$, $\varphi(x) = 0$ $(x \in S\backslash O)$.

Proof. By Proposition 0.2.15 we may assume that $\bar{O}$ is compact, and that there is an open set $O_{\frac{1}{2}}$ such that

$$C \subset O_{\frac{1}{2}} \subset \bar{O}_{\frac{1}{2}} \subset O .$$

Since $\bar{O}_{\frac{1}{2}}$ is compact (by 0.2.8), this procedure can be repeated. It follows that there exist open sets $O_{\frac{1}{4}}$ and $O_{\frac{3}{4}}$ such that

$$C \subset O_{\frac{3}{4}} \subset \bar{O}_{\frac{3}{4}} \subset O_{\frac{1}{2}} \subset \bar{O}_{\frac{1}{2}} \subset O_{\frac{1}{4}} \subset \bar{O}_{\frac{1}{4}} \subset O .$$

Continuing in this way we obtain for each $p \in D := \{k \cdot 2^{-n} \mid n \in \mathbb{N}, k = 1,2,\ldots,2^n-1\}$ an open set O_p such that

$$C \subset O_p \subset \bar{O}_p \subset O_q \subset \bar{O}_q \subset O$$

if $0 < q < p < 1$. Now define φ on S by

$$\varphi(x) := \begin{cases} \sup\{p \in D \mid x \in O_p\} & \text{if } x \in \bigcup_{p \in D} O_p , \\ 0 & \text{otherwise.} \end{cases}$$

It is obvious that $\varphi(x) = 1$ if $x \in C$, and $\varphi(x) = 0$ if $x \in S\backslash O$. So we need only show that φ is continuous.

For every $\alpha \in \mathbb{R}$, put $U_\alpha := \{x \in S \mid \varphi(x) < \alpha\}$ and $V_\alpha := \{x \in S \mid \varphi(x) > \alpha\}$. It is sufficient to show that U_α and V_α are open (Exercise 0.2-11). As to U_α, we may assume that $0 < \alpha \leq 1$. Let $x \in U_\alpha$, and take $p \in D$, $q \in D$ with $\varphi(x) < q < p < \alpha$. Then $x \notin O_q$, so $x \notin \bar{O}_p$, and if $y \notin \bar{O}_p$, then $y \notin O_p$, so $\varphi(y) < p < \alpha$. This shows that $S\backslash\bar{O}_p$ is a neighborhood of x, and that it is contained in U_α. Hence U_α is open. As to V_α, we may assume that $0 \leq \alpha < 1$. Let $x \in V_\alpha$, and take $p \in D$ with $\alpha < p < \varphi(x)$. Obviously, $x \in O_p$. Also, if $y \in O_p$, then $\varphi(y) \geq p > \alpha$, so $y \in V_\alpha$. Hence O_p is a neighborhood of x, and it is contained in V_α. Therefore V_α is open. □

0.2.17. Urysohn's lemma is a special case of the following result, which provides what is called a *partition of unity*. The existence of these partitions will be important in the treatment of integrals on product spaces.

Theorem. Let (S,T) be a locally compact Hausdorff space, C a compact subset of S, and let $O_1,\ldots,O_n$ be open subsets of S that together cover C. Then there exist continuous real-valued functions $\varphi_1,\ldots,\varphi_n$ on S with compact support, and such that

$$0 \le \varphi_k(x) \le 1 \quad (x \in S,\ 1 \le k \le n), \qquad \varphi_k(x) = 0 \quad (x \in S\backslash O_k,\ 1 \le k \le n),$$

$$\sum_{k=1}^{n} \varphi_k(x) = 1 \quad (x \in C).$$

Proof. We first show that there exist compact sets $C_1,\ldots,C_n$ such that $C = \bigcup_{k=1}^n C_k$, $C_k \subset O_k$ $(k = 1,\ldots,n)$. For every $x \in C$ there exists an open neighborhood U_x such that $\bar{U}_x$ is contained in some O_k (see 0.2.14). Now $\{U_x \mid x \in C\}$ is an open covering of C, which contains a finite subcovering $\{U_x \mid x \in E\}$, where E is a finite subset of C. For $k = 1,2,\ldots,n$, let V_k be the union of those U_x's with $x \in E$, $U_x \subset O_k$. Then $C_k := C \cap \bar{V}_k$ is compact (0.2.8), $C_k \subset O_k$ $(1 \le k \le n)$, and $C = \bigcup_{k=1}^n C_k$.

Now use Urysohn's lemma 0.2.16 to see that there exist continuous real-valued functions ψ_k $(k = 1,\ldots,n)$ such that $0 \le \psi_k(x) \le 1$ $(x \in S)$, $\psi_k(x) = 1$ $(x \in C_k)$, $\psi_k(x) = 0$ $(x \in S\backslash O_k)$. Put $\vartheta(t) := 1$ if $0 \le t \le 1$, $\vartheta(t) := t^{-1}$ if $t > 1$, $\psi(x) := \sum_{k=1}^n \psi_k(x)$ $(x \in S)$, and $\varphi_k(x) := \psi_k(x)\vartheta(\psi(x))$ $(x \in S,\ k = 1,\ldots,n)$. These φ_k satisfy our conditions. □

0.2.18. Let (S,T) and (S',T') be topological spaces. The collection of all sets $U \times V$, where $U \in T$, $V \in T'$, is a basis for a topology on $S \times S'$, which is called the *product topology*. The *projections* π_1 and π_2 defined by $\pi_1(x,y) := x$, $\pi_2(x,y) := y$ are now continuous. In fact, the product topology is the smallest topology on the product which makes π_1 and π_2 continuous.

0.2.19. Theorem. Let (S,T) and (S',T') be topological spaces. Let $S'' := S \times S'$, and let T'' be the product topology.

(i) If (S,T) and (S',T') are Hausdorff, then so is (S'',T'').

(ii) If (S,T) and (S',T') are compact, then so is (S'',T'').

(iii) If (S,T) and (S',T') are locally compact, then so is (S'',T'').

Proof. (i) Exercise 0.2-18.

(ii) Let (S,T) and (S',T') be compact. Let $\{W_\alpha \mid \alpha \in A\}$ be an open covering of S''. For every $s \in S$, $\alpha \in A$, let $W_\alpha^s := \pi_2(W_\alpha \cap (\{s\} \times S'))$, then W_α^s is open (Exercise 1.2-17(i)). Now let $s \in S$. It is easy to see that the collection $\{W_\alpha^s \mid \alpha \in A\}$

covers S'. Hence, there exists a finite subcovering $\{W_\alpha^s \mid \alpha \in E_s\}$ where $E_s \subset A$ is finite. Put $U_s := \bigcap_{\alpha \in E_s} \pi_1(W_\alpha)$, then U_s is open (by Exercise 0.2-17(i)), and obviously contains s. The collection $\{U_s \mid s \in S\}$ thus obtained is an open covering for the compact space S, so it contains a finite subcovering $\{U_s \mid s \in E\}$, where $E \subset S$ is finite. It is now not difficult to check that the finite family $\{W_\alpha \mid \alpha \in E_s$ for some $s \in E\}$ still covers S".

(iii) Exercise 0.2-18. ⊓

Exercises Section 0.2

1. Consider $\mathbb{R}$ with

$$d_1(x,y) := \frac{|x - y|}{1 + |x - y|} ,$$

$$d_2(x,y) := \arctan|x - y| ,$$

$$d_3(x,y) := |\arctan x - \arctan y| .$$

Show that d_1, d_2, and d_3 are metrics for $\mathbb{R}$ and that each generates the usual topology. Show that $(\mathbb{R}, d_1)$ and $(\mathbb{R}, d_2)$ are complete, while $(\mathbb{R}, d_3)$ is not.

2. If S is a set, then $\{\emptyset, S\}$ and $\{E \mid E \subset S\}$ are topologies for S (the *trivial* and the *discrete* topology, respectively).

3. Let S be a set and let $\{T_\alpha \mid \alpha \in A\}$ be a collection of topologies for A. Show that $\bigcap_{\alpha \in A} T_\alpha$ is a topology for S.

4. If S is a topological space and $A \subset B \subset S$, then $\bar{A} \subset \bar{B}$ and $A^\circ \subset B^\circ$.

5. Give the proofs of the assertions in 0.2.4.

6. Show that $\mathbb{Q}$ is dense in $\mathbb{R}$ (with the usual topology).

7. Show that every open set in $\mathbb{R}$ is the union of at most countably many pairwise disjoint open intervals.

8. (i) In a Hausdorff space the limit of a convergent sequence is unique.

(ii) Let S contain at least two elements, let $a \in S$, and call a subset of S open if it is empty or if it contains a as a member. Show that there is a sequence $(x_n)_{n \in \mathbb{N}}$ in S that converges to two different points.

9. Prove Proposition 0.2.8.

10. Show that a subset E of $\mathbb{R}$ is compact if and only if it is closed and bounded. Use the following steps.

(i) If E is compact, show that E is closed and bounded.

(ii) Let E be closed and bounded, and let $\{O_\alpha \mid \alpha \in A\}$ be an open covering of E containing no finite subcovering. Assume $E \subset [0,1]$ and put $E_{k,n} := E \cap [(k-1)\cdot 2^{-n}, k\cdot 2^{-n}]$ for $n \in \mathbb{N}$, $k = 1,\ldots,2^n-1$. Show that there is a sequence $(k_n)_{n\in\mathbb{N}}$ in $\mathbb{N}$ with $1 \le k_n \le 2^n-1$ such that $E_{k_{n+1},n+1} \subset E_{k_n,n}$ and such that $\{O_\alpha \mid \alpha \in A\}$ does not contain a finite subcovering of $E_{k_n,n}$ for any n. Show there is an $x \in E$ such that $x \in E_{k_n,n}$ for all n. Choose $\alpha \in A$ with $x \in O_\alpha$ and derive a contradiction by showing that $E_{k_n,n} \subset O_\alpha$ for large n.

11. Let (S,T) be a topological space. Show that $\varphi : S \to \mathbb{R}$ is continuous if and only if $\{s \in S \mid \varphi(s) < \alpha\}$ and $\{s \in S \mid \varphi(s) > \alpha\}$ are open for every $\alpha \in \mathbb{R}$.

12. Let (S,T) and (S',T') be topological spaces and let $\varphi : S \to S'$ be continuous. If $C \subset S$ is compact, then $\varphi(C)$ is compact.

13. Let (S,T) be a compact topological space, and let $\varphi : S \to \mathbb{R}$ be continuous. Show that φ is bounded and that it attains its upper and lower bounds.

14. Let (S,T) be a topological space, and let $(f_n)_{n\in\mathbb{N}}$ be a sequence of continuous real-valued functions defined on S that *converges uniformly on* S to the function f, i.e., $\sup_{s\in S}|f_n(s) - f(s)| \to 0$ as $n \to \infty$. Show that f is continuous.

15. Let (S,T) be a topological space, and let V be the class of bounded continuous real-valued functions defined on S. Define

$$d(f,g) := \sup_{s\in S} |f(s) - g(s)| \qquad (f,g \in V) .$$

Show that (V,d) is a complete metric space.

16. Let (S,d) and (S',d') be metric spaces. A function $\varphi : S \to S'$ is called *uniformly continuous* on S if for every $\varepsilon > 0$ there exists a $\delta > 0$ such that $d'(\varphi(s),\varphi(t)) < \varepsilon$ whenever $d(s,t) < \delta$. Show that if S is compact and $\varphi : S \to S'$ is continuous then φ is uniformly continuous on S.

17. Let (S,T) and (S',T') be topological spaces. Let S", T", π_1 and π_2 be as in 0.2.18. Show the following.

(i) If $E \subset S''$ is open, then $\pi_1(E)$, $\pi_2(E)$ and $\{y \mid (x,y) \in E\}$ are open for all $x \in S$.

(ii) If $C \subset S''$ is compact, then $\pi_1(C)$ and $\pi_2(C)$ are compact.

(iii) If $C \subset S''$ is compact, then there exist compact sets A and B in S and S' such that $C \subset A \times B$.

Is (i) true if "open" is replaced by "closed"?

18. Prove 0.2.19 (i) and (iii).

0.3. Normed spaces and inner product spaces

In the familiar Euclidean vector spaces we have the concepts of length and of angle. Generalization of these notions leads to the concepts of a normed linear space and of an inner product space, respectively.

0.3.1. Let V be a vector space over the scalar field K (equal to $\mathbb{R}$ or $\mathbb{C}$), and assume that there is a mapping $\|\cdot\| : V \to \mathbb{R}^+$ such that for all $x,y \in V$, $\alpha \in K$, one has

$$\|x + y\| \leq \|x\| + \|y\| ,$$

$$\|\alpha x\| = |\alpha| \cdot \|x\| ,$$

$$\|x\| = 0 \text{ if and only if } x = 0 .$$

Then V is called a *normed linear space*, and $\|x\|$ is called the *norm* of x. (In Section 1.1 we shall meet another concept of norm, related to integration.) If the distance between two points $x,y \in V$ is defined by $d(x,y) := \|x - y\|$, then d is a metric for V, so any normed linear space can also be regarded as a metric space and therefore also as a topological space. If the resulting metric space is complete, then V is called a *Banach space*.

Examples of Banach spaces are $\mathbb{R}^n$ (with $K = \mathbb{R}$) and $\mathbb{C}^n$ (with $K = \mathbb{C}$), where $\|x\| := \{\sum_{k=1}^n |x_k|^2\}^{\frac{1}{2}}$ for $x = (x_1,\ldots,x_n) \in \mathbb{R}^n$ and $\mathbb{C}^n$, respectively. Another example appears in Exercise 0.2-15 if one puts $\|f - g\| := d(f,g)$.

0.3.2. Let V be a real or complex vector space, K its scalar field. Assume there is a mapping $(\cdot,\cdot) : V \times V \to K$ such that for all $x,y,z \in V$ and $\alpha,\beta \in K$ one has

$$(\alpha x+\beta y,z) = \alpha(x,z) + \beta(y,z) ,$$

$$(x,y) = \overline{(y,x)} ,$$

$$(x,x) > 0 \text{ if } x \neq 0 .$$

Then V is called an *inner product space*, and (x,y) is called the *inner product* of x and y. If we define $\|x\| := \sqrt{(x,x)}$ for $x \in V$, then $\|\cdot\|$ is a norm on V (see Exercise 0.3-2), called the inner product norm. If V is a Banach space with this norm, then V is called a *Hilbert space*.

Examples of Hilbert spaces are $\mathbb{R}^n$ (with $K = \mathbb{R}$) and $\mathbb{C}^n$ (with $K = \mathbb{C}$) where the inner product in both cases is defined by $(x,y) := \sum_{k=1}^{n} x_k \bar{y}_k$ if $x = (x_1,\ldots,x_n)$, $y = (y_1,\ldots,y_n)$ are in $\mathbb{R}^n$ (or $\mathbb{C}^n$).

Exercises Section 0.3

1. Let S be a non-empty set and let B be the class of bounded real-valued functions defined on S. Define

$$\|\varphi\| := \sup_{s\in S} |\varphi(s)| \qquad (\varphi \in B) .$$

Show that B with $\|\cdot\|$ is a Banach space. Show that the space V of Exercise 0.2-15 with $\|\cdot\|$ is a Banach space, too.

2. Let $(\cdot,\cdot)$ be an inner product on the vector space V. Show that $\|x\| := \sqrt{(x,x)}$ gives a norm on V. (Hint. Assume $K = \mathbb{R}$ and let $x,y \in V$. Show that $\alpha^2\|x\|^2 + 2\alpha\beta(x,y) + \beta^2\|y\|^2 \geq 0$ for all $\alpha,\beta \in \mathbb{R}$, and conclude that $|(x,y)| \leq \|x\|\cdot\|y\|$.)

3. Let ℓ_e^2 be the space of all mappings $f : \mathbb{N} \to \mathbb{R}$ such that there is an $N \in \mathbb{N}$ with $f(x) = 0$ $(n \geq N)$. Show that ℓ_e^2 is a function space over $\mathbb{R}$. Put $(f,g) := \sum_{n=1}^{\infty} f(n)g(n)$. Show that this makes ℓ_e^2 into an inner product space, but not into a Hilbert space.

4. Let ℓ^2 be the space of all mappings $f : \mathbb{N} \to \mathbb{R}$ that satisfy $\sum_{n=1}^{\infty} |f(n)|^2 < \infty$.

(i) Show that ℓ^2 is a linear space. (Hint. $2|ab| \leq a^2 + b^2$.)

(ii) Let $f,g \in \ell^2$. Then $\sum_{n=1}^{\infty} f(n)g(n)$ is absolutely convergent. Define $(f,g) := \sum_{n=1}^{\infty} f(n)g(n)$. Show that $(\cdot,\cdot)$ is an inner product.

(iii) Show that ℓ^2 is a Hilbert space with this inner product as follows. Let $(f_k)_{k\in\mathbb{N}}$ be a fundamental sequence in ℓ^2.

(a) Show that $\lim_{k\to\infty} f_k(n)$ exists for $n \in \mathbb{N}$ and put $f(n) := \lim_{k\to\infty} f_k(n)$.

(b) Let $\varepsilon > 0$ and take $L \in \mathbb{N}$ such that $\|f_k - f_\ell\| < \varepsilon/2$ for $k,\ell \geq L$ ($\|\cdot\|$ denotes the inner product norm.) Let $N \in \mathbb{N}$ be such that $\sum_{n=N}^{\infty} |f_L(n)|^2 < \frac{1}{4}\varepsilon^2$. Use the inequality

$$\left(\sum_{n=N}^{M} |f_k(n)|^2\right)^{\frac{1}{2}} \le \left(\sum_{n=N}^{M} |f_k(n) - f_L(n)|^2\right)^{\frac{1}{2}} + \left(\sum_{n=N}^{M} |f_L(n)|^2\right)^{\frac{1}{2}}$$

to show that $\sum_{n=N}^{\infty} |f(n)|^2 \le \varepsilon^2$. Hence $f \in \ell^2$. Finally show that $\|f - f_k\| \to 0$ as $k \to \infty$.

0.4. Summation

In this section we treat unordered summation. This concept will be useful for illustrative purposes in Chapter 1, and it provides a simple example of a non-trivial Riesz function space. Moreover, Proposition 0.4.7 on change of order of summation will often be used in this book.

0.4.1. Let S be a set, and let $f : S \to \mathbb{R}$. Denote by F (or $F(S)$) the class of all finite subsets of S, and put

$$I(f;A) := \sum_{s \in A} f(s) \qquad (A \in F) .$$

Thus the numbers $I(f;A)$ are sums of finitely many real numbers.

Definition. f is *summable over* S if there exists a number $I(f)$ with the following property: if $\varepsilon > 0$, then there exists an $A \in F$ such that $B \in F$ and $B \supset A$ implies $|I(f;B) - I(f)| < \varepsilon$.

The number $I(f)$ of the definition is denoted by $\sum_{s \in S} f(s)$; it is uniquely determined (see Exercise 0.4-1 below). Instead of saying that f is summable over S, we may also say that $\sum_{s \in S} f(s)$ exists. It is convenient to allow S to be empty and to put $\sum_{s \in S} f(s)$ equal to 0 in this case.

0.4.2. Using the definition is often not the best way to show that $\sum_{s \in S} f(s)$ exists, because one first must have a candidate $I(f)$ to compare the numbers $I(f;A)$ with. The way out is well-known from elementary analysis where one meets the same difficulty in establishing the convergence of sequences and series: we prove a Cauchy criterion for convergence.

Proposition. Let $f : S \to \mathbb{R}$. The following three statements are equivalent.

(i) f is summable over S.

(ii) For every $\varepsilon > 0$ there is an $A \in F$ such that if $B, C \in F$ and $B \supset A$, $C \supset A$, then $|I(f;B) - I(f;C)| < \varepsilon$.

(iii) For every $\varepsilon > 0$ there is an $A \in F$ such that if $B \in F$ and $B \cap A = \emptyset$, then $|I(f;B)| < \varepsilon$.

Proof. Equivalence of (ii) and (iii) is obvious, so we only show that (i) and (ii) are equivalent.

First assume (i), and let $\varepsilon > 0$. Choose a set $A \in F$ such that $|I(f;B) - I(f)| < \frac{\varepsilon}{2}$ for every $B \in F$ with $B \supset A$. Now if $B \in F$, $C \in F$, $B \supset A$ and $C \supset A$, then $|I(f;B) - I(f;C)| < \varepsilon$ by the triangle inequality.

Next assume (ii). This is really the hard part of the proof, for we must now somehow find the number $I(f)$. This is done as follows. For each $n \in \mathbb{N}$, take $A_n \in F$ such that $|I(f;B) - I(f;C)| < 1/n$ if $B \in F$, $C \in F$, $B \supset A_n$, $C \supset A_n$. The sequence $(I(f;A_n))_{n\in\mathbb{N}}$ is a fundamental sequence in $\mathbb{R}$, since

$$(*)\qquad \begin{aligned} &|I(f;A_n) - I(f;A_m)| \le \\ &\le |I(f;A_n) - I(f;A_n \cup A_m)| + |I(f;A_n \cup A_m) - I(f;A_m)| \le \frac{1}{n} + \frac{1}{m}, \end{aligned}$$

for $n \in \mathbb{N}$, $m \in \mathbb{N}$, so the sequence has a limit, α say. This α is the number we are after. To show this, let $\varepsilon > 0$, and fix $n \in \mathbb{N}$ such that $n > 2\varepsilon^{-1}$. Let $m \to \infty$ in $(*)$ to obtain $|I(f;A_n) - \alpha| \le 1/n < \varepsilon/2$. Now if $A \in F$, $A \supset A_n$, then

$$|I(f;A) - \alpha| \le |I(f;A) - I(f;A_n)| + |I(f;A_n) - \alpha| < \varepsilon .$$

Hence, f is summable and $I(f) = \alpha$. □

0.4.3. Here is a first application of the preceding result.

Proposition. Let $f : S \to \mathbb{R}$ be summable over S. Then so is $|f|$.

Proof. Let $\varepsilon > 0$, and take $A \in F$ such that $|I(f;B)| < \varepsilon/2$ whenever $B \in F$, $A \cap B = \emptyset$. If $B \in F$, $B \cap A = \emptyset$, and $B_+ := \{s \in B \mid f(s) \ge 0\}$, $B_- := \{s \in B \mid f(s) < 0\}$, then

$$|I(|f|;B)| = I(f;B_+) - I(f;B_-) < \varepsilon .$$

So $|f|$ is summable by Proposition 0.4.2. □

0.4.4. Let L (or $L(S)$) denote the class of summable functions on S.

Proposition. L is a Riesz function space, and I is a positive linear functional on L.

Proof. Let $f,g \in L$. Let $\varepsilon > 0$, and take $A_1, A_2 \in F$ such that $|I(f;B) - I(f)| < \varepsilon/2$ whenever $B \in F$, $B \supset A_1$, and $|I(g;B) - I(g)| < \varepsilon/2$ whenever $B \in F$, $B \supset A_2$. If $A := A_1 \cup A_2$, and $B \in F$, $B \supset A$, then

$$|I(f+g;B) - I(f) - I(g)| = |I(f;B) + I(g;B) - I(f) - I(g)| < \varepsilon .$$

Hence $f + g \in L$, and $I(f + g) = I(f) + I(g)$. The proof that $\alpha f \in L$ and $I(\alpha f) = \alpha I(f)$ if $f \in L$, $\alpha \in \mathbb{R}$, is similar. Hence L is a function space, and I is a linear functional on L.

Proposition 0.4.3 and the second paragraph of 0.1.6 imply that L is a Riesz function space, and it is easy to see that I is positive, that is, $I(f) \geq 0$ whenever $f \geq 0$. □

0.4.5. Let $\{S_j \mid j \in J\}$ be a partition of S. By this we mean that J is a non-empty set, and the S_j's are pairwise disjoint subsets of S with union S. We relate the summation over S to summations over the S_j's.

Proposition. Let $f : S \to \mathbb{R}$.

(a) If f is summable over S, then

(i) $\sum_{s \in S_j} f(s)$ exists for every $j \in J$,

(ii) $\sum_{j \in J} \left(\sum_{s \in S_j} f(s) \right)$ exists.

Moreover, $\sum_{j \in J} (\sum_{s \in S_j} f(s)) = \sum_{s \in S} f(s)$.

(b) Conversely, if (i) and (ii) of (a) hold and $f \geq 0$, then f is summable over S.

Proof. (a) Let f be summable over S. The proof of (i) is left to the reader as Exercise 0.4-4.

To prove (ii), let $\varepsilon > 0$, and take $A \in F(S)$ such that $|I(f;B)| < \varepsilon$ if $B \in F(S)$ and $B \cap A = \emptyset$. Then

$$(*) \qquad |I(f) - I(f;A)| \leq \varepsilon .$$

Let $P := \{j \in J \mid A \cap S_j \neq \emptyset\}$. Obviously, P is finite, that is, $P \in F(J)$. Let $Q \in F(J)$ be such that $Q \cap P = \emptyset$. Then $|\sum_{j \in Q} (\sum_{s \in S_j} f(s))| \leq \varepsilon$. For, if this is not true, there are finite subsets B_j of S_j for $j \in Q$ such that $|\sum_{j \in Q} (\sum_{s \in B_j} f(s))| > \varepsilon$. If $B := \bigcup_{j \in Q} B_j$, then $B \in F(S)$ and $B \cap A = \emptyset$, which implies $|I(f;B)| = |\sum_{j \in Q} (\sum_{s \in B_j} f(s))| > \varepsilon$, contrary to the assumption. It follows from Proposition 0.4.2(iii) that $\sum_{j \in J} (\sum_{s \in S_j} f(s))$ exists, and

$$(**) \qquad \left| \sum_{j \in P} \left(\sum_{s \in S_j} f(s) \right) - \sum_{j \in J} \left(\sum_{s \in S_j} f(s) \right) \right| \leq \varepsilon .$$

We show next that

$$(***) \qquad \left| I(f;A) - \sum_{j \in P} \left(\sum_{s \in S_j} f(s) \right) \right| \leq 2\varepsilon ,$$

arguing by contradiction. So assume (***) is not true.

Take finite sets A_j with $A \cap S_j \subset A_j \subset S_j$ such that $|\sum_{j\in P}(\sum_{s\in S_j} f(s) - \sum_{s\in A_j} f(s))| < \varepsilon$, and let $B := \bigcup_{j\in P} A_j \setminus A$. Then $B \cap A = \emptyset$, whence $|I(f;B)| < \varepsilon$, while also

$$|I(f;B)| = |\sum_{j\in P} \sum_{s\in A_j} f(s) - \sum_{j\in P} \sum_{s\in S_j\cap A} f(s)| =$$

$$= |\sum_{j\in P} \sum_{s\in A_j} f(s) - I(f;A)| \geq |\sum_{j\in P} \sum_{s\in S_j} f(s) - I(f;A)| - \varepsilon > \varepsilon .$$

Combination of (*), (**) and (***) gives $|I(f) - \sum_{j\in J} \sum_{s\in S_j} f(s)| \leq 4\varepsilon$. Let $\varepsilon > 0$.

(b) Assume that (i) and (ii) hold, and that $f \geq 0$. Let $\varepsilon > 0$, and take $P \in F(J)$ such that $|\sum_{j\in Q}(\sum_{s\in S_j} f(s))| < \varepsilon/2$ if $Q \in F(J)$ and $Q \cap P = \emptyset$. Let q be the number of elements of P. For each $j \in P$, let $A_j \in F(S_j)$ be such that $\sum_{s\in B} f(s) < \varepsilon(2q + 1)^{-1}$ if $B \in F(S_j)$, $B \cap A_j = \emptyset$. Let $A := \bigcup_{j\in P} A_j$; then $A \in F(S)$. Now let $B \in F(S)$, $B \cap A = \emptyset$. If $B_j := B \cap S_j$ for $j \in J$, and $Q := \{j \in J\setminus P \mid B_j \neq \emptyset\}$, then $\sum_{s\in B_j} f(s) \leq \sum_{s\in S_j} f(s)$ because $f \geq 0$, and

$$I(f;B) = \sum_{j\in P} \left(\sum_{s\in B_j} f(s)\right) + \sum_{j\in Q} \left(\sum_{s\in B_j} f(s)\right) \leq q \cdot \frac{\varepsilon}{2q + 1} + \frac{\varepsilon}{2} < \varepsilon .$$

Now Proposition 0.4.2(iii) shows that f is summable. □

0.4.6. Unordered summation as defined in 0.4.1 and summation of series as taught in elementary calculus are different concepts. For example, there are series that are convergent but not absolutely convergent (consider $\sum_{n=1}^{\infty}(-1)^n n^{-1}$), whereas the corresponding phenomenon does not occur with unordered summation (see Proposition 0.4.3). The precise situation is as follows. Let $\sum_{n=1}^{\infty} u(n)$ be a series with real terms. The function u is summable over $\mathbb{N}$ if and only if $\sum_{n=1}^{\infty}|u(n)|$ is convergent, and if u is summable, then $\sum_{n\in\mathbb{N}} u(n) = \sum_{n=1}^{\infty} u(n)$. A result of the same type is this: u is summable over $\mathbb{N}$ (hence $\sum_{n=1}^{\infty}|u(n)| < \infty$) if and only if $\sum_{n=1}^{\infty} u(\pi(n))$ is convergent for every permutation π of $\mathbb{N}$. The proofs of these facts are left to the reader.

0.4.7. Combining the remarks in 0.4.6 with Proposition 0.4.5 we get the following useful result.

Proposition. Let $u : \mathbb{N} \times \mathbb{N} \to \mathbb{R}$, and assume that $\sum_{n=1}^{\infty}(\sum_{m=1}^{\infty}|u(n,m)|) < \infty$. Then $|u|$ and u are summable over $\mathbb{N} \times \mathbb{N}$. Moreover, $\sum_{(n,m)\in\mathbb{N}\times\mathbb{N}} u(n,m)$ can be summed in any order, always with the same sum. In particular,

$$\sum_{n=1}^{\infty} \left(\sum_{m=1}^{\infty} u(n,m)\right) = \sum_{m=1}^{\infty} \left(\sum_{n=1}^{\infty} u(n,m)\right) .$$

Exercises Section 0.4

1. Show that $I(f)$ in Definition 0.4.1 is uniquely determined.

2. Let L and S be as in 0.4.4. If $f \in L^+$, $g : S \to \mathbb{R}$ and $|g| \le f$, then $g \in L$ and $|I(g)| \le I(f)$.

3. Let L and S be as in 0.4.4. Show that $\|f\| := I(|f|)$ defines a norm on L that makes L into a Banach space.

4. If f is summable over S and $S_1 \subset S$, then f is summable over S_1.

5. Let f be summable over S. Show that $\{s \in S \mid f(s) \neq 0\}$ is countable.

6. Let $(f_n)_{n\in\mathbb{N}}$ be a sequence in L. Let $g \in L$ and assume $|f_n| \le g$ for all $n \in \mathbb{N}$. Assume that $f_n(s) \to f(s)$ for $s \in S$. Show that $f \in L$, that $I(f_n) \to I(f)$, and that even $I(|f_n - f|) \to 0$.

7. Show by an example that the condition "$f \ge 0$" in 0.4.5(b) may not be left out. Show that it can be replaced by "$f \ge g$ with g summable over S".

8. Prove the assertions in 0.4.6.

0.5. The Riemann and the Riemann-Stieltjes integral

Although we assume that the reader is more or less familiar with the theory of Riemann and Riemann-Stieltjes integration on the real line, we give the main features of this theory in this section, mainly to fix notation and definitions. The treatment is not very general, but some supplementary material is contained in the exercises at the end of this section.

0.5.1. In all of this section $[a,b]$, where $a \le b$, is a fixed interval in $\mathbb{R}$. A *partition* P of $[a,b]$ is a finite set of numbers $\{x_0,x_1,\ldots,x_n\}$ in $[a,b]$, where $a = x_0 < x_1 < \ldots < x_n = b$; we also write $P = [x_0,x_1,\ldots,x_n]$. The *width* of the partition P is defined as $\max_{1\le i\le n}(x_i - x_{i-1})$. If P and Q are partitions of $[a,b]$, we say that Q is *finer* than P (or that Q is a *refinement* of P) if every point of P occurs in Q.

Let $g : [a,b] \to \mathbb{R}$ be non-decreasing, let $f : [a,b] \to \mathbb{R}$ be bounded, and let $P := [x_0,x_1,\ldots,x_n]$ be a partition of $[a,b]$. The *upper sum of* f *with respect to* g *and the partition* P is

$$S(P,f,g) := \sum_{i=1}^{n} \{g(x_i) - g(x_{i-1})\} \sup_{x_{i-1} \le x \le x_i} f(x) .$$

The *lower sum of* f *with respect to* g *and the partition* P is

$$s(P,f,g) := \sum_{i=1}^{n} \{g(x_i) - g(x_{i-1})\} \inf_{x_{i-1} \le x \le x_i} f(x) .$$

(The reference to g or P may be left out if no confusion can arise.)

0.5.2. The following result will often be useful.

Proposition. Let $f : [a,b] \to \mathbb{R}$ be bounded and let $g : [a,b] \to \mathbb{R}$ be non-decreasing. Let P and Q be partitions of [a,b] with Q finer than P. Then

$$s(P,f,g) \le s(Q,f,g) \le S(Q,f,g) \le S(P,f,g) .$$

Proof. It is sufficient to consider the case that Q contains just one point more than P, say $P = [x_0,x_1,\ldots,x_n]$, $Q = [y_0,y_1,\ldots,y_{n+1}]$, with $x_0 = y_0,\ldots,x_j = y_j$, $x_j < y_{j+1} < x_{j+1}$, $x_{j+1} = y_{j+2},\ldots,x_n = y_{n+1}$. Then

$$\begin{aligned} S(Q,f,g) - S(P,f,g) = &\{g(y_{j+1}) - g(y_j)\} \sup_{y_j \le x \le y_{j+1}} f(x) + \\ &+ \{g(y_{j+2}) - g(y_{j+1})\} \sup_{y_{j+1} \le x \le y_{j+2}} f(x) + \\ &- \{g(x_{j+1}) - g(x_j)\} \sup_{x_j \le x \le x_{j+1}} f(x) \le 0 . \end{aligned}$$

That $S(Q,f,g) \ge s(Q,f,g)$ is obvious from the definition, and the proof of the inequality for the lower sums is similar to the one for the upper sums. □

0.5.3. Corollary. Let f and g be as in 0.5.2. Let P and Q be partitions of [a,b]. Then $s(P,f,g) \le S(Q,f,g)$.

Proof. Let R be the partition of [a,b] that consists of all points of P together with all points of Q. Application of 0.5.2 to the pair P,R gives $s(P,f,g) \le s(R,f,g)$. Another application of 0.5.2, now to the pair Q,R, gives $s(R,f,g) \le S(R,f,g) \le$ $\le S(Q,f,g)$. Hence the result. □

0.5.4. Let f and g be as at the end of 0.5.1. The *upper integral of* f *with respect to* g *over* [a,b] is

$$\overline{\int_a^b} f(x)dg(x) := \inf\{S(P,f,g) \mid P \text{ partition of } [a,b]\} ,$$

and the *lower integral of* f *with respect to* g *over* [a,b] is

$$\underline{\int_a^b} f(x)dg(x) := \sup\{s(P,f,g) \mid P \text{ partition of } [a,b]\} .$$

The preceding corollary shows that the lower integral is at most equal to the upper integral.

0.5.5. Here are some useful formulas for upper and lower integrals.

Proposition. Let f and k be bounded real-valued functions on [a,b]; let g and h be non-decreasing real-valued functions defined on [a,b], and let $\lambda \in \mathbb{R}$. Then

(i) $$\overline{\int_a^b} f(x)dg(x) = \overline{\int_a^c} f(x)dg(x) + \overline{\int_c^b} f(x)dg(x) \quad \text{if } a \le c \le b ,$$

(ii) $$\overline{\int_a^b} f(x)d(g(x) + h(x)) = \overline{\int_a^b} f(x)dg(x) + \overline{\int_a^b} f(x)dh(x) ,$$

(iii) $$\overline{\int_a^b} (f(x) + k(x))dg(x) \le \overline{\int_a^b} f(x)dg(x) + \overline{\int_a^b} k(x)dg(x) ,$$

(iv) $$\overline{\int_a^b} \lambda f(x)dg(x) = \begin{cases} \lambda \overline{\int_a^b} f(x)dg(x) & \text{if } \lambda \ge 0 , \\ \lambda \underline{\int_a^b} f(x)dg(x) & \text{if } \lambda < 0 , \end{cases}$$

(v) $$\int_a^b f(x)d(\lambda g(x)) = \overline{\int_a^b} f(x)dg(x) \quad \text{if } \lambda \ge 0 .$$

There are similar results for lower integrals (with the inequality sign in (iii) reversed).

Proof. (i) If P is a partition of [a,c], Q one of [c,b], and R the partition of [a,b] obtained by taking the points in P and Q together, then obviously $S(P,f,g) + S(Q,f,g) = S(R,f,g) \ge \overline{\int_a^b} f(x)dg(x)$. Now let P and Q vary to obtain

$$\overline{\int_a^c} f(x)dg(x) + \overline{\int_c^b} f(x)dg(x) \ge \overline{\int_a^b} f(x)dg(x) .$$

To prove the reverse inequality, let $\varepsilon > 0$, and choose a partition R of [a,b] such that $S(R,f,g) < \overline{\int_a^b} f(x)dg(x) + \varepsilon$. It follows from Proposition 0.5.2 that we may assume that c occurs in the partition R. Let P be the partition of [a,c] consisting of the points of R that belong to [a,c], and let Q be the partition of [c,b] consisting of the points of R that belong to [c,b]. Then

$$\overline{\int_a^c} f(x)dg(x) + \overline{\int_c^b} f(x)dg(x) \leq S(P,f,g) + S(Q,f,g) =$$

$$= S(R,f,g) < \overline{\int_a^b} f(x)dg(x) + \varepsilon \, .$$

Now let $\varepsilon \to 0$.

The remaining proofs, which are much easier, are left to the reader. □

0.5.6. Definition. Let f and g be as at the end of 0.5.1. If $\underline{\int_a^b} f(x)dg(x) = \overline{\int_a^b} f(x)dg(x)$, then this common value is denoted by $\int_a^b f(x)dg(x)$ and f is said to be *integrable with respect to* g *over* [a,b].

The integral thus defined is called a *Riemann-Stieltjes integral*. The special case where $g = \Upsilon_{x\in[a,b]} x$ gives rise to the ordinary *Riemann integral*, denoted by $\int_a^b f(x)dx$. The terminology is often simplified. For instance, instead of saying that f is integrable with respect to g over [a,b], we also say that $\int_a^b f(x)dg(x)$ exists, and in the Riemann case that $\int_a^b f(x)dx$ exists, or that f is (Riemann) integrable over [a,b].

0.5.7. Theorem. Let f and k be bounded real-valued functions defined on [a,b]. Let g and h be non-decreasing real-valued functions defined on [a,b]. Let $\lambda \in \mathbb{R}$ and $c \in [a,b]$. Assume that f is integrable with respect to g and h, and that k is integrable with respect to g (integration over [a,b]). Then the following assertions hold.

(i) f is integrable with respect to g over [a,c] and over [c,b], and

$$\int_a^b f(x)dg(x) = \int_a^c f(x)dg(x) + \int_c^b f(x)dg(x) \, .$$

(ii) f is integrable with respect to g + h over [a,b], and

$$\int_a^b f(x)d(g+h)(x) = \int_a^b f(x)dg(x) + \int_a^b f(x)dh(x) \, .$$

(iii) $f + k$ is integrable with respect to g over $[a,b]$, and

$$\int_a^b (f(x) + k(x))dg(x) = \int_a^b f(x)dg(x) + \int_a^b k(x)dg(x) .$$

(iv) $$\int_a^b \lambda f(x)dg(x) = \lambda \int_a^b f(x)dg(x) .$$

(v) If $\lambda \geq 0$, then f is integrable with respect to g over $[a,b]$, and

$$\int_a^b f(x)d(\lambda g(x)) = \lambda \int_a^b f(x)dg(x) .$$

Proof. Everything follows immediately from Proposition 0.5.5. □

0.5.8. Which functions f are integrable with respect to a non-decreasing function g? There is a large class of functions for which things are simple.

Proposition. Let $f : [a,b] \to \mathbb{R}$ be continuous, and let $g : [a,b] \to \mathbb{R}$ be non-decreasing. Then $\int_a^b f(x)dg(x)$ exists.

Proof. We must show that the upper and the lower integral of f with respect to g are equal, and by Exercise 0.5-2 it is enough to show that for every $\varepsilon > 0$ there exists a partition P of $[a,b]$ such that $S(P,f,g) - s(P,f,g) < \varepsilon$. Let $\varepsilon > 0$, and define $\eta := \varepsilon\{g(b) - g(a) + 1\}^{-1}$. Since $[a,b]$ is compact, f is uniformly continuous on $[a,b]$ (Exercise 0.2-16). Hence, there exists $\delta > 0$ such that $u \in [a,b]$, $v \in [a,b]$ and $|u - v| < \delta$ implies $|f(u) - f(v)| < \eta$. Let $P := [x_0,\ldots,x_n]$ be any partition of $[a,b]$ with width less than δ. Then

$$S(P,f,g) - s(P,f,g) =$$

$$= \sum_{i=1}^{n} \{g(x_i) - g(x_{i-1})\}\left\{ \sup_{x_{i-1} \leq x \leq x_i} f(x) - \inf_{x_{i-1} \leq x \leq x_i} f(x)\right\} \leq$$

$$\leq \sum_{i=1}^{n} \{g(x_i) - g(x_{i-1})\}\eta = \eta\{g(b) - g(a)\} < \varepsilon .$$ □

0.5.9. The next result is a first example of a convergence theorem for integrals. We shall meet many more of these.

Theorem. Let $g : [a,b] \to \mathbb{R}$ be non-decreasing. Let $(f_n)_{n \in \mathbb{N}}$ be a sequence of bounded real-valued functions, all integrable with respect to g. Assume that the sequence

$(f_n)_{n\in\mathbb{N}}$ converges to f, uniformly on [a,b]. Then f is integrable with respect to g, and

$$\int_a^b f(x)dg(x) = \lim_{n\to\infty} \int_a^b f_n(x)dg(x) .$$

<u>Proof</u>. Let $\varepsilon > 0$. Choose $N \in \mathbb{N}$ so large that $|f(x) - f_n(x)| < \varepsilon$ for $x \in [a,b]$, $n \geq N$. Since f_N is bounded on [a,b], by M say, we conclude that $|f(x)| < M + \varepsilon$ for $x \in [a,b]$, that is, f is bounded. Moreover, since f_N is integrable with respect to g on [a,b], there exists a partition P of [a,b] such that $S(P,f_N,g) - s(P,f_N,g) < \varepsilon$ (one half of the result of Exercise 0.5-2). It is easy to see that this implies that $S(P,f,g) - s(P,f,g) < \varepsilon + 2\varepsilon\{g(b) - g(a)\}$. Since $\varepsilon > 0$ is arbitrary, this last inequality implies that ${}_a\int^b f(x)dg(x)$ exists (the other half of the result of Exercise 0.5-2).

For the remaining part of the proof, let $\varepsilon > 0$ and let N be as before. If P is any partition of [a,b], and $n \geq N$, then $S(P,f,g) - S(P,f_n,g) \leq \varepsilon\{g(b) - g(a)\}$. Hence

$$\underline{\int_a^b} f(x)dg(x) \leq S(P,f,g) < S(P,f_n,g) + \varepsilon\{g(b) - g(a)\} .$$

Varying P, we obtain

$$\underline{\int_a^b} f(x)dg(x) = \overline{\int_a^b} f(x)dg(x) \leq \overline{\int_a^b} f_n(x)dg(x) + \varepsilon\{g(b) - g(a)\} =$$

$$= \int_a^b f_n(x)dg(x) + \varepsilon\{g(b) - g(a)\} .$$

Interchanging the roles of f_n and f in the last reasoning, we get the reverse inequality. Hence

$$\left|\int_a^b f(x)dg(x) - \int_a^b f_n(x)dg(x)\right| \leq \varepsilon\{g(b) - g(a)\}$$

if $n \geq N$. □

0.5.10. The notion of integral can be usefully extended by allowing more general functions behind the d-sign. We take only one step in this direction, admitting functions of bounded variation, which we now define.

Let $g : [a,b] \to \mathbb{R}$. For any partition $P := [x_0,x_1,\ldots,x_n]$ of [a,b] put

$$\mathrm{var}_P\, g := \sum_{i=1}^{n} |g(x_i) - g(x_{i-1})| .$$

If $\sup\{\mathrm{var}_P\, g \mid P \text{ partition of } [a,b]\}$ is finite, then we say that g *has bounded variation on* $[a,b]$. The value of the supremum is denoted by $\mathrm{var}(g;[a,b])$ and called the *total variation of* g *on* $[a,b]$.

0.5.11. Obviously, every monotone function g on $[a,b]$ has bounded variation on $[a,b]$ with $\mathrm{var}(g;[a,b]) = |g(b) - g(a)|$. Moreover, any linear combination of functions of bounded variation has bounded variation. In particular, the difference of two non-decreasing functions has bounded variation. The next result shows a converse to the last assertion.

<u>Theorem</u> (Jordan). Suppose $g : [a,b] \to \mathbb{R}$ has bounded variation on $[a,b]$. Then there are non-decreasing real-valued functions g_1 and g_2, defined on $[a,b]$, such that $g = g_1 - g_2$.

<u>Proof</u>. It is clear that g has bounded variation on each subinterval of $[a,b]$, so $g_1 := \Upsilon_{x\in[a,b]}\, \mathrm{var}(g;[a,x])$ is well-defined, and obviously non-decreasing. Let $g_2 := g - g_1$. By Exercise 0.5-6, $\mathrm{var}(g;[a,x]) + \mathrm{var}(g;[x,b]) = \mathrm{var}(g;[a,b])$ for every $x \in [a,b]$. This means that g_2 is non-decreasing too. So we have a decomposition of g as desired. □

0.5.12. The preceding result gives a simple way to extend the notion of integral.

<u>Definition</u>. Let $f : [a,b] \to \mathbb{R}$ be bounded, and suppose $g : [a,b] \to \mathbb{R}$ has bounded variation on $[a,b]$. Assume that g has a decomposition $g = g_1 - g_2$ into non-decreasing functions such that ${}_a\!\int^b f(x)dg_1(x)$ and ${}_a\!\int^b f(x)dg_2(x)$ both exist. Then f is said to be *integrable with respect to* g *over* $[a,b]$, and the integral is defined as

$$\int_a^b f(x)dg(x) := \int_a^b f(x)dg_1(x) - \int_a^b f(x)dg_2(x) .$$

Of course we have to check that the value of the integral does not depend on the particular decomposition chosen in defining it. This is easy. For, let $g = g_1' - g_2'$ be another decomposition for which the integrals exist. Then $g_1 + g_2' = g_1' + g_2$, and now Theorem 0.5.7 (ii) shows that

$$\int_a^b f(x)dg_1(x) + \int_a^b f(x)dg_2'(x) = \int_a^b f(x)dg_1'(x) + \int_a^b f(x)dg_2(x) ,$$

which is what we need.

0.5.13. Theorem 0.5.7, Proposition 0.5.8 and Theorem 0.5.9 hold if the functions behind the d-sign have bounded variation only. Most of the proofs consist in using the definition of the integral together with 0.5.7, 0.5.8, and 0.5.9. (There are some complications with 0.5.7(iii) and 0.5.9, see Exercise 0.5-14.)

0.5.14. Additional limit processes may be used to extend the integral still further. It will be sufficient to consider two examples that are typical for the kind of generalization we have in mind. We restrict ourselves to the Riemann case.

First assume that the interval of integration is unbounded, $[a,\infty)$ say. Let $f : [a,\infty) \to \mathbb{R}$ be bounded on every interval $[a,b]$, assume that ${}_a\int^b f(x)dx$ exists for every $b \in [a,\infty)$, and that $\lim_{b\to\infty} {}_a\int^b f(x)dx$ exists. Then f is *(improperly) integrable over* $[a,\infty)$ and ${}_a\int^b f(x)dx$ is defined to be equal to the limit.

In the second place, let $[a,b]$ be a finite interval. Let $f : [a,b] \to \mathbb{R}$ have the property that f is bounded on every interval $[a,c]$ and integrable over $[a,c]$ for every $c \in [a,b]$, while $\lim_{c\uparrow b} {}_a\int^c f(x)dx$ exists. Then we say that f is *integrable over* $[a,b]$, and ${}_a\int^b f(x)dx$ is defined to be equal to the limit. In this case it is possible that ${}_a\int^b f(x)dx$ was already defined by Definition 0.5.6, and a conflict might arise as to the value to be attributed to the integral. There are no problems, however (Exercise 0.5-16).

Exercises Section 0.5

1. Complete the proof of Proposition 0.5.5.

2. Let $f : [a,b] \to \mathbb{R}$ be bounded, $g : [a,b] \to \mathbb{R}$ non-decreasing. Show that ${}_a\int^b f(x)dg(x)$ exists if and only if for every $\varepsilon > 0$ there is a partition P of $[a,b]$ such that $S(P,f,g) - s(P,f,g) < \varepsilon$. Show that $S(P,f,g) \ge {}_a\int^b f(x)dg(x) \ge s(P,f,g)$ if ${}_a\int^b f(x)dg(x)$ exists and P is a partition of $[a,b]$.

3. Let $f : [a,b] \to \mathbb{R}$ be bounded. We define the *oscillation* of f over $[\alpha,\beta]$ by

$$\Delta(f,[\alpha,\beta]) := \sup_{\alpha\le x\le\beta} f(x) - \inf_{\alpha\le x\le\beta} f(x) .$$

(i) Show that

$$\Delta(f,[\alpha,\beta]) = \sup\{|f(x) - f(y)| \mid \alpha \le x \le \beta,\ \alpha \le y \le \beta\} .$$

(ii) Show that f is Riemann integrable over $[a,b]$ if and only if for every $\sigma > 0$, $\eta > 0$, there is a partition $P = [x_0,x_1,\dots,x_n]$ of $[a,b]$ such that the sum of the lengths of the intervals $[x_{i-1},x_i]$ where $\Delta(f,[x_{i-1},x_i]) > \sigma$, is less than η.

4. Let $f : [a,b] \to \mathbb{R}$ be bounded. Show that f is Riemann integrable over $[a,b]$ if and only if there exists $I \in \mathbb{R}$ with the following property: for every $\varepsilon > 0$ there is a $\delta > 0$ such that for any partition $P = [x_0, x_1, \ldots, x_n]$ of $[a,b]$ with width less than δ and every choice $\xi_1, \ldots, \xi_n$ with $\xi_i \in [x_{i-1}, x_i]$ we have $|\sum_{i=1}^n (x_i - x_{i-1}) f(\xi_i) - I| < \varepsilon$. (This is Riemann's original definition; I is just the integral.)

5. Let $f : [a,b] \to \mathbb{R}$ be bounded and Riemann integrable over $[a,b]$. Let $\varphi : \mathbb{R} \to \mathbb{R}$ be continuous. Show that $\varphi \circ f$ is Riemann integrable over $[a,b]$. Apply this to prove that f^2, $|f|$, $\max(f,0)$ are Riemann integrable. Use the relation $4f_1 f_2 = (f_1 + f_2)^2 + - (f_1 - f_2)^2$ to prove that the product of two bounded Riemann integrable functions is Riemann integrable.

6. Let $g : [a,b] \to \mathbb{R}$ have bounded variation on $[a,b]$. Let $c \in [a,b]$. Show that $\mathrm{var}(f;[a,c]) + \mathrm{var}(f;[c,b]) = \mathrm{var}(f;[a,b])$. (Hint. Argue with partitions P, Q and R as in the proof of 0.5.5(i).)

7. Let $g : [a,b] \to \mathbb{R}$ be non-decreasing. Then $g(x + 0) := \lim_{y \downarrow x} g(y)$ and $g(x - 0) := \lim_{y \uparrow x} g(y)$ exist for $x \in [a,b)$ and $(a,b]$, respectively. For $x \in (a,b)$ we have g discontinuous at x if and only if the *jump* $g(x + 0) - g(x - 0)$ is positive. Show that g has at most countably many discontinuities on $[a,b]$. This result also holds if g has bounded variation.

8. Let $g : [a,b] \to \mathbb{R}$ have bounded variation. Let $x \in [a,b)$ and assume that g is continuous from the right at x. Let $v(y) := \mathrm{var}(g;[a,y])$ for $y \in [a,b]$. Show that v is continuous from the right at x. (Hint. Assume $x = a$ and let $\varepsilon > 0$. There is a partition $[a = x_0, x_1, \ldots, x_n = b]$ of $[a,b]$ such that $|g(x_0) - g(x_1)| < \varepsilon/2$, $v(b) < \sum_{i=1}^n |g(x_i) - g(x_{i-1})| + \varepsilon/2$. Now show that $v(y) < \varepsilon$ if $x_0 \le y \le x_1$.)

9. Let $f : [a,b] \to \mathbb{R}$ be bounded and Riemann integrable over $[a,b]$. Let $g : [a,b] \to \mathbb{R}$ be continuously differentiable. Show that g has bounded variation, that f is integrable with respect to g over $[a,b]$, and that $\int_a^b f(x)dg(x) = \int_a^b f(x)g'(x)dx$. (Hint. Use Exercises 4 and 5.)

10. Let $f : [a,b] \to \mathbb{R}$, $g : [a,b] \to \mathbb{R}$ be continuously differentiable. Show that

$$\int_a^b f(x)dg(x) + \int_a^b g(x)df(x) = f(b)g(b) - f(a)g(a) .$$

(Together with Exercise 9 this gives the familiar formula for *integration by parts* from calculus.)

11. Let $g : [a,b] \to \mathbb{R}$ have bounded variation and let $f : [a,b] \to \mathbb{R}$ be bounded and integrable with respect to g. Show that

$$\left| \int_a^b f(x)dg(x) \right| \leq \sup_{a \leq x \leq b} |f(x)| \cdot \mathrm{var}(g;[a,b]) .$$

Assume in addition that g is continuous. Show that $\forall_{c \in [a,b]}\ {}_a\int^c f(x)dg(x)$ is continuous.

12. Let $f : [a,b] \to \mathbb{R}$ be bounded, let $g : [a,b] \to \mathbb{R}$ be non-decreasing, and assume that g is not continuous from the right at a point c. Show that f is continuous from the right at c if ${}_a\int^b f(x)dg(x)$ exists.

13. Let f be a bounded real-valued function, and g and h non-decreasing functions defined on [a,b]. Assume that $|h(x) - h(y)| \leq |g(x) - g(y)|$ for all x,y in [a,b]. If f is integrable with respect to g, then f is also integrable with respect to h.

14. Let $f : [a,b] \to \mathbb{R}$ be bounded and let $g : [a,b] \to \mathbb{R}$ have bounded variation. Let $v := \forall_{x \in [a,b]}\ \mathrm{var}(g;[a,x])$. Show that ${}_a\int^b f(x)dg(x)$ exists if and only if ${}_a\int^b f(x)dv(x)$ exists. (Hint. Take any decomposition $g = g_1 - g_2$ used in defining ${}_a\int^b f(x)dg(x)$. Let $k := g_1 + g_2$, compare v and k, and use Exercise 13.) Now let $p := \frac{1}{2}(g + v)$, $n := \frac{1}{2}(v - g)$. If f is integrable with respect to g, then it is integrable with respect to p and to n.

15. Prove 0.5.7(iii) and 0.5.9 for a g of bounded variation. (Hint. Use Exercise 14 to get a decomposition of g that works for all functions involved at the same time.)

16. Prove the last statement of 0.5.14.

17. Let R be the space consisting of all continuous real-valued functions on [a,b]. Then R is a Riesz function space, and

$$\|f\| := \int_0^1 |f(x)|dx \qquad (f \in R)$$

defines a norm on R. Show that R is not a Banach space with this norm.

CHAPTER ONE

THE FUNDAMENTAL STATIONS AND THEIR CONNECTIONS

Essential for any integral, besides being positive and linear, is that it should behave nicely under limit operations. What we want is: If $(f_n)_{n\in\mathbb{N}}$ is a sequence of integrable functions that tends to a function f, then f is integrable and its integral is the limit of the integrals of the f_n. No theory of integration attains this end completely, but the theory initiated by Lebesgue around 1900 gets remarkably far. The fundamental theorem in integration theory, the dominated convergence theorem, is due to Lebesgue and named after him. In our set-up this important result is derived right at the start, which is only possible, of course, by choosing the point of departure cleverly.

This point of departure is called station $\mathcal{L}$. It consists of a Riesz function space L of real-valued functions on a set S together with a positive linear functional I on L. The functional I is assumed to have a certain continuity property that is really a special case of Lebesgue's theorem. From station $\mathcal{L}$ the theory of integration can be developed further. For instance, measurability of functions can be introduced, and some functions that are not defined on all of S can be assigned integrals. This is done in the next chapter.

At present our problem is how to get at station $\mathcal{L}$. Always in practice one reaches $\mathcal{L}$ from a situation we call station $\mathcal{N}$. In $\mathcal{N}$ one has a Riesz function space B on the set S with a positive linear functional I on B. In addition one has a norm on Φ, the class of $\mathbb{R}^*$-valued functions defined on S. The norm and the functional I are connected in the following way: if $\|\varphi\|$ denotes the norm of a $\varphi \in \Phi$, then

$$(*) \qquad I(b) = \|b\|$$

for all non-negative members b of B. This relation leads to the fundamental inequality

$$(**) \qquad |I(b)| \leq \|b\| \qquad (b \in B) .$$

The norm induces a sort of topological structure on Φ, and (**) can be interpreted as saying that I is continuous on B with respect to this structure. This enables us to derive a class L of integrable functions on S together with an integral that extends I in such a way that we obtain $\mathcal{L}$. The procedure, denoted by $\mathcal{N} \to \mathcal{L}$, consists in taking the closure of B in Φ, and extending the integral by continuity. A station $\mathcal{L}$ constructed in this way has a slightly richer structure than an abstract one; as a result we obtain a form of Lebesgue's theorem which is easier to use.

Positive linear functionals I on Riesz function spaces B are not hard to get. For instance, on the Riesz function space of continuous real-valued functions on [0,1] the Riemann integral is both positive and linear. If we want to integrate more functions than the members of B, then the problem is how to obtain norms on Φ that are compatible with I in the sense of (*). To solve this problem, another class A of auxiliary functions is introduced, together with a functional J on A. The classes A and B, and the functionals I and J, must be properly connected if we are to expect anything useful; the desired situation is called station $\mathcal{A}$. From $\mathcal{A}$ it is easy to get to $\mathcal{N}$; we indicate the procedure by $\mathcal{A} \to \mathcal{N}$. Since the pair (A,J) determines the norm, and the norm in turn determines the result of the extension process for the integral, the choice of A and J is critical. It is precisely at this point that the various theories of integration differ most.

We describe two paths to $\mathcal{A}$, starting from stations $\mathcal{D}$ and $\mathcal{D}'$, respectively. Both stations consist of a Riesz function space B and a positive linear functional I on B, where I is assumed to have an appropriate continuity property. Station $\mathcal{D}$ is the point of departure needed to develop the theory according to the method of Daniell as it was modified by Stone. Station $\mathcal{D}'$ is a strengthened form of $\mathcal{D}$; it is an abstraction of the approach to integration on locally compact Hausdorff spaces worked out by Bourbaki. The two paths are denoted by $\mathcal{D} \to \mathcal{A}$ and $\mathcal{D}' \to \mathcal{A}$.

It turns out that every situation $\mathcal{L}$ is also a situation $\mathcal{D}$. This enables us to close the circuit, and we obtain the circle line in its simplest form:

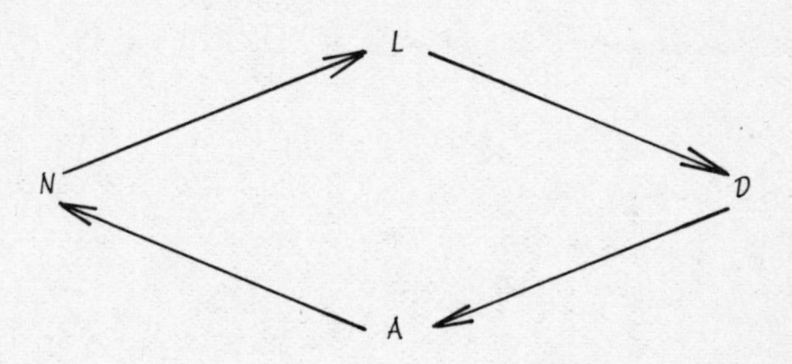

(The name circle line is borrowed from the London underground system.) The picture leads to an obvious question: If we start from $\mathcal{L}$ and make a tour in the circle line, what has happened to the class of integrable functions and the integral when we return to $\mathcal{L}$? It will turn out that after this tour we may be able to integrate more functions, but not many more, and that the new integral is an extension of the original one. A second tour will yield no more gains, however.

The reader may have noticed that we develop the theory in order of decreasing abstractness. It is more common to start from a concrete situation and to construct the integral by some extension procedure; and, in general, different procedures give different theories of integration. Our aim is to combine these procedures as far as possible, to show where they are similar, and to establish their differences.

1.1. Basic functions, norms, and the extension process

This section starts with two equivalent versions of the continuity property that an integral should have. We show that in a system with this property a form of Lebesgue's dominated convergence theorem holds. Next we describe the basic extension process by which integrals are constructed; the starting point is a positive linear functional on a Riesz function space on a set S together with a norm on the $\mathbb{R}^*$-valued functions on S. The additional structure induced by the norm leads to a more useful form of Lebesgue's theorem; here null functions and null sets play an important part. Of course, the result of the extension process depends heavily on the norm used; we deal with this dependence at the end of the section.

1.1.1. Let S be a non-empty set, let L be a Riesz function space of real-valued functions defined on S, and let I be a positive linear functional defined on L (definitions of these concepts are in 0.1.6). We write $\mathcal{L}(L,I)$ (shortened to $\mathcal{L}$ if it is clear which pair (L,I) is meant) if the pair (L,I) satisfies the following condition $\mathcal{L}1$.

$\mathcal{L}1$. If $(f_n)_{n\in\mathbb{N}}$ is a sequence in L^+, if $f = \sum_{n=1}^{\infty} f_n$ is finite-valued, and if $\sum_{n=1}^{\infty} I(f_n) < \infty$, then $f \in L^+$ and $I(f) = \sum_{n=1}^{\infty} I(f_n)$.

We have chosen the word "finite-valued" instead of "real-valued" in the formulation of $\mathcal{L}1$ to emphasize that $\sum_{n=1}^{\infty} f_n$ is required everywhere to converge to a finite real number. In a situation $\mathcal{L}(L,I)$ we shall occasionally call I an *integral*, and the members of L *integrable functions*. The meaning of "integrable" will be extended in Section 2.2.

We write $\mathcal{L}_\sigma(L,I)$ (or $\mathcal{L}_\sigma$) if (L,I) satisfies $\mathcal{L}1$ and the additional condition $\mathcal{L}2$.

$\mathcal{L}2$. There exists a sequence $(f_n)_{n\in\mathbb{N}}$ in L^+ such that $\sum_{n=1}^{\infty} f_n > 0$.

If we are in an $\mathcal{L}_\sigma$-case, we call the system *σ-finite*.

1.1.2. Example. Let S be a non-empty set. Let L consist of the real-valued functions that are summable over S in the sense of 0.4.1, and put

$$I(f) := \sum_{s\in S} f(s) \qquad (f \in L) .$$

Proposition 0.4.4 tells us that L is a Riesz function space, and that I is a positive linear functional on L. We leave it to the reader to show that $\mathcal{L}1$ is satisfied. (Start with a sequence $(f_n)_{n\in\mathbb{N}}$ in L^+ as stipulated in $\mathcal{L}1$. Show first that the sum function is summable over S, proceeding roughly as in the proof of 0.4.5, and then use the result of Exercise 0.4.6.) Obviously, the system is σ-finite if and only if S is countable.

1.1.3. *L*1 is formulated for series with non-negative terms. There is an equivalent formulation in terms of non-decreasing sequences which is occasionally easier to handle. It reads as follows.

*L*1'. If $(f_n)_{n\in\mathbb{N}}$ is a sequence in L^+, if $f_n \leq f_{n+1}$ $(n \in \mathbb{N})$, if $f := \lim_{n\to\infty} f_n$ is finite-valued, and if the sequence $(I(f_n))_{n\in\mathbb{N}}$ is bounded, then $f \in L^+$ and $I(f) = \lim_{n\to\infty} I(f_n)$.

The proof of the equivalence of *L*1 and *L*1' is left to the reader.

1.1.4. In a situation *L*(L,I) we can already prove a version of *Lebesgue's dominated convergence* theorem. Several more general versions of this highly important result will be treated later.

Theorem (Dominated convergence theorem). Assume *L*(L,I). Let $(f_n)_{n\in\mathbb{N}}$ be a sequence in L that converges pointwise to a real-valued function f, and assume that there exists $h \in L^+$ such that $|f_n| \leq h$ $(n \in \mathbb{N})$. Then $f \in L$, and $I(f_n) \to I(f)$ as $n \to \infty$.

Proof. First consider the case that the sequence $(f_n)_{n\in\mathbb{N}}$ is pointwise increasing, that is, $f_m \leq f_{m+1}$ for all $m \in \mathbb{N}$. For each $m \in \mathbb{N}$, put $g_m := f_m - f_1$, then $g_m \in L^+$. Obviously, $(g_m)_{m\in\mathbb{N}}$ increases to the finite-valued function $f - f_1$, and the sequence $(I(g_m))_{m\in\mathbb{N}}$ is bounded because

$$|I(g_m)| = |I(f_m - f_1)| \leq |I(f_m)| + |I(f_1)| \leq 2I(h) \qquad (m \in \mathbb{N}).$$

Hence, by *L*1', $f - f_1 \in L^+$ and $I(g_m) \to I(f - f_1) = I(f) - I(f_1)$. Therefore, $f \in L$ and $I(f_m) \to I(f)$.

Next suppose that the sequence $(f_n)_{n\in\mathbb{N}}$ is pointwise decreasing. Then $(-f_n)_{n\in\mathbb{N}}$ is pointwise increasing. Hence, by the preceding special case, $-f \in L$ and $-I(f_n) = I(-f_n) \to I(-f) = -I(f)$.

To treat the general case, for each pair $(m,n) \in \mathbb{N} \times \mathbb{N}$ with $n \geq m$ put $g_{mn} := \sup(f_m, f_{m+1}, \ldots, f_n)$. First keep m fixed. Clearly $g_{mn} \leq g_{m,n+1}$, $g_{mn} \in L$, and $|g_{mn}| \leq h$ for every $n \geq m$, while $g_m := \lim_{n\to\infty} g_{mn}$ is a finite-valued function on S. The first case already treated shows that $g_m \in L$. Now the sequence $(g_m)_{m\in\mathbb{N}}$ decreases to f, while $|g_m| \leq h$ $(m \in \mathbb{N})$. Hence, by the second case, $f \in L$ and $I(f) = \lim_{m\to\infty} I(g_m)$. Let $\varepsilon > 0$. Since $g_m \geq f_m$, we have $I(g_m) \geq I(f_m)$ for all m, and since $I(g_m) \to I(f)$ as $m \to \infty$, it follows that $I(f_m) \leq I(f) + \varepsilon$ for all large m. Now apply this result to the sequence $(-f_m)_{m\in\mathbb{N}}$ to get $-I(f_m) = I(-f_m) \leq I(-f) + \varepsilon = -I(f) + \varepsilon$ for all large m. The conclusion is that $I(f) - \varepsilon \leq I(f_m) \leq I(f) + \varepsilon$ for all large m, that is, $I(f_m) \to I(f)$. □

The function h of the theorem is sometimes called a *majorant* of the sequence $(f_n)_{n\in\mathbb{N}}$, and the elements of the sequence are said to be *dominated* by h (hence the name of the theorem).

1.1.5. In almost all interesting cases L arises from a situation N, which we describe presently. Let S be a non-empty set, let B be a Riesz function space consisting of real-valued functions defined on S, and let I be a positive linear functional defined on B. Let Φ denote the class of $\mathbb{R}^*$-valued functions defined on S. Assume that a norm has been introduced on Φ that extends I, that is, there is a mapping $\|\cdot\|$ from Φ into $\mathbb{R}^*_+$ such that the following conditions are satisfied.

$N1$. If $\varphi \in \Phi$, $\alpha \in \mathbb{R}$, then $\|\alpha\varphi\| = |\alpha|\cdot\|\varphi\|$.

$N2$. If $\varphi \in \Phi$, $\psi \in \Phi$, $|\varphi| \le |\psi|$, then $\|\varphi\| \le \|\psi\|$.

$N3$. If $(\varphi_n)_{n\in\mathbb{N}}$ is a sequence in Φ, then $\|\sum_{n=1}^{\infty} |\varphi_n|\| \le \sum_{n=1}^{\infty} \|\varphi_n\|$.

$N4$. If $b \in B^+$, then $\|b\| = I(b)$.

We write $N(B,I,\|\cdot\|)$ (abbreviated N) if the triple $(B,I,\|\cdot\|)$ satisfies the conditions $N1$ through $N4$. The elements of B will be called *basic functions*. If the pair (B,I) satisfies the condition $L2$ of 1.1.1, then we write $N_\sigma(B,I,\|\cdot\|)$, or N_σ for short, and we call the system σ-*finite*.

For later use, note that $N2$ immediately implies $\|\varphi\| = \||\varphi|\|$ for $\varphi \in \Phi$. Moreover, if $b \in B$, then $|I(b)| \le \|b\|$. To prove this, start from the obvious inequality $b \le \sup(b,0)$ to get $I(b) \le I(\sup(b,0)) = \|\sup(b,0)\| \le \|b\|$, by the positivity of I, $N4$, and $N2$. Now apply this result to $-b$ to get $-I(b) = I(-b) \le \|-b\| = \|b\|$.

Though we call $\|\cdot\|$ a norm, it should be noted that it is not a norm in the sense of 0.3.1. For while $\|0\| = 0$ (take $\alpha = 0$ in $N1$), it may well occur that $\|\varphi\| = 0$ without $\varphi = 0$. Moreover, norms usually take finite values only, whereas our norms may be infinite.

1.1.6. Example. Let S be a non-empty set and let α be an $\mathbb{R}^+$-valued function on S. Let B_α be the class of real-valued functions b defined on S such that $\alpha\cdot b$ is summable over S. Take

$$I_\alpha(b) := \sum_{s\in S} \alpha(s)b(s) \qquad (b \in B_\alpha) .$$

Then B_α is a Riesz function space, and I_α is a positive linear functional on B_α; this follows easily from Proposition 0.4.4. Let Φ have the usual meaning. Put

$$\|\varphi\|_\alpha := \inf\{I_\alpha(b) \mid b \in B^+, |\varphi| \le b\} \qquad (\varphi \in \Phi) ,$$

where $\inf \emptyset = \infty$, as usual. We leave it as an exercise to show that $(B_\alpha, I_\alpha, \|\cdot\|_\alpha)$ satisfies the conditions N. The system is σ-finite if and only if $\{s \in S \mid \alpha(s) > 0\}$ is countable.

1.1.7. Starting from a station $N(B,I,\|\cdot\|)$ as described in 1.1.5, we propose to extend I to a large subclass of Φ in such a way that the extension is linear and positive, and also satisfies $L1$. However, Φ is not a linear space if the algebraic operations are defined pointwise (what to do with expressions like $\infty - \infty$?), and it is not clear what a linear functional on Φ might be. Therefore we introduce the set $\Phi_F := \{f \in \Phi \mid f \text{ finite-valued and } \|f\| < \infty\}$. If addition and scalar multiplication are defined pointwise (as described in 0.1.5), then Φ_F is a Riesz function space (check this). The requirement that the elements of Φ_F have finite norms could be left out, but this would not give greater generality. The restriction to finite-valued functions will be partially removed later (Section 2.2); this will make practical manipulation of integrals much easier.

1.1.8. <u>Definition</u>. Assume $N(B,I,\|\cdot\|)$. A function $\varphi \in \Phi_F$ is called *integrable* if for every $\varepsilon > 0$ there exists $b \in B$ with $\|b - \varphi\| < \varepsilon$. The class of integrable functions in Φ_F is denoted by L.

Thus the integrable functions are those members of Φ_F that can be approximated as closely as we please by basic functions, where the degree of the approximation is measured by the norm of the difference. It will turn out that L is a Riesz function space, and that I can be extended to a positive functional I on L such that $L(L,I)$ holds. So there is no danger of confusion with the usage of "integrable" introduced in 1.1.1.

1.1.9. <u>Proposition</u>. Assume $N(B,I,\|\cdot\|)$. Then L is a Riesz function space which contains B.

<u>Proof</u>. First we show that L is a linear space. Let $f_1, f_2 \in L$, $\alpha_1, \alpha_2 \in \mathbb{R}$ and $\varepsilon > 0$. Obviously $\alpha_1 f_1 + \alpha_2 f_2 \in \Phi_F$. And if $b_1 \in B$, $b_2 \in B$ satisfy

$$\|b_i - f_i\| < \varepsilon(1 + 2|\alpha_i|)^{-1} \qquad (i = 1,2) ,$$

then

$$\|\alpha_1 b_1 + \alpha_2 b_2 - (\alpha_1 f_1 + \alpha_2 f_2)\| \le |\alpha_1|\cdot\|b_1 - f_1\| + |\alpha_2|\cdot\|b_2 - f_2\| < \varepsilon .$$

Since $\alpha_1 b_1 + \alpha_2 b_2 \in B$, this shows that $\alpha_1 f_1 + \alpha_2 f_2 \in L$. Hence L is a linear space.

If $f \in L$ and $\varepsilon > 0$, then there exists $b \in B$ such that $\|b - f\| < \varepsilon$. The inequality $||f| - |b|| \le |f - b|$ implies $\||f| - |b|\| \le \|f - b\| < \varepsilon$. Since $|b| \in B$, and $|f| \in \Phi_F$ is obvious, it follows that $|f| \in L$. Together with the result of the preceding paragraph this shows that L is a Riesz function space (see the second paragraph of 0.1.6).

The inclusion $B \subset L$ is trivial. □

1.1.10. <u>Theorem</u>. Assume $N(B,I,\|\cdot\|)$. There exists exactly one linear functional $\bar{I}$ on L that satisfies (i) and (ii), where

(i) $\bar{I}(b) = I(b) \quad (b \in B)$, that is, $\bar{I}$ extends I,

(ii) $|\bar{I}(f)| \leq \|f\| \quad (f \in L)$.

This $\bar{I}$ is positive, and $\bar{I}(f) = \|f\|$ whenever $f \in L^+$. Moreover, the pair $(L,\bar{I})$ satisfies $L1$.

<u>Proof</u>. To get started, assume that $\bar{I}$ is a linear functional on L that satisfies the conditions (i) and (ii). Let $f \in L$, and let $(b_n)_{n \in \mathbb{N}}$ be any sequence in B such that $\|b_n - f\| \to 0$. Since

$$|I(b_n) - \bar{I}(f)| = |\bar{I}(b_n) - \bar{I}(f)| = |\bar{I}(b_n - f)| \leq \|b_n - f\| ,$$

it follows that $I(b_n) \to \bar{I}(f)$. In the first place this implies that there is at most one possible extension. In the second place it follows that if the extension exists, it will be positive. For if $f \in L^+$, and $(b_n)_{n \in \mathbb{N}}$ is a sequence in B with $\|b_n - f\| \to 0$, then also $\||b_n| - f\| \to 0$ because $||b_n| - f| \leq |b_n - f|$. Therefore $\bar{I}(f) = \lim_{n\to\infty} I(|b_n|)$, and since all $|b_n| \in B^+$, it follows that $\bar{I}(f) \geq 0$. Finally, most important of all, we get a clue how to obtain the extension.

Now we turn to the construction of $\bar{I}$. Let $f \in L$, take a sequence $(b_n)_{n \in \mathbb{N}}$ in B with $\|b_n - f\| \to 0$, and put

$$\bar{I}(f) := \lim_{n\to\infty} I(b_n) .$$

We shall show that this limit exists, and that its value does not depend on the particular sequence $(b_n)_{n \in \mathbb{N}}$ chosen. Let $(c_n)_{n \in \mathbb{N}}$ be any sequence in B with $\|c_n - f\| \to 0$. Then

$$(*) \qquad |I(b_n) - I(c_m)| = |I(b_n - c_m)| \leq \|b_n - c_m\| \leq \|b_n - f\| + \|c_m - f\| \to 0$$

as $n \to \infty$, $m \to \infty$. Conclusion: if $\lim_{n\to\infty} I(b_n)$ exists, then so does $\lim_{m\to\infty} I(c_m)$ and the limits are equal. Now take $c_m = b_m$ in (*). Then (*) shows that $(I(b_n))_{n \in \mathbb{N}}$ is a fundamental sequence in $\mathbb{R}$. Therefore it is convergent. So the value of $\bar{I}(f)$ is unambiguously defined. (The technical term is: $\bar{I}$ is well-defined on L.)

To show that $\bar{I}$ is linear, let $f_1, f_2 \in L$ and $\alpha_1, \alpha_2 \in \mathbb{R}$. Choose sequences $(b_{1n})_{n \in \mathbb{N}}$ and $(b_{2n})_{n \in \mathbb{N}}$ in B such that $\|b_{1n} - f_1\| \to 0$ and $\|b_{2n} - f_2\| \to 0$. Then

$$\|\alpha_1 b_{1n} + \alpha_2 b_{2n} - (\alpha_1 f_1 + \alpha_2 f_2)\| \leq |\alpha_1| \cdot \|b_{1n} - f_1\| + |\alpha_2| \cdot \|b_{2n} - f_2\| \to 0 ,$$

and therefore

$$\bar{I}(\alpha_1 f_1 + \alpha_2 f_2) = \lim_{n\to\infty} I(\alpha_1 b_{1n} + \alpha_2 b_{2n})$$

$$= \alpha_1 \lim_{n\to\infty} I(b_{1n}) + \alpha_2 \lim_{n\to\infty} I(b_{2n}) = \alpha_1 \bar{I}(f_1) + \alpha_2 \bar{I}(f_2) .$$

That $\bar{I}$ satisfies (i) is trivial, and that it satisfies (ii) follows easily from the definition. In fact, if $f \in L$ and if $(b_n)_{n\in\mathbb{N}}$ is a sequence in B with $\|b_n - f\| \to 0$, then the inequality $|\|b_n\| - \|f\|| \le \|b_n - f\|$ shows that $\|b_n\| \to \|f\|$. We have already noted in 1.1.5 that $|I(b_n)| \le \|b_n\|$, so $|\bar{I}(f)| = \lim_{n\to\infty} |I(b_n)| \le \lim_{n\to\infty} \|b_n\| = \|f\|$. If, in addition, $f \in L^+$, then by the reasoning in the first paragraph of the proof

$$\bar{I}(f) = \lim_{n\to\infty} \bar{I}(|b_n|) = \lim_{n\to\infty} I(|b_n|) = \lim_{n\to\infty} \||b_n|\| = \lim_{n\to\infty} \|b_n\| = \|f\| ,$$

where we also used *N4*.

Finally we show that *L1* holds for the pair $(L,\bar{I})$. Let $(f_n)_{n\in\mathbb{N}}$ be a sequence in L^+ such that $f := \sum_{n=1}^{\infty} f_n$ is everywhere finite, while $\sum_{n=1}^{\infty} \bar{I}(f_n)$ is convergent. Let $\varepsilon > 0$, and choose for each $n \in \mathbb{N}$ a $b_n \in B$ with $\|b_n - f_n\| \le \varepsilon\cdot 2^{-n-1}$. Use *N3* and the equality $\bar{I}(f_n) = \|f_n\|$ to see that

$$\|f - \sum_{n=1}^{N} b_n\| \le \|f - \sum_{n=1}^{N} f_n\| + \|\sum_{n=1}^{N} (b_n - f_n)\|$$

$$\le \sum_{n=N+1}^{\infty} \|f_n\| + \sum_{n=1}^{N} \|b_n - f_n\|$$

$$\le \sum_{n=N+1}^{\infty} \bar{I}(f_n) + \frac{\varepsilon}{2} < \varepsilon$$

if N is sufficiently large. So $f \in L^+$. Moreover, since $|\bar{I}(g)| \le \|g\|$ for all $g \in L$, it follows that

$$|\bar{I}(f) - \sum_{n=1}^{N} \bar{I}(f_n)| = |\bar{I}(f - \sum_{n=1}^{N} f_n)|$$

$$\le \|f - \sum_{n=1}^{N} f_n\| = \|\sum_{n=N+1}^{\infty} f_n\| \le \sum_{n=N+1}^{\infty} \|f_n\| \to 0$$

as $N \to \infty$, that is, $\bar{I}(f) = \sum_{n=1}^{\infty} \bar{I}(f_n)$. □

Now that the proof is finished, there is no longer any reason to distinguish between I and $\bar{I}$: we drop the bar and simply write I for the extension, which is called an *integral* on L.

1.1.11. The process described in 1.1.8, 1.1.9 and 1.1.10, leading from N to L, is a standard transition, which will often occur in the sequel. It is denoted by $N(B,I,\|\cdot\|) \to L(L,I)$, which is shortened to $N \to L$ if there is no danger of confusion. If one starts with a situation N_σ, then the process leads to a σ-finite system (L,I), so there is also a transition $N_\sigma \to L_\sigma$.

1.1.12. Example. Consider Example 1.1.6, and follow the line $N \to L$. Then $L = B_\alpha$, and $I(f) = \sum_{s\in S} \alpha(s)f(s)$ for $f \in L$, so we do not obtain an extension in this case.

However, things are different if one starts with B', where B' is the set of $b \in B$ such that $b(s) \neq 0$ for only finitely many points of S. Obviously, the Riesz function space B' is contained in B_α, and if we put $I'(b) := I_\alpha(b)$ for $b \in B'$, then I' is a positive linear functional on B'. Follow the line $N(B',I',\|\cdot\|_\alpha) \to L(L',I')$. It is an exercise to show that $L' = B_\alpha$ and that $I' = I_\alpha$.

1.1.13. A station L derived by the process $N \to L$ has additional structure. This fact leads to a first extension of Lebesgue's theorem. For the formulation of the extension some new concepts are needed.

Definition. Assume $N(B,I,\|\cdot\|)$. A function $\varphi \in \Phi$ is called a *null function* if $\|\varphi\| = 0$. A subset E of S is called a *null set* if its characteristic function χ_E is a null function. If P is a proposition involving points of S, then P is said to hold *almost everywhere* in S (abbreviated a.e.), if there exists a null set $E \subset S$ such that $P(s)$ holds for all $s \in S\backslash E$.

Later we shall have occasion to consider various kinds of null sets at the same time, and then we shall also have to add a qualification to the "almost everywhere" clause.

1.1.14. Example. Consider the situation of Example 1.1.6. Here $\varphi \in \Phi$ is a null function if and only if $\varphi(s) = 0$ whenever $\alpha(s) \neq 0$. And $E \subset S$ is a null set if and only if $\alpha(s) = 0$ whenever $s \in E$.

1.1.15. The contents of the following simple proposition is used frequently.

Proposition. Assume $N(B,I,\|\cdot\|)$.

(i) If $\varphi \in \Phi$ and $\|\varphi\| < \infty$, then φ is finite (a.e.).

(ii) If $\varphi \in \Phi$, then $\|\varphi\| = 0$ if and only if $\varphi = 0$ (a.e.).

(iii) If $\varphi \in \Phi$, $\psi \in \Phi$, $|\psi| \leq |\varphi|$ (a.e.), then $\|\psi\| \leq \|\varphi\|$.

(iv) If $(\varphi_n)_{n\in\mathbb{N}}$ is a sequence of non-negative null functions, then $\sum_{n=1}^{\infty} \varphi_n$ is a null function.

(v) A union of countably many null sets is a null set.

(vi) Any subset of a null set is a null set.

(vii) If φ is a finite-valued null function, then φ is integrable and $I(\varphi) = 0$.

Proof. We prove only (i) and (ii); the proofs of the remaining assertions are left to the reader.

(i) Let $N := \{s \in S \mid |\varphi(s)| = \infty\}$. Then $0 \le n\chi_N \le |\varphi|$, and therefore $n\|\chi_N\| \le \|\varphi\|$ for all $n \in \mathbb{N}$, by $N2$ and $N3$. Hence $\|\chi_N\| = 0$.

(ii) Suppose $\|\varphi\| = 0$. Let $N := \{s \in S \mid \varphi(s) \ne 0\}$. Since $|\chi_N| \le \sum_{n=1}^{\infty} |\varphi|$ (the sum is just $|\varphi| + |\varphi| + \ldots$), it follows that $\|\chi_N\| \le \sum_{n=1}^{\infty} \|\varphi\| = 0$, that is, $\|\chi_N\| = 0$. Conversely, if $\varphi = 0$ (a.e.), then $N := \{s \in S \mid \varphi(s) \ne 0\}$ is a null set, that is $\|\chi_N\| = 0$. Now $|\varphi| \le \sum_{n=1}^{\infty} \chi_N$, and therefore $\|\varphi\| \le \sum_{n=1}^{\infty} \|\chi_N\| = 0$. Hence $\|\varphi\| = 0$. □

We point out an important consequence of (vii): if one modifies the values of an integrable function on a null set, then the modified function is still integrable, and it has the same integral as the original function. This fact will prove useful many times in the sequel.

1.1.16. We arrive at our first generalization of Lebesgue's theorem 1.1.4: instead of integrable majorants we allow majorants in Φ^+ with finite norm (Φ^+ is the class of those $\varphi \in \Phi$ with $\varphi(s) \ge 0$ for all $s \in S$), and instead of convergence at all points of S we require only convergence almost everywhere.

Theorem (Dominated convergence theorem). Assume $N(B, I, \|\cdot\|)$ and $L(L, I)$ derived from it. Let $(f_n)_{n \in \mathbb{N}}$ be a sequence in L, let $h \in \Phi^+$ with $\|h\| < \infty$ and assume $|f_n| \le h$ (a.e.) for $n \in \mathbb{N}$. Let f be a finite-valued function on S such that $f = \lim_{n\to\infty} f_n$ (a.e.). Then $f \in L$ and $I(f) = \lim_{n\to\infty} I(f_n)$.

Proof. There are several null sets that play a role in this proof. First there is one, M_1 say, such that $h(s) < \infty$ for $s \in S \backslash M_1$. Then there is a null set M_2 such that $f(s) = \lim_{n\to\infty} f_n(s)$ for $s \in S\backslash M_2$. And finally, for each $n \in \mathbb{N}$ there is a null set N_n such that $|f_n(s)| \le h(s)$ for $s \in S\backslash N_n$. The union N of all these null sets is still a null set (by 1.1.15(v)), and outside N we have h finite, the f_n's dominated by h, and pointwise convergence of the sequence $(f_n)_{n\in\mathbb{N}}$ to f. For $n \in \mathbb{N}$, let

$$g_n(s) := \begin{cases} f_n(s) & \text{if } s \notin N\ , \\ 0 & \text{if } s \in N\ . \end{cases}$$

Then $g_n \in L$ and $I(g_n) = I(f_n)$ by the remark made at the end of 1.1.15. Also, $(g_n)_{n\in\mathbb{N}}$ converges everywhere on S to a finite-valued function g, and $g = f$ (a.e.).

For each $n \in \mathbb{N}$, put $k_n := \sup\{|g_1|,\dots,|g_n|\}$. The sequence $(k_n)_{n\in\mathbb{N}}$ is contained in L^+, it increases to a finite-valued function k, while $I(k_n) = \|k_n\| \le \|k\| \le \|h\| < \infty$ for each $n \in \mathbb{N}$. So *L1'* implies that $k \in L^+$. Since $|g_n| \le k$ for all n, Lebesgue's theorem 1.1.4 shows that $g \in L$, and $I(g) = \lim_{n\to\infty} I(g_n)$. Since $f = g$ (a.e.) and $f_n = g_n$ (a.e.), the remark at the end of 1.1.15 implies that $f \in L$ and $I(f) = \lim_{n\to\infty} I(f_n)$. □

1.1.17. Corollary (Beppo Levi theorem). Assume $N(B,I,\|\cdot\|)$ and $L(L,I)$ derived from it. Let $(f_n)_{n\in\mathbb{N}}$ be a sequence in L^+ such that $\sum_{n=1}^{\infty} I(f_n) < \infty$. Then $\sum_{n=1}^{\infty} f_n$ is finite almost everywhere, there exists $f \in L^+$ such that $f = \sum_{n=1}^{\infty} f_n$ (a.e.), and $I(f) = \sum_{n=1}^{\infty} I(f_n)$.

The proof of the corollary is left to the reader as an exercise. The wording of the corollary reminds one of *L1*, and indeed there is an equivalent form in terms of sequences which is comparable to *L1'*. In this form the result is known as the monotone convergence theorem.

1.1.18. Corollary (Monotone convergence theorem). Assume $N(B,I,\|\cdot\|)$ and $L(L,I)$ derived from it. Let $(f_n)_{n\in\mathbb{N}}$ be a sequence in L^+ with $f_n \le f_{n+1}$ (a.e.) for each $n \in \mathbb{N}$, and let $\lim_{n\to\infty} I(f_n)$ be finite. Then $\lim_{n\to\infty} f_n$ is finite almost everywhere, there exists $f \in L^+$ with $f = \lim_{n\to\infty} f_n$ (a.e.), and $I(f) = \lim_{n\to\infty} I(f_n)$.

It is not difficult to see that Lebesgue's theorem, the Beppo Levi theorem, and the monotone convergence theorem are equivalent. However, each of these various forms comes in handy sometimes.

1.1.19. The transition $N \to L$ described in Subsection 1.1.11 works with a space B, a functional I, and a norm $\|\cdot\|$, all properly related. In many cases B and I appear naturally at the outset (for instance, B can be the class of continuous real-valued functions on [0,1] and I the Riemann integral), and to extend the integral with our theory only a norm is needed. Occasionally, there are several ways to define a norm, which may lead to different extensions of the integral. Obviously, it is then of interest to compare these extensions. For instance, we might want to know which extension is defined on the largest class of functions. In the remainder of this section we look into this question for the case of two comparable norms.

First we derive a result concerning approximation of integrable functions by means of basic functions.

Theorem. Assume $N(B,I,\|\cdot\|)$ and $L(L,I)$ derived from it. Let $f \in L$.

(i) There exists a sequence $(b_n)_{n\in\mathbb{N}}$ in B such that $\|f - b_n\| \to 0$ and $b_n \to f$ (a.e.).

(ii) There exists a sequence (b'_n) in B such that $\|f - \sum_{n=1}^{N} b'_n\| \to 0$ and $f = \sum_{n=1}^{\infty} b'_n$ (a.e.).

Proof. Only (i) needs proof, since (ii) immediately follows from (i) by putting $b'_1 := b_1$, $b'_{n+1} := b_{n+1} - b_n$ $(n \geq 1)$. There exists for each $n \in \mathbb{N}$ a $b_n \in B$ such that $\|b_n - f\| < 2^{-n}$, by the definition of integrability. Since $\|\sum_{n=1}^{\infty} |b_n - f|\| \leq$ $\leq \sum_{n=1}^{\infty} \|b_n - f\| < \infty$, Proposition 1.1.15(i) shows that $\sum_{n=1}^{\infty} |b_n - f|$ is a.e. finite. Hence $b_n \to f$ (a.e.). □

1.1.20. Now let B be a Riesz function space on S, let I be a positive linear functional defined on B. Assume that there are two norms $\|\cdot\|_1$ and $\|\cdot\|_2$ on Φ, and that $N(B,I,\|\cdot\|_1)$ and $N(B,I,\|\cdot\|_2)$ hold. In each case make the transition $N \to L$. This gives the systems $L(L_1,I_1)$ and $L(L_2,I_2)$, respectively, and these are to be compared.

Theorem. Under the conditions described, assume that

$$(*) \qquad \|\varphi_1\| \leq \|\varphi_2\| \qquad (\varphi \in \Phi) .$$

Then

(i) $L_2 \subset L_1$, and I_1 extends I_2, that is, $I_1(f) = I_2(f)$ for all $f \in L_2$;

(ii) if $f \in L_1$, then there exists $g \in L_2$ such that $\|f - g\|_1 = 0$.

Proof. (i) Let $f \in L_2$. Choose a sequence $(b_n)_{n\in\mathbb{N}}$ in B such that $\|b_n - f\|_2 \to 0$. By (*), $\|b_n - f\|_1 \to 0$, which implies $f \in L_1$. Moreover, $I_1(f) = \lim_{n\to\infty} I(b_n) = I_2(f)$ (compare the construction of the extended functional in the proof of 1.1.10).

(ii) Let $f \in L_1$. Let $(b_n)_{n\in\mathbb{N}}$ be a sequence in B such that $\|b_n - f\|_1 < 2^{-n}$. Let E_1 be the set of those $s \in S$ where $(b_n(s))_{n\in\mathbb{N}}$ does not converge to $f(s)$. We have seen in the proof of Theorem 1.1.19 that E_1 is a $\|\cdot\|_1$-null set. Moreover, the inequalities

$$I(|b_{n+1} - b_n)| = \|b_{n+1} - b_n\| \leq \|b_{n+1} - f\| + \|b_n - f\| < 2^{-n+1} \qquad (n \in \mathbb{N})$$

imply $\sum_{n=1}^{\infty} I(|b_{n+1} - b_n|) < \infty$. Hence, by the Beppo Levi theorem 1.1.17, there exists $h \in L_2^+$ such that $h = \sum_{n=1}^{\infty} |b_{n+1} - b_n|$ a.e. with respect to $\|\cdot\|_2$. Now apply Lebesgue's theorem 1.1.16 to the series $\sum_{n=1}^{\infty} (b_{n+1} - b_n)$ with I_2 to conclude that there exists $g \in L_2$ such that $b_1 + \sum_{n=1}^{\infty} (b_{n+1} - b_n) = g$ a.e. with respect to $\|\cdot\|_2$. Let E_2 be the $\|\cdot\|_2$-null set outside of which $b_n(s) \to g(s)$. Since E_2 is also a $\|\cdot\|_1$-null set, it follows that $g(s) = f(s)$ outside the $\|\cdot\|_1$-null set $E_1 \cup E_2$, that is, $\|g - f\|_1 = 0$. □

The theorem shows that using a smaller norm in the extension process gives more integrable functions. However, all one gains are just null functions with respect to the smaller norm.

Exercises Section 1.1

1. Let the pair (L,I) satisfy condition $\mathcal{L}1$ of 1.1.1. Let F be a real-valued function defined on $S \times \mathbb{R}$ which is continuous in the second variable, that is $\lim_{t\to t_0} F(s,t) = F(s,t_0)$ for $(s,t_0) \in S \times \mathbb{R}$. Assume that

(i) $F_t := \Upsilon_{s\in S} F(s,t) \in L$ for every $t \in \mathbb{R}$,

(ii) there exists $h \in L^+$ such that $|F_t| \le h$ for every $t \in \mathbb{R}$.

Prove that $g := \Upsilon_{t\in\mathbb{R}} I(F_t)$ is continuous. (Hint. The function g is continuous at t_0 if and only if for every sequence $(t_n)_{n\in\mathbb{N}}$ with $t_n \to t_0$ we have $I(F_{t_n}) \to I(F_{t_0})$.)

2. Let the pair (L,I) satisfy condition $\mathcal{L}1$ of 1.1.1. Let F be a real-valued function defined on $S \times \mathbb{R}$ which is differentiable with respect to the second variable (i.e. $\lim_{h\to 0} h^{-1}(F(s,t+h) - F(s,t))$ exists for every $(s,t) \in S \times \mathbb{R}$). Let $G_t := \Upsilon_{s\in S} \lim_{h\to 0} h^{-1}(F(s,t+h) - F(s,t))$ for $t \in \mathbb{R}$. Assume that

(i) $F_t = \Upsilon_{s\in S} F(s,t) \in L$ for every $t \in \mathbb{R}$,

(ii) there exists an $h \in L^+$ such that $|G_t| \le h$ for every $t \in \mathbb{R}$.

Prove that $G_t \in L$ for every $t \in \mathbb{R}$, and that $\Upsilon_{t\in\mathbb{R}} I(F_t)$ is differentiable with derivative $\Upsilon_{t\in\mathbb{R}} I(G_t)$. (Hint. Use the mean value theorem to estimate $h^{-1}(F(s,t+h) - F(s,t))$.)

3. Let the pair (L,I) satisfy condition $\mathcal{L}1$ of 1.1.1.

(i) Let $(f_n)_{n\in\mathbb{N}}$ be a sequence in L^+ with $\sum_{n=1}^{\infty} I(f_n) < \infty$.

(a) Let $g = \sum_{n=1}^{\infty} f_n$, and let A be the subset of S where g is infinite. Show that $h\cdot\chi_A \in L^+$ and that $I(h\cdot\chi_A) = 0$ if $h \in L^+$.
(Hint. We have $h\cdot\chi_A = \lim_{n\to\infty} \lim_{m\to\infty} \inf(h, \frac{1}{n} \sum_{k=1}^{m} f_k)$; use Lebesgue's theorem twice.)

(b) Let $f = g\cdot\chi_{S\setminus A}$, where $0\cdot\infty = 0$ as usual. Show that $f \in L^+$ and that $I(f) = \sum_{n=1}^{\infty} I(f_n)$.

(ii) Let $(f_n)_{n\in\mathbb{N}}$ be a sequence in L with $\sum_{n=1}^{\infty} I(|f_n|) < \infty$. Let A be the set where $\sum_{n=1}^{\infty} |f_n|$ is infinite. Show that the function f defined by

$$f(s) := \begin{cases} \sum_{n=1}^{\infty} f_n(s) & (s \notin A), \\ 0 & (s \in A), \end{cases}$$

belongs to L and that $I(f) = \sum_{n=1}^{\infty} I(f_n)$.

4. Let the pair (L,I) satisfy condition $L1$ of 1.1.1. Let $f \in L^+$ and let $A := \{s \in S \mid f(s) \neq 0\}$. If $g \in L^+$, then $g \cdot \chi_A \in L^+$.

5. Let the pair (L,I) satisfy the conditions $L1$ and $L2$ of 1.1.1. Show that there is an $f \in L^+$ with $f(s) > 0$ for $s \in S$. (Hint. Use Exercise 4 to show that we can find a sequence $(f_n)_{n \in \mathbb{N}}$ as in $L2$ with $f_n \cdot f_m = 0$ if $n \neq m$.)

6. Let L be the set of Riemann integrable functions defined on $[0,1]$ and let I be the Riemann integral. Show that (L,I) does not satisfy condition $L1$. (Hint. Consider χ_V where $V = \mathbb{Q} \cap [0,1]$.)

7. Let $S = [0,1]$, and let B be the set of all functions of the form $\sum_{k=1}^{n} \alpha_k \chi_{A_k}$ where $n \in \mathbb{N}$, $\alpha_k \in \mathbb{R}$, $A_1 = [a_0, a_1]$, $A_k = (a_{k-1}, a_k]$ $(k = 2,\dots,n)$ and where $[a_0, a_1, \dots, a_n]$ is a partition of $[0,1]$; for such a b put $I(b) := \sum_{k=1}^{n} \alpha_k (a_k - a_{k-1})$. Also, let $\|\varphi\| := \inf\{I(b) \mid |\varphi| \leq b,\ b \in B^+\}$ for $\varphi \in \Phi$, where, by convention, $\inf \emptyset = \infty$. Show that $f \in \Phi$ is Riemann integrable if and only if for every $\varepsilon > 0$ there is a $b \in B$ such that $\|f - b\| < \varepsilon$, and that the Riemann integral of such an f can be found by mimicing the extension procedure of Theorem 1.1.10. Which of the properties $N1$ - $N4$ of 1.1.5 is not satisfied by $\|\cdot\|$?

8. Let the triple $(B, I, \|\cdot\|)$ satisfy conditions $N1$ - $N4$ of 1.1.5. If $(b_n)_{n \in \mathbb{N}}$ is a sequence in B^+ then $\|\sum_{n=1}^{\infty} b_n\| = \sum_{n=1}^{\infty} \|b_n\|$. Is it true that $\|\varphi_1 + \varphi_2\| = \|\varphi_1\| + \|\varphi_2\|$ for all $\varphi_1, \varphi_2 \in \Phi^+$?

9. Elaborate Examples 1.1.2, 1.1.6, 1.1.12, and 1.1.14.

10. Prove Corollaries 1.1.17 and 1.1.18.

1.2. Auxiliary functions

We shall now describe the construction of norms, starting with objects that are available quite often. These objects are a class of auxiliary functions and a certain kind of functional defined on it. Of the examples illustrating the theory, two treat the Lebesgue integral on the real line. These examples should be studied carefully, not only because of the importance of the Lebesgue integral in analysis, but also because they give a preview of the main lines of two of the following chapters. In fact, the construction in Example 1.2.4, which takes as its starting point the ordinary Riemann integral on $\mathbb{R}$, will reappear in Chapter 4 in the more general context of integrals on locally compact Hausdorff spaces. And the length of an interval, which is fundamental in Example 1.2.5, will turn out to be an example of the measures that we shall study in Chapter 3.

In all cases of interest the norm can be described in terms of integrable functions. At the end of the section we give a condition on the auxiliary functions for this to be the case, and explore some consequences of this further relation between norms and integrals. This part of the section does not belong to the main line of the development, however.

1.2.1. Let S be a non-empty set, let B be a Riesz function space of real-valued functions defined on S, and let I be a positive linear functional defined on B. The elements of B will be called *basic functions*. Let A be a non-empty class of functions defined on S with values in $\mathbb{R}_+^*$ (the non-negative extended real numbers), and let J be an $\mathbb{R}_+^*$-valued functional defined on A. We write $\mathcal{A}(B,I,A,J)$, or briefly $\mathcal{A}$, to indicate that the quadruple (B,I,A,J) satisfies the following three conditions.

$\mathcal{A}$1. If $h \in A$, $0 \le \alpha < \infty$, then $\alpha h \in A$, and $J(\alpha h) = \alpha J(h)$.

$\mathcal{A}$2. If $h_1 \in A$, $h_2 \in A$, $h_1 \le h_2$, then $J(h_1) \le J(h_2)$.

$\mathcal{A}$3. If $b \in B^+$, then

$$I(b) = \inf\left\{\sum_{n=1}^{\infty} J(h_n) \mid h_n \in A \ (n \in \mathbb{N}),\ b \le \sum_{n=1}^{\infty} h_n\right\}.$$

The members of A will be called *auxiliary functions*. In connection with $\mathcal{A}$1 we recall our convention that $0\cdot\infty = 0$, which implies that $0 \in A$, and also that $J(0) = 0$.

We write $\mathcal{A}_\sigma(B,I,A,J)$, or $\mathcal{A}_\sigma$, and call the system *σ-finite*, if besides the conditions $\mathcal{A}$1, $\mathcal{A}$2 and $\mathcal{A}$3 the following additional condition is satisfied.

$\mathcal{A}$4. There exists a sequence $(b_n)_{n\in\mathbb{N}}$ in B^+ such that $\sum_{n=1}^{\infty} b_n > 0$.

1.2.2. Starting with a situation $\mathcal{A}(B,I,A,J)$ we shall now construct a norm on Φ, the class of $\mathbb{R}^*$-valued functions defined on S. For each $\varphi \in \Phi$, let

$$\|\varphi\| := \inf\left\{\sum_{n=1}^{\infty} J(h_n) \mid h_n \in A \ (n \in \mathbb{N}),\ |\varphi| \le \sum_{n=1}^{\infty} h_n\right\},$$

where $\inf \emptyset = \infty$ by convention. It is not difficult to see that $\mathcal{N}$1, $\mathcal{N}$2 and $\mathcal{N}$4 are satisfied by the triple $(B,I,\|\cdot\|)$, so we only check $\mathcal{N}$3.

Let $(\varphi_n)_{n\in\mathbb{N}}$ be a sequence in Φ, and assume $\sum_{n=1}^{\infty}\|\varphi_n\| < \infty$, for otherwise there is nothing to prove. Let $\varepsilon > 0$. According to the definition, there exists for each $n \in \mathbb{N}$ a sequence $(h_{nm})_{m\in\mathbb{N}}$ in A such that $|\varphi_n| \le \sum_{m=1}^{\infty} h_{nm}$ and $\sum_{m=1}^{\infty} J(h_{nm}) \le$ $\le \|\varphi_n\| + \varepsilon\cdot 2^{-n}$. The double sequence $(h_{nm})_{n\in\mathbb{N},\, m\in\mathbb{N}}$ can be arranged into a single sequence $(h_k)_{k\in\mathbb{N}}$, and then $\sum_{n=1}^{\infty}|\varphi_n| \le \sum_{k=1}^{\infty} h_k$. Since the series $\sum_{k=1}^{\infty} J(h_k)$ is a rearrangement of the double series $\sum_{n=1}^{\infty} \sum_{m=1}^{\infty} J(h_{nm})$, which has non-negative terms, Proposition 0.4.7 shows

$$\| \sum_{n=1}^{\infty} |\varphi_n| \| \le \sum_{k=1}^{\infty} J(h_k) = \sum_{n=1}^{\infty} \sum_{m=1}^{\infty} J(h_{nm}) \le$$

$$\le \sum_{n=1}^{\infty} (\|\varphi_n\| + \varepsilon \cdot 2^{-n}) = \varepsilon + \sum_{n=1}^{\infty} \|\varphi_n\| .$$

Since $\varepsilon > 0$ is arbitrary, $N3$ follows.

The transition described in this subsection is denoted by $A \to N$, or $A(B,I,A,J) \to N(B,I,\|\cdot\|)$, if we want to be more specific. Since $A4$ implies $L2$, there is also a transition $A_\sigma \to N_\sigma$.

1.2.3. We shall now study some examples to illustrate the situations for which our approach is intended. The first one is about summation, and is quite simple. The next two are about Lebesgue integration on the real line; they are already quite complicated.

Example. Let S be a non-empty set. Take for B the class of real-valued functions that vanish outside a finite set, and let $I(b) := \sum_{s\in S} b(s)$ for $b \in B$. A possible choice for A and the functional J is $A := B^+$, $J(h) := I(h)$ for $h \in A$. It is not difficult to see that the conditions A are satisfied. The system is σ-finite if and only if S is countable.

1.2.4. Example. In this example the reader must be familiar with the Riemann integral for continuous real-valued functions on a bounded interval of $\mathbb{R}$ and with its most elementary properties. Section 0.5 contains already much more on the Riemann integral than needed here.

Let $S := \mathbb{R}$, and let B be the class of continuous real-valued functions defined on S that have compact support. This amounts to requiring that for each $b \in B$ the set $\{s \in S \mid b(s) \neq 0\}$ is contained in a bounded interval. For I take the ordinary Riemann integral, so $I(b) := {}_{-\infty}\int^{\infty} b(x)dx$ for $b \in B$. (The integral is improper in the sense of Subsection 0.5.14. Because each member of B vanishes off a bounded set, this does not give any trouble.) Let $A := B^+$, and $J(h) := I(h)$ for $h \in A$.

The class B is a Riesz function space (Exercise 0.5.17), and I is a positive linear functional on B (see the developments in Section 0.5), so $A1$ and $A2$ are obviously satisfied. The proof of $A3$ will depend on Dini's theorem 0.2.12. For this proof, let $b \in B^+$. Write

$$\rho := \inf \left\{ \sum_{n=1}^{\infty} J(h_n) \mid h_n \in A \ (n \in \mathbb{N}),\ b \le \sum_{n=1}^{\infty} h_n \right\} .$$

Since $b \in A$ and $0 \in A$, it is clear that $\rho \le I(b)$. To prove the reverse inequality, let $(h_n)_{n\in\mathbb{N}}$ be any sequence in A such that $b \le \sum_{n=1}^{\infty} h_n$. For each $N \in \mathbb{N}$, put $g_N := \inf\{\sum_{n=1}^{N} h_n, b\}$. Obviously, $g_N \in B^+$, $g_N \le g_{N+1}$, and $I(g_N) \le \sum_{n=1}^{N} I(h_n)$ for every $N \in \mathbb{N}$. Since $g_N \to b$ pointwise on $\mathbb{R}$, Theorem 0.2.12 can be applied to the sequence $(b - g_N)_{N\in\mathbb{N}}$. This shows that $g_N \to b$ uniformly on each closed interval containing the support of b. (In fact, the convergence is uniform on $\mathbb{R}$, but we do not need this.) Now each g_N has its support contained in that of b, so Theorem 0.5.9 shows that $I(g_N) \to I(b)$. Together with the inequalities $I(g_N) \le \sum_{n=1}^{N} J(h_n)$ $(N \in \mathbb{N})$ this gives $I(b) \le \sum_{n=1}^{\infty} J(h_n)$. Hence $\rho \ge I(b)$, so $A3$ holds.

The system is clearly σ-finite: if $b_n(x) = \max\{n^2 - x^2, 0\}$ for $x \in \mathbb{R}$, $n \in \mathbb{N}$, then the b_n are in B^+, and $\sum_{n=1}^{\infty} b_n = \infty$.

1.2.5. Example. In the preceding example we started with the Riemann integral, a notion which depends on a limit process, even though it is a simple one. We shall now proceed in a more elementary way: it will suffice to know what the "length" of an interval is. We have to pay for using this simple starting point, for a direct proof of $A3$ turns out to be quite hard now. We will not attempt such a proof at present, because it probably would complicate matters too much. Instead, we make a connection with the preceding example, which greatly eases our problems. In addition we will be able to show that the integral furnished by the method of the preceding example is identical to the present one. After reading Chapter 3, the reader will be able to construct his own direct proof of $A3$.

Let B_1 be the class of functions that can be written as $b = \sum_{n=1}^{N} \alpha_n \chi_{A_n}$, where $N \in \mathbb{N}$, $\alpha_n \in \mathbb{R}$, A_n is an interval $(x_n, y_n]$ with $-\infty < x_n < y_n < \infty$ $(n = 1,2,\dots,N)$, and $A_n \cap A_m = \emptyset$ for different n and m. The members of B_1 are commonly called *step functions*. For such a b, put $I_1(b) := \sum_{n=1}^{N} \alpha_n \mu(A_n)$, where $\mu(A_n)$ is the ordinary length $y_n - x_n$ of A_n. It is easy to see that B_1 is a Riesz function space, and that I_1 is well defined. The class A_1 of auxiliary functions consists of the functions h that have the form $h = \alpha\chi_C$, where $0 \le \alpha < \infty$, and C is an interval $(x,y]$ with $-\infty < x < y < \infty$. Put $J_1(h) := I_1(h)$ for $h \in A_1$ (note that $A_1 \subset B_1^+$).

It is not difficult to show that $A1$, $A2$ and $A4$ are satisfied. To prove $A3$, we use the results of the preceding example. Let B, I, A and J have the meaning of 1.2.4. Let $\|\cdot\|$ denote the norm, and let L be the class of integrable functions from this quadruple (the integral on L is again denoted by I). The connection between the two systems is made by the following assertions.

(i) For every $b \in B_1$ and every $\varepsilon > 0$ there exist $b_1, b_2 \in B$ such that $b_1 \le b \le b_2$ and $I(b_1) + \varepsilon \ge I_1(b) \ge I(b_2) - \varepsilon$.

(ii) If $b \in B_1$, then $b \in L$ and $I_1(b) = I(b)$.

The first of these is obvious (b_1 and b_2 can be taken piecewise linear). The second one is proved as follows. Let $b \in B_1$, and $\varepsilon > 0$. Choose b_1 and b_2 as in (i), then

(*) $\quad \|b - b_1\| \le \|b_2 - b_1\| = I(b_2 - b_1) < 2\varepsilon$.

Since ε is arbitrary, this shows that $b \in L$. To prove the remaining part of (ii), let ε, b_1 and b_2 be as before. The inequalities for the functionals in (i) give first that $I_1(b) - I(b_1) < \varepsilon$, and also that $I(b_1) - I_1(b) \le I(b_2) - I_1(b) < \varepsilon$. Hence $|I_1(b) - I(b_1)| < \varepsilon$. Combining this with $|I(b) - I(b_1)| \le \|b - b_1\| < 2\varepsilon$, which follows from (*), we conclude that $|I(b) - I_1(b)| < 3\varepsilon$. Hence $I(b) = I_1(b)$.

The proof of A3 is easy now. Let $b \in B_1^+$, and let $(h_n)_{n\in\mathbb{N}}$ be a sequence in A_1 such that $b \le \sum_{n=1}^\infty h_n$. Apply (ii) above, together with property A3 for $\|\cdot\|$ to get

$$I_1(b) = I(b) = \|b\| \le \sum_{n=1}^\infty \|h_n\| = \sum_{n=1}^\infty I(h_n) = \sum_{n=1}^\infty I_1(h_n) = \sum_{n=1}^\infty J_1(h_n) .$$

This gives one half of A3, and the remaining part follows from the fact that $A_1 \subset B_1^+$.

The method described in the beginning of this section gives a norm $\|\cdot\|_1$, and thus we obtain a space of integrable functions L_1 with an integral I_1 by means of the process $N \to L$. It is an important fact that $L = L_1$ and $I = I_1$. To prove this, note first that $\|\varphi\| \le \|\varphi\|_1$ if $\varphi \in \Phi$. For let $\varphi \in \Phi$, and let $(h_n)_{n\in\mathbb{N}}$ be a sequence in A_1 such that $|\varphi| \le \sum_{n=1}^\infty h_n$. Since $J_1(h_n) = I_1(h_n) = I(h_n) = \|h_n\|$ for all n, we have $\|\varphi\| \le \sum_{n=1}^\infty \|h_n\| = \sum_{n=1}^\infty J_1(h_n)$. Hence every sum $\sum_{n=1}^\infty J_1(h_n)$ occurring in the definition of $\|\varphi\|_1$ exceeds $\|\varphi\|$, and so $\|\varphi\| \le \|\varphi\|_1$. Now let $f \in L_1$ and $\varepsilon > 0$. There exists $b \in B_1$ such that $\|f - b\|_1 < \varepsilon$, and by (ii) above there exists $b_1 \in B$ with $\|b - b_1\| < \varepsilon$. Hence $\|f - b_1\| \le \|f - b\| + \|b - b_1\| \le \|f - b\|_1 + \varepsilon < 2\varepsilon$. Therefore $f \in L$. We know already that $b \in L$ and $I_1(b) = I(b)$, so $|I(f) - I_1(f)| = |I(f) - I(b)| + |I_1(b) - I_1(f)| \le \|f - b\| + \|f - b\|_1 < 2\varepsilon$. Hence $I(f) = I_1(f)$, which proves half of our assertion. For the proof of the remaining half we need something like (i) and (ii) above with the roles of B and B_1, L and L_1 interchanged. Those assertions can be obtained by considering upper and lower sums approximating the Riemann integral. Details are left to the reader as Exercise 1.2-11.

The integral we have obtained in two ways now is called the *Lebesgue integral*, and the integrable functions are called *Lebesgue integrable*. We note that the present example would also have worked with more general "length functions" μ, constructed from non-decreasing real-valued functions on $\mathbb{R}$. This would have led to Lebesgue-Stieltjes integrals, which generalize the Riemann-Stieltjes integrals of Section 0.5; we shall meet these in Example 3.1.12.

1.2.6. Although in the process $A \to N \to L$ the norm is determined from the auxiliary functions, in all cases arising in practice the norm can also be described by means of integrable functions. For instance, in the three examples treated above the auxiliary functions with finite J-value turn out to be integrable, and

(*) $\quad \|\varphi\| = \inf \left\{ \sum_{n=1}^\infty I(f_n) \;\middle|\; f_n \in L^+ \ (n \in \mathbb{N}),\ \sum_{n=1}^\infty f_n \ge |\varphi| \right\}$

for every $\varphi \in \Phi$ with $\|\varphi\| < \infty$. As a sideline we shall now discuss a condition on the auxiliary functions that implies (*), and explore some other consequences of that condition. The main line of the text will be picked up again in the next section.

Suppose we have a non-empty set S and a situation $A(B,I,A,J)$. Follow the route $A \to N$ described in 1.2.2, and then $N \to L$ as described in 1.1.11. The condition we have in mind is the following one.

A5. For every $h \in A$ with $J(h) < \infty$, and every $\varepsilon > 0$ there exists a sequence $(f_n)_{n\in\mathbb{N}}$ in L^+ such that $h \le \sum_{n=1}^{\infty} f_n$ and $\sum_{n=1}^{\infty} I(f_n) \le J(h) + \varepsilon$.

1.2.7. Proposition. Assume $A(B,I,A,J)$ and A5. Then (*) of 1.2.6 holds for every $\varphi \in \Phi$ with $\|\varphi\| < \infty$.

Proof. Let $\varphi \in \Phi$ have finite norm. Denote the expression at the right hand side of (*) by R. If $(f_n)_{n\in\mathbb{N}}$ is a sequence in L^+ such that $|\varphi| \le \sum_{n=1}^{\infty} f_n$, then by N3 and Theorem 1.1.10, $\|\varphi\| \le \sum_{n=1}^{\infty} \|f_n\| = \sum_{n=1}^{\infty} I(f_n)$. So $\|\varphi\| \le R$.

To prove the reverse inequality, let $\varepsilon > 0$ and let $(h_n)_{n\in\mathbb{N}}$ be a sequence in A such that $\sum_{n=1}^{\infty} J(h_n) < \|\varphi\| + \varepsilon$ and $|\varphi| \le \sum_{n=1}^{\infty} h_n$. The condition A5 implies that for each $n \in \mathbb{N}$ there is a sequence $(f_{nm})_{m\in\mathbb{N}}$ in L^+ such that $h_n \le \sum_{m=1}^{\infty} f_{nm}$ and $\sum_{m=1}^{\infty} I(f_{nm}) \le J(h_n) + \varepsilon\cdot 2^{-n}$. Now proceed as in 1.2.2: rearrange the double sum $\sum_{n=1}^{\infty}\sum_{m=1}^{\infty} f_{nm}$ into a single one, $\sum_{k=1}^{\infty} f_k$, which dominates $|\varphi|$. The numbers $I(f_{nm})$ are all non-negative, so $\sum_{n=1}^{\infty}\sum_{m=1}^{\infty} I(f_{nm}) = \sum_{k=1}^{\infty} I(f_k)$. Hence $\sum_{k=1}^{\infty} I(f_k) \le \|\varphi\| + 2\varepsilon$. Now let $\varepsilon \to 0$ to deduce $R \le \|\varphi\|$. □

1.2.8. Corollary. Assume $A(B,I,A,J)$ and A5. Let $\varphi \in \Phi$ and $\|\varphi\| < \infty$. Then there exists $f \in L^+$ such that $\|\varphi\| = I(f)$ and $|\varphi| \le f$ (a.e.).

Proof. According to the proposition there exists for each $n \in \mathbb{N}$ a sequence $(f_{nm})_{m\in\mathbb{N}}$ in L^+ such that $|\varphi| \le \sum_{m=1}^{\infty} f_{nm}$, while

$$(**)\qquad \|\varphi\| \le \sum_{m=1}^{\infty} I(f_{nm}) \le \|\varphi\| + n^{-1} .$$

By the Beppo Levi theorem 1.1.17 there exists for each $n \in \mathbb{N}$ an $f_n \in L^+$ such that $I(f_n) = \sum_{m=1}^{\infty} I(f_{nm})$ and $f_n = \sum_{m=1}^{\infty} f_{nm}$ (a.e.). Let $f := \inf_{n\in\mathbb{N}} f_n$. Then $f \in L^+$ by Lebesgue's theorem 1.1.4. Since $|\varphi| \le f_n$ (a.e.) for each n, it follows that $|\varphi| \le f$ (a.e.), which immediately implies $\|\varphi\| \le \|f\| = I(f)$. On the other hand, $I(f) \le I(f_n) = \sum_{m=1}^{\infty} I(f_{nm})$ for each n. Therefore, by (**), $I(f) \le \|\varphi\|$. □

1.2.9. The property in the theorem below is called the *Fatou property* for the norm. The name indicates that this theorem does for norms what Fatou's lemma (to be derived in 2.2.12) does for integrals.

<u>Theorem</u>. Assume $A(B,I,A,J)$ and $A5$. Let $(\varphi_n)_{n\in\mathbb{N}}$ be a sequence in Φ^+ such that $\varphi_n \le \varphi_{n+1}$ $(n \in \mathbb{N})$, and let $\varphi := \lim_{n\to\infty} \varphi_n$. Then $\|\varphi_n\| \to \|\varphi\|$.

<u>Proof</u>. Obviously, $\|\varphi_n\| \le \|\varphi_{n+1}\| \le \|\varphi\|$ for each $n \in \mathbb{N}$, so $\lim_{n\to\infty}\|\varphi_n\|$ exists, and $\lim_{n\to\infty}\|\varphi_n\| \le \|\varphi\|$. Therefore we need only show that $\|\varphi\| \le \lim_{n\to\infty}\|\varphi_n\|$.

We may as well assume that $\lim_{n\to\infty}\|\varphi_n\|$ is finite. Corollary 1.2.8 shows that for each $n \in \mathbb{N}$ there exists $f_n \in L^+$ such that $\|\varphi_n\| = I(f_n)$ and $\varphi_n \le f_n$ (a.e.). Put $g_n := \sup_{1\le k\le n} f_k$. Then $g_n \in L^+$ and $g_n \le g_{n+1}$ for all n. Moreover, it is easy to see that $g_n = f_n$ (a.e.), and therefore

(†) $$\lim_{n\to\infty} I(g_n) = \lim_{n\to\infty} I(f_n) = \lim_{n\to\infty} \|\varphi_n\| < \infty .$$

So the monotone convergence theorem 1.1.18 implies that there exists $g \in L^+$ such that $g = \lim_{n\to\infty} g_n$ (a.e.) and

(††) $$I(g) = \lim_{n\to\infty} I(g_n) .$$

Since $0 \le \varphi_n \le f_n = g_n \le g$ (a.e.) for all n, we have $\varphi \le g$ (a.e.) and therefore $\|\varphi\| \le \|g\| = I(g)$, which, when combined with (†) and (††), gives what we want. □

<u>Exercises Section 1.2</u>

1. Let the quadruple (B,I,A,J) satisfy the conditions A of 1.2.1. Is it true that $J(h_1 + h_2) = J(h_1) + J(h_2)$ for $h_1 \in A$, $h_2 \in A$, with $h_1 + h_2 \in A$?

2. Let $S = [0,1]$ and take $B = C([0,1])$, $I(b) = \int_0^1 b(x)dx$ $(b \in B)$ where the integral is the ordinary Riemann integral. Let $g(x) := x^{-\frac{1}{2}}$ $(0 < x \le 1)$, $g(0) = 0$. Take $A = \{b + \alpha g \mid b \in B^+, 0 \le \alpha < \infty\}$, $J(h) = I(b) + 3\alpha$ if $h = b + \alpha g$ with $b \in B^+$, $0 \le \alpha < \infty$. Show that the conditions $A1 - A5$ of 1.2.1 and 1.2.6 are satisfied.

3. Start from station $N(B,I,\|\cdot\|)$ and take $A = \Phi^+$, $J(h) = \|h\|$ $(h \in A)$. Show that the conditions A of 1.2.1 are satisfied by (B,I,A,J) and that the norm obtained by following the route $A \to N$ equals $\|\cdot\|$. Is condition $A5$ always satisfied?

In Exercises 4 - 11 we consider the Lebesgue integral for functions defined on $\mathbb{R}$.

4. For $\varphi \in \Phi$, $t \in \mathbb{R}$ define $\varphi_t := \Upsilon_{s\in\mathbb{R}}\, \varphi(s + t)$. Show that $\|\varphi_t\| = \|\varphi\|$, and that $I(\varphi_t) = I(\varphi)$ if φ is Lebesgue integrable. This property of the Lebesgue integral is called *translation invariance*.

5. (i) Let $a \in \mathbb{R}$, $b \in \mathbb{R}$, $a < b$. Let $f : [a,b] \to \mathbb{R}$ be Riemann integrable. Show that f is Lebesgue integrable and that its Riemann integral and Lebesgue integral are equal (we have to extend f to a function $\tilde{f}$ defined on $\mathbb{R}$ by $\tilde{f}(x) = f(x)$ or 0 according as $x \in [a,b]$ or $x \notin [a,b]$).

(ii) Let $f : \mathbb{R} \to \mathbb{R}$ be Riemann integrable over every interval $[a,b]$ and assume that $\lim_{a\to-\infty, b\to\infty} \int_a^b |f(t)|dt < \infty$. Show that f is Lebesgue integrable and that its Lebesgue integral equals $\lim_{a\to-\infty, b\to\infty} \int_a^b f(t)dt$.

(iii) Show that $\int_{-\infty}^{\infty} t^{-1} \sin t\, dt$ exists as an improper Riemann integral but not as a Lebesgue integral.

6. If g is Lebesgue integrable, then $\forall_{t\in\mathbb{R}} \int_0^t g(s)ds$ is a continuous function.

7. (i) Let $a \in \mathbb{R}$, $b \in \mathbb{R}$, $a < b$. Show that $\chi_{\{a\}}$, $\chi_{(a,b]}$, $\chi_{(a,b)}$, $\chi_{[a,b]}$ are integrable, with integrals $0, b - a, b - a, b - a$, respectively.

(ii) Show that $\mathbb{Q}$ is a Lebesgue null set.

(iii) Let $(I_n)_{n\in\mathbb{N}}$ be a sequence of pairwise disjoint intervals in $\mathbb{R}$ and assume that $\sum_{n=1}^{\infty}$ length $(I_n) < \infty$. Show that the characteristic function of $\bigcup_{n=1}^{\infty} I_n$ is Lebesgue integrable, with integral $\sum_{n=1}^{\infty}$ length (I_n).

(iv) If O is a bounded open set in $\mathbb{R}$, then χ_O is Lebesgue integrable.

8. Let $N \subset \mathbb{R}$ be a Lebesgue null set, and let $\varepsilon > 0$. Show that there is a sequence $(I_n)_{n\in\mathbb{N}}$ of disjoint open intervals such that $N \subset \bigcup_{n=1}^{\infty} I_n$, $\sum_{n=1}^{\infty}$ length $(I_n) < \varepsilon$ by using the following steps.

(i) Let $(b_n)_{n\in\mathbb{N}}$ be a sequence of non-negative compactly supported continuous functions such that $\chi_N \le \sum_{n=1}^{\infty} b_n$, $\sum_{n=1}^{\infty} I(b_n) < \varepsilon/2$. Show that the set $O_\alpha := \{x \in \mathbb{R} \mid \sum_{n=1}^{\infty} b_n(x) > \alpha\}$ is open for $\alpha \in \mathbb{R}$.

(ii) Show that $N \subset O_\alpha$, $\chi_{O_\alpha} \le \alpha^{-1} \sum_{n=1}^{\infty} b_n$ for $\alpha \in (0,1)$.

(iii) Complete the proof by using Exercise 7.

9. Prove the following characterization of Lebesgue null sets: $N \subset S$ is a null set if and only if there is a sequence $(b_n)_{n\in\mathbb{N}}$ in B^+ such that $\sum_{n=1}^{\infty} I(b_n) < \infty$, $\sum_{n=1}^{\infty} b_n(s) = \infty$ $(s \in N)$. Here B is either one of the spaces of basic functions of 1.2.4 and 1.2.5.

10. Let $f : \mathbb{R} \to \mathbb{R}$ be continuously differentiable.

(i) Let J be a bounded interval in $\mathbb{R}$. Show that there is a $K > 0$ such that for every interval $I \subset J$ we have $\|\chi_{f(I)}\| \le K\|\chi_I\|$.

(ii) Show that $f(A)$ is a null set if A is a null set. (Hint. Use Exercise 8.)

11. Elaborate the last part of Example 1.2.5.

Exercises 12 - 16 deal with the condition $\mathcal{A}5$ of 1.2.6.

12. Give an example of a quadruple (B,I,A,J) satisfying the conditions $\mathcal{A}$ of 1.2.1 but not condition $\mathcal{A}5$ of 1.2.6. (Hint. One can give an example where S consists of two elements.)

13. Assume that (B,I,A,J) satisfies $\mathcal{A}$ and that the norm obtained in the process $\mathcal{A} \to \mathcal{N}$ satisfies $(*)$ of 1.2.6. Show that $\mathcal{A}5$ is satisfied.

14. Let $S = [0,1]$, $B = \{\alpha\chi_S \mid \alpha \in \mathbb{R}\}$, $I(b) = b(0)$ $(b \in B)$, $A = \{\alpha\chi_E \mid E \subset S,\ \alpha \geq 0\}$, $J(h) = J(\alpha\chi_E) = 0$ or α according as E is finite or infinite. Show that the conditions $\mathcal{A}1 - \mathcal{A}5$ are satisfied. Show that there is an $h \in A$ such that $\|h\| < \infty$, $J(h) < \infty$, $\|h\| \neq J(h)$.

15. Give an example where $\mathcal{A}1 - \mathcal{A}4$ is satisfied, $\|h\| = J(h)$ for all $h \in A$ with $J(h) < \infty$, and where $\mathcal{A}5$ does not hold.

16. Assume that $\mathcal{A}1 - \mathcal{A}5$ is satisfied and follow the line $\mathcal{A} \to \mathcal{N} \to \mathcal{L}$. Let $f_n \in L^+$ $(n \in \mathbb{N})$ and assume that $f_n \cdot f_m = 0$ (a.e.) if $n \neq m$. Also, let $E_n := \{s \in S \mid f_n(s) > 0\}$ for $n \in \mathbb{N}$ and put $E := \bigcup_{n=1}^{\infty} E_n$. Show that

$$\|\varphi\cdot\chi_E\| = \sum_{n=1}^{\infty} \|\varphi\cdot\chi_{E_n}\|$$

for $\varphi \in \Phi$. (Hint. To prove the inequality $\|\varphi\cdot\chi_E\| \geq \sum_{n=1}^{\infty}\|\varphi\cdot\chi_{E_n}\|$ we can assume that $\|\varphi\cdot\chi_E\| < \infty$. Now take $f \in L^+$ with $I(f) = \|\varphi\cdot\chi_E\|$, $f \geq |\varphi\cdot\chi_E|$ (a.e.) and use Exercise 1.1-4 to finish the proof.)

1.3. <u>The Daniell approach</u>

We now turn to the problem of obtaining auxiliary functions. We treat two methods, in both of which the starting point is a Riesz function space B with a positive linear functional defined on B that satisfies a certain continuity condition. The first method is in essence due to Daniell (he wrote about 1920); it was made popular by Stone in a famous series of papers (1948-1950). The second one is an abstract version of the Bourbaki approach to integration, where one constructs integrals on locally compact Hausdorff spaces. The two methods are not very different, and in many cases lead to the same results, which explains why only Daniell's name occurs in the heading of this section.

1.3.1. Let B be a Riesz function space of functions defined on a non-empty set S, and let I be a positive linear functional defined on B. The members of B are called *basic functions*. Assume that the following condition is satisfied.

$\mathcal{D}1$. If $(b_n)_{n\in\mathbb{N}}$ is a sequence in B^+, and $b_n \downarrow 0$, then $I(b_n) \to 0$.

We denote this situation by $\mathcal{D}(B,I)$, or $\mathcal{D}$. We write $\mathcal{D}_\sigma(B,I)$, or $\mathcal{D}_\sigma$, and call the system σ-*finite*, if the following additional condition $\mathcal{D}2$ is satisfied.

$\mathcal{D}2$. There exists a sequence $(b_n)_{n\in\mathbb{N}}$ in B^+ such that $\sum_{n=1}^\infty b_n > 0$.

1.3.2. Example. Take $S := [0,1]$, $B := C([0,1])$ and $I(b) := {}_0\!\int^1 b(s)ds$ $(b \in B)$. Now $\mathcal{D}1$ follows from Dini's theorem 0.2.12 together with Theorem 0.5.9. The system is clearly σ-finite.

1.3.3. Start with a situation $\mathcal{D}(B,I)$ as described above. It is easy to obtain auxiliary functions: just put $A := B^+$ and $J(h) := I(h)$ for $h \in A$. It is obvious that the conditions $A1$ and $A2$ are satisfied, so again only $A3$ needs a proof. Let $b \in B^+$. In the first place it is clear that

$$I(b) \geq \inf\left\{\sum_{n=1}^{\infty} J(h_n) \mid h_n \in A\ (n \in \mathbb{N}),\ b \leq \sum_{n=1}^{\infty} h_n\right\},$$

for $I(b)$ is one of the numbers occurring in the set on the right hand side. On the other hand, let $(h_n)_{n\in\mathbb{N}}$ be a sequence in A such that $b \leq \sum_{n=1}^\infty h_n$. We have to show that $I(b) \leq \sum_{n=1}^\infty J(h_n)$. For each $n \in \mathbb{N}$, let $g_n := b - \inf(\sum_{k=1}^n h_k, b)$. Then $g_n \downarrow 0$, and therefore $I(g_n) \to 0$ by $\mathcal{D}1$. Since I is positive and $g_n \geq b - \sum_{k=1}^n h_k$, it follows that $I(g_n) \geq I(b) - \sum_{k=1}^n I(h_k)$ for every $n \in \mathbb{N}$. This shows that $0 \geq I(b) - \sum_{k=1}^\infty I(h_k)$, that is, $I(b) \leq \sum_{k=1}^\infty J(h_k)$. So $A3$ is satisfied.

The transition from station $\mathcal{D}$ to station A described in the preceding paragraph is denoted by $\mathcal{D}(B,I) \to A(B,I,A,J)$, or just $\mathcal{D} \to A$. If the original system satisfies $\mathcal{D}2$, the resulting system is also σ-finite, which means that we also have a transition $\mathcal{D}_\sigma \to A_\sigma$.

So we can obtain integrable functions and an integral by means of the transition $\mathcal{D} \to A$, treated here, and the transition $A \to N \to L$, treated in the preceding two sections. Conversely, any pair (L,I) that satisfies $L1$ of 1.1.1 also satisfied $\mathcal{D}1$. Otherwise stated: every station L is a station $\mathcal{D}$. This immediately raises questions like: what happens if the procedure of the present subsection is applied to a pair (L,I) satisfying L? In the next section we deal with this kind of questions.

1.3.4. In the process $\mathcal{D}(B,I) \to \mathcal{A}(B,I,A,J)$ the auxiliary functions are basic functions. So the norm could have been described in terms of B right away, and the reference to auxiliary functions is really superfluous. The motive for this presentation is to exhibit the relation between this method of obtaining norms and the one described below in 1.3.8.

To make this relation more obvious, the following observation may be useful. Assume that B is a Riesz function space on a set S and that I is a positive linear functional on B. Then the pair (B,I) satisfies $\mathcal{D}1$ if and only if the following condition (*) is satisfied.

(*) If V is a countable non-empty subset of B^+ and $\inf_{b\in V} b = 0$, then for every $\varepsilon > 0$ there exists a finite subset $W \subset V$ such that $I(\inf_{b\in W} b) < \varepsilon$.

The proof of the equivalence of $\mathcal{D}1$ and (*) is left to the reader.

1.3.5. Once again, let B be a Riesz function space on the non-empty set S, and let I be a positive linear functional on B. Obviously, $\mathcal{D}1$ is the weakest condition on the pair (B,I) that enables us to use the procedures of Sections 1.1 and 1.2. In fact, in the end B will be a subspace of a space of integrable functions with an integral that extends I, and then $\mathcal{D}1$ appears as a consequence of the simple form of Lebesgue's theorem in 1.1.4. However, in the important case of continuous functions of compact support on a locally compact Hausdorff space, there is a property much stronger than $\mathcal{D}1$. Our abstraction of it is the following condition $\mathcal{D}1'$.

$\mathcal{D}1'$. If V is a non-empty subset of B^+ and $\inf_{b\in V} b = 0$, then for every $\varepsilon > 0$ there exists a finite subset $W \subset V$ such that $I(\inf_{b\in W} b) < \varepsilon$.

If the pair (B,I) satisfies $\mathcal{D}1'$, then we denote this by $\mathcal{D}'(B,I)$, or $\mathcal{D}'$; if $\mathcal{D}2$ of 1.3.1 is also satisfied, then we write $\mathcal{D}'_\sigma(B,I)$, or $\mathcal{D}'_\sigma$, and call the system σ-*finite*.

1.3.6. Example. Take $\mathbb{R}$ in the role of S. Let B be the class of continuous real-valued functions defined on $\mathbb{R}$ that have compact support, and put $I(b) := \int_{-\infty}^{\infty} b(s)ds$ for $b \in B$. To show that (B,I) satisfies $\mathcal{D}'$, we need only check $\mathcal{D}1'$. Let V be a non-empty subset of B^+ such that $\inf_{b\in V} b = 0$. Let $\varepsilon > 0$. Take any function $b_0 \in V$, denote its support by C, and let $[\alpha,\beta]$ be an interval containing C. For every $s \in C$ there exists $b_s \in V$ such that $b_s(s) < \frac{1}{2}\varepsilon(\beta - \alpha)^{-1}$, and by continuity there is an open interval I_s around s such that $b_s(t) < \varepsilon(\beta - \alpha)^{-1}$ for $t \in I_s$. Now C is compact, and it can therefore be covered by finitely many of these intervals, $I_{s_1}, I_{s_2}, \ldots, I_{s_n}$, say. If $h := \inf(b_0, b_{s_1}, \ldots, b_{s_n})$, then $0 \le h < \varepsilon(\beta - \alpha)^{-1}$ on C and $h = 0$ outside C. Hence $I(h) = \int_{-\infty}^{\infty} h(x)dx < \varepsilon$ as required.

1.3.7. Let (B,I) be a system that satisfies $\mathcal{D}1'$. Then it also satisfies $\mathcal{D}1$. This is obvious from the equivalence of $\mathcal{D}1$ and (*) mentioned in 1.3.4. It follows that auxiliary functions for the $\mathcal{D}'$-system can be obtained in the manner of 1.3.3; the procedure is denoted by $\mathcal{D}' \to \mathcal{D} \to \mathcal{A}$. But there is another possibility, too, which often leads to a smaller norm, and hence to a larger class of integrable functions (see Theorem 1.1.20). We turn to this now.

1.3.8. Start with a situation $\mathcal{D}'(B,I)$. For A take the real-valued functions on S that can be written as $\sup_{b\in V} b$ for some non-empty subset V of B^+, and put $J(h) := \sup\{I(b) \mid b \in B^+, b \le h\}$ for $h \in A$.

It is easy to see that $\mathcal{A}1$ and $\mathcal{A}2$ of 1.2.1 are satisfied, but $\mathcal{A}3$ causes some difficulties. Let $b \in B^+$. Let $(h_n)_{n\in\mathbb{N}}$ be a sequence in A with $b \le \sum_{n=1}^{\infty} h_n$. We shall show that $I(b) \le \sum_{n=1}^{\infty} J(h_n)$. Let U be the collection of all finite sums $\sum_{k=1}^{n} b_k$, where $n \in \mathbb{N}$, $b_k \in B^+$, $b_k \le h_k$ $(k = 1,2,\ldots,n)$ and $\sum_{k=1}^{n} b_k \le b$. Then $\sup_{c\in U} c = b$, so if $V := \{b - c \mid c \in U\}$, then $\inf_{d\in V} d = 0$. Let $\varepsilon > 0$. Apply $\mathcal{D}1'$ to the family V. It follows that there is a finite subset W of U such that $I(\inf_{c\in W}(b - c)) < \varepsilon$, that is $I(\sup_{c\in W} c) > I(b) - \varepsilon$. Suppose $W = \{c_1,c_2,\ldots,c_m\}$. Each c_k can be written as $c_k = \sum_{n=1}^{N} b_{kn}$, where $b_{kn} \in B^+$, $b_{kn} \le h_n$ $(n = 1,2,\ldots,N)$, and it is very convenient that N be may chosen independently of k (include zero functions in the sums, if necessary). Now

$$\sup_{c\in W} c = \sup_{1\le k\le m}\left(\sum_{n=1}^{N} b_{kn}\right) \le \sum_{n=1}^{N} \sup_{1\le k\le m} b_{kn} ,$$

and therefore

$$I(b) - \varepsilon < I(\sup_{c\in W} c) \le \sum_{n=1}^{N} I(\sup_{1\le k\le m} b_{kn}) \le \sum_{n=1}^{N} J(h_n) \le \sum_{n=1}^{\infty} J(h_n) .$$

Hence $I(b) \le \sum_{n=1}^{\infty} J(h_n)$. Now combine this with the remark that $b \in A$ and $J(b) = I(b)$ to complete the proof of $\mathcal{A}3$.

This way of obtaining auxiliary functions is denoted by $\mathcal{D}' \to \mathcal{A}$.

1.3.9. Each of the transitions $\mathcal{D} \to \mathcal{A}$ and $\mathcal{D}' \to \mathcal{A}$ leads to a situation where the condition $\mathcal{A}5$ of 1.2.6 is satisfied. This is obvious for $\mathcal{D} \to \mathcal{A}$ (in this case the auxiliary functions are basic functions, so they become integrable in the process), but not quite so for $\mathcal{D}' \to \mathcal{A}$. The latter case can be handled by means of the following result, which is also interesting in its own right.

1.3.10. <u>Proposition</u>. Assume $\mathcal{D}'(B,I)$, and follow $\mathcal{D}' \to \mathcal{A} \to \mathcal{N}$. Let $h \in A$ and $J(h) < \infty$. Then there exists a sequence $(b_n)_{n\in\mathbb{N}}$ in B^+ and a non-negative null function $\varphi \in \Phi$ such that $h = \varphi + \sum_{n=1}^{\infty} b_n$, while $J(h) = \sum_{n=1}^{\infty} I(b_n)$. Moreover, after the transition $\mathcal{N} \to \mathcal{L}$ there is an integrable function f such that $f = h$ (a.e.) and $I(f) = J(h)$.

Proof. By the definition of $J(h)$ there is a sequence $(c_n)_{n\in\mathbb{N}}$ in B^+ such that $c_n \le h$ $(n \in \mathbb{N})$, $I(c_n) \to J(h)$. We may assume that the sequence is increasing (replace c_n by $\sup(c_1,c_2,\ldots,c_n)$ if necessary), so if $b_1 := c_1$, $b_{n+1} := c_{n+1} - c_n$ $(n \in \mathbb{N})$, then $(b_n)_{n\in\mathbb{N}}$ is a sequence in B^+ that satisfies $J(h) = \sum_{n=1}^{\infty} I(b_n)$. Since $c_n \le h$ for all n, we have $\sum_{n=1}^{\infty} b_n \le h$. Hence, if $\varphi : S \to \mathbb{R}$ is defined by

$$\varphi(s) := \begin{cases} h(s) - \sum_{n=1}^{\infty} b_n(s) & \text{if } \sum_{n=1}^{\infty} b_n(s) < \infty , \\ 0 & \text{otherwise,} \end{cases}$$

then φ is non-negative, and $h = \varphi + \sum_{n=1}^{\infty} b_n$.

To show that φ is a null function, for all $n \in \mathbb{N}$ put $h_n := h - c_n$ and $V_n := \{d - c_n \mid d \in B^+,\ c_n \le d \le h\}$. Then $h_n := \sup_{b\in V_n} b \in A$, and it is not difficult to see that

$$J(h_n) = \sup\{I(d) - I(c_n) \mid d \in B^+,\ c_n \le d \le h\} = J(h) - I(c_n) .$$

Hence $J(h_n) \to 0$. Since $0 \le \varphi \le h_n$, it follows that $\|\varphi\| \le \|h_n\| \le J(h_n)$, so $\|\varphi\| = 0$.

The remaining assertion follows immediately from Beppo Levi's theorem 1.1.17. □

1.3.11. Suppose that the pair (B,I) satisfies the conditions $\mathcal{D}'$ of 1.3.4, and introduce the auxiliary functions by means of $\mathcal{D}' \to \mathcal{A}$. If $\emptyset \neq V \subset B^+$ and $h := \sup_{b\in V} b$, then h is an auxiliary function. Why don't we define $J(h)$ to be $\sup_{b\in V} I(b)$? The answer is that $J(h)$ (as we did define it) may well exceed $\sup_{b\in V} I(b)$, so use of the latter quantity as $J(h)$ might cause $J(h)$ to depend on V (see Exercise 1.3-4). If V is "large enough", however, this cannot occur. (We have already seen this in a particular case, namely in the proof of 1.3.10, where we used that $J(h) = \sup\{I(d) \mid d \in B^+,$ $c_n \le d \le h\}$.) The technical term for what is needed is that V should be directed (see 0.1.6 for the definition).

Proposition. Assume $\mathcal{D}'(B,I)$ and follow $\mathcal{D}' \to \mathcal{A}$. Let V be a non-empty directed subset of B^+, and $h := \sup_{b\in V} b$. Then $J(h) = \sup_{b\in V} I(b)$.

Proof. Since $\sup_{b\in V} I(b) \le J(h)$ is obvious, we need only prove the reverse inequality.

Let $b_0 \in B^+$, $b_0 \le h$. Then $\sup_{b\in V} b \ge b_0$. Let $V_0 := \{\inf(b,b_0) \mid b \in V\}$, then $\sup_{c\in V_0} c = b_0$, that is, $\inf_{c\in V_0}(b_0 - c) = 0$. Let $\varepsilon > 0$. By $\mathcal{D}1'$ there exists a finite set $W_0 \subset V_0$ such that $I(\inf_{c\in W_0}(b_0 - c)) < \varepsilon$. Hence $I(\sup_{c\in W_0} c) \ge I(b_0) - \varepsilon$. For each $c \in W_0$ choose a $b \in V$ with $c = \inf(b,b_0)$, and denote the set of all those b's by W. Then $I(\sup_{b\in W} b) \ge I(b_0) - \varepsilon$. Since V is directed, and W is a finite subset of V, there exists $b_1 \in V$ such that $\sup_{b\in W} b \le b_1$. Hence $\sup_{b\in V} I(b) \ge I(b_1) \ge I(b_0) - \varepsilon$. It follows that $\sup_{b\in V} I(b) \ge I(b_0)$, and since b_0 may be chosen to be any member of B^+ that satisfies $b_0 \le h$, we conclude that $\sup_{b\in V} I(b) \ge J(h)$. □

1.3.12. The space of auxiliary functions obtained by means of $\mathcal{D}' \to A$ has a nice structure which enables us to reformulate the norm definition.

Proposition. Assume $\mathcal{D}'(B,I)$, and follow $\mathcal{D}' \to A \to N$. Then the following assertions hold.

(i) If $(h_n)_{n\in\mathbb{N}}$ is a sequence in A, then $\sum_{n=1}^{\infty} h_n \in A$ and $J(\sum_{n=1}^{\infty} h_n) = \sum_{n=1}^{\infty} J(h_n)$.

(ii) If $\varphi \in \Phi$, then $\|\varphi\| = \inf\{J(h) \mid h \in A, |\varphi| \le h\}$.

(iii) If $h \in A$, then $J(h) = \|h\|$.

Proof. (i) Let $(h_n)_{n\in\mathbb{N}}$ be a sequence in A, and put $h := \sum_{n=1}^{\infty} h_n$. Let

$$V := \{b_1 + b_2 + \ldots + b_m \mid b_n \in B^+, b_n \le h_n, 1 \le n \le m \text{ and } m \in \mathbb{N}\}.$$

Obviously, $h = \sup_{b\in V} b$, so $h \in A$. Since V is directed, 1.3.11 implies that $J(h) = \sup_{b\in V} I(b)$.

First let $m \in \mathbb{N}$ be fixed. If $b_n \in B^+$, $b_n \le h_n$ $(1 \le n \le m)$, then $b := \sum_{n=1}^{m} b_n \in V$, so

$$I(b_1) + I(b_2) + \ldots + I(b_m) = I(b) \le J(h) .$$

Now take the suprema over $b_1, b_2, \ldots, b_m$, successively, to obtain

$$J(h_1) + J(h_2) + \ldots + J(h_m) \le J(h) .$$

Hence $\sum_{n=1}^{\infty} J(h_n) \le J(h)$.

To prove the reverse inequality, choose $\alpha \in \mathbb{R}$, $\alpha < J(h)$. By 1.3.11 there exists $b \in V$ such that $I(b) > \alpha$. This b can be written as $b = b_1 + b_2 + \ldots + b_m$, where $b_n \in B^+$, $b_n \le h_n$ for $1 \le n \le m$, where m is some natural number. Therefore

$$\sum_{n=1}^{\infty} J(h_n) \ge \sum_{n=1}^{m} J(h_n) \ge \sum_{n=1}^{m} I(b_n) = I\left(\sum_{n=1}^{m} b_n\right) = I(b) > \alpha .$$

Hence $\sum_{n=1}^{\infty} J(h_n) \ge J(h)$, which completes the proof of (i).

The second assertion follows immediately from (i) and the norm definition in 1.2.2, while (iii) is obvious because of (ii). □

Exercises Section 1.3

1. Let B be a Riesz function space of functions defined on S and let I be a positive linear functional defined on B. Show that (B,I) satisfy the condition $\mathcal{D}1$ of 1.3.1 if and only if $I(b) \le \sum_{n=1}^{\infty} I(b_n)$ whenever $b \in B^+$, $b_n \in B^+$ $(n \in \mathbb{N})$, $b \le \sum_{n=1}^{\infty} b_n$.

2. Give an example of a pair (B,I) that satisfies the conditions $\mathcal{D}'$ of 1.3.5 such that the processes $\mathcal{D}' \to \mathcal{D} \to A \to N \to L$ and $\mathcal{D}' \to A \to N \to L$ give different results. (Hint. Take S uncountable, B the class of real-valued functions that are zero except for finitely many points, I as simple as possible.)

3. Let $S_0 := [0,1]$ and let $S = \mathbb{Q} \cap S_0$. Take for B the set $\{\prod_{x\in S} b_0(x) \mid b_0 : S_0 \to \mathbb{R}$ is continuous$\}$, and put $I(b) = {}_0\int^1 b_0(x)dx$ (ordinary Riemann integral) where $b \in B$ and $b_0 : S_0 \to \mathbb{R}$ is continuous with $b_0(x) = b(x)$ $(x \in S)$. Are the conditions $\mathcal{D}$ or $\mathcal{D}'$ satisfied by (B,I)?

4. Give an example of a pair (B,I) satisfying the conditions $\mathcal{D}'$ of 1.3.5 such that there exists an $h \in A$ and a set $V \in B^+$ with $h = \sup_{b\in V} b$, $J(h) > \sup_{b\in V} I(b)$. (Hint. Consider the Lebesgue integral on [0,1].)

5. Take S, B and I as in Example 1.3.6. Show that the routes $\mathcal{D}' \to \mathcal{D} \to A \to N \to L$ and $\mathcal{D}' \to A \to N \to L$ give the same results. (Hint. This is a difficult exercise. In Section 4.2 a problem like this is solved under more general conditions.)

6. Start from $\mathcal{D}(B,I)$ and let O be the class of $\mathbb{R}^*$-valued functions f for which there exists a sequence $(b_n)_{n\in\mathbb{N}}$ in B such that $b_n \uparrow f$ and $\lim_{n\to\infty} I(b_n) < \infty$.

(i) Show that $f + g \in O$, $\alpha f \in O$, $\sup(f,g) \in O$, $\inf(f,g) \in O$ if $f \in O$, $g \in O$, $\alpha \ge 0$.

(ii) Let $f \in O$ and let $(b_n)_{n\in\mathbb{N}}$ be a sequence in B such that $b_n \uparrow f$. Show that $\lim_{n\to\infty} I(b_n) < \infty$ and that $J'(f) := \lim_{n\to\infty} I(b_n)$ is well-defined. (Hint. Use Exercise 1.)

(iii) Follow the line $\mathcal{D} \to A \to N \to L$. Show that $f \in L$ if and only if for every $\varepsilon > 0$ there is g_1, g_2 with $g_1 \in O$, $-g_2 \in O$ such that $g_2 \le f \le g_1$ and $J'(g_1) \le -J'(-g_2) + \varepsilon$. (This is almost literally *Daniell*'s approach to integration; note the similarity with upper and lower sums in the theory of Riemann integration.)

7. Follow the route $\mathcal{D} \to A \to N$. Show that we have the same characterization of null sets as in Exercise 1.2-9.

8. Follow the route $\mathcal{D}'(B,I) \to A(B,I,A,J) \to N \to L(L,I)$. Show that for every $f \in L$ and every $\varepsilon > 0$ there is a $b \in B$ and an $h \in A$ such that $f \leq b + h$, $I(b) + J(h) \leq$ $\leq I(f) + \varepsilon$.

9. Prove the assertion of 1.3.9 that the direct route $\mathcal{D}' \to A$ leads to a system where A5 holds.

1.4. The circle line

In the preceding sections we successively constructed an integral I on a class of integrable functions L, starting with N; then a norm $\|\cdot\|$, starting with A; and finally auxiliary functions with a J-functional, starting with $\mathcal{D}$ or $\mathcal{D}'$. So $\mathcal{D}$ and $\mathcal{D}'$ may act as starting points for the procedure, which furnishes a system $L(L,I)$. An L-system, in turn, is also a $\mathcal{D}$-system (see 1.3.3), so the procedure can be iterated. Since the real interest is in integrating functions, the proper question is whether more functions get integrable by such a repetition. It will turn out that in the cases occurring in practice nothing is gained by going through the movements a second time.

1.4.1. The development up to now can be nicely couched in terms of *stations* together with their connections. The stations are L, N, A, $\mathcal{D}$ and $\mathcal{D}'$, and the connections $L \to \mathcal{D}$, $\mathcal{D} \to A$, $A \to N$, $N \to L$, $\mathcal{D}' \to \mathcal{D}$, and $\mathcal{D}' \to A$. Together they form the *circle line*. (See figure; the numbers indicate the subsections where the stations and their connections have been described. Some further stations will be added later.)

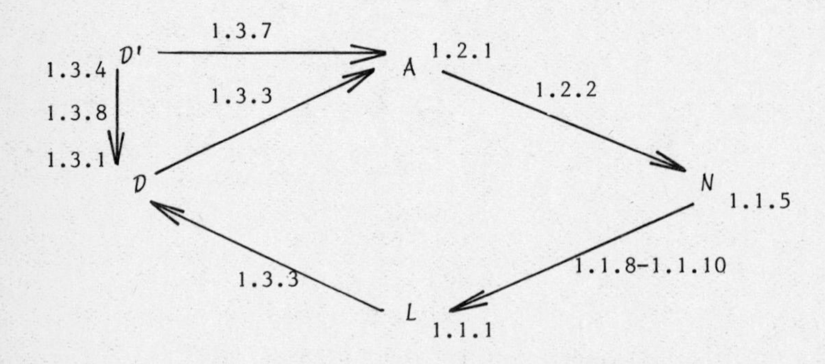

The station $\mathcal{D}'$ is located somewhat beside the line; the circle line may be entered at $\mathcal{D}'$, but there is no way to get at $\mathcal{D}'$ from any point in the circle line. Note that there is one-way traffic everywhere.

1.4.2. First consider what happens if one starts from station $L(L_0,I_0)$ and then follows the line $L \to \mathcal{D} \to A \to N \to L$ to arrive at a set L_1 of integrable functions with an integral I_1. Question: how are the pairs (L_0,I_0) and (L_1,I_1) related? The anwer is simple.

Proposition. $L_0 \subset L_1$, and I_1 extends I_0. Moreover, for every $f \in L_1$ there exists a sequence $(f_n)_{n\in\mathbb{N}}$ in L_0 such that $f = \sum_{n=1}^{\infty} f_n$ (a.e. with respect to $\|\cdot\|_1$) and $I_1(f) = \sum_{n=1}^{\infty} I_0(f_n)$.

Proof. L_0 serves as the class of basic functions in the extension process. Hence 1.1.9 and 1.1.10 imply that $L_0 \subset L_1$ and that I_1 extends I_0. The remaining assertion follows immediately from 1.1.19(ii). □

It is not difficult to construct an example where L_0 is a proper subset of L_1 (see Exercise 1.4-2).

1.4.3. In the circle line itself, L can only be reached through N. As a consequence the class of integrable functions and the integral are both definitely determined at the first passage through L. To prove this we need the following lemma.

Lemma. Assume $N(B_0,I_0,\|\cdot\|_0)$, and follow the route $N(B_0,I_0,\|\cdot\|_0) \to L(L_0,I_0) \to \mathcal{D} \to A \to$ $\to N(B_1,I_1,\|\cdot\|_1)$. Let $\varphi \in \Phi$. Then $\|\varphi\|_0 \le \|\varphi\|_1$.

Proof. We may assume that $\|\varphi\|_1 < \infty$. The definition of $\|\varphi\|_1$ in 1.2.2, together with $N4$, shows

$$\|\varphi\|_1 = \inf\left\{\sum_{n=1}^{\infty} \|f_n\|_0 \;\middle|\; f_n \in L_0^+ \ (n \in \mathbb{N}),\ |\varphi| \le \sum_{n=1}^{\infty} f_n\right\}.$$

By $N3$ and $N2$, each of the numbers $\sum_{n=1}^{\infty}\|f_n\|_0$ is at least $\|\varphi\|_0$. □

1.4.4. Theorem. Assume $N(B_0,I_0,\|\cdot\|_0)$, and follow the route $N(B_0,I_0,\|\cdot\|_0) \to L(L_0,I_0) \to$ $\to \mathcal{D} \to A \to N(B_1,I_1,\|\cdot\|_1) \to L(L_1,I_1)$. Then $L_1 = L_0$, and $I_1(f) = I_0(f)$ for $f \in L_0$.

Proof. First, let $f \in L_1$ and $\varepsilon > 0$. There exists $b \in B_1 = L_0$ such that $\|f - b\|_1 < \varepsilon/2$, and $b_0 \in B_0$ such that $\|b - b_0\|_0 < \varepsilon/2$. By Lemma 1.4.3 and $N3$, $\|f - b_0\|_1 < \varepsilon$, which shows that $f \in L_0$. Hence $L_1 \subset L_0$. On the other hand, Proposition 1.4.2 implies that $L_0 \subset L_1$, and that L_1 extends L_0. Therefore $L_1 = L_0$ and $I_1 = I_0$. □

1.4.5. Lemma 1.4.3 leaves open the possibility that the norm increases in the circle line. Here is an example where this actually occurs.

Example. Let $S := [0,1)$, let B_0 be the class of constant real-valued functions on $[0,1)$, and let $I_0(b) := {}_0\int^1 b(x)dx$ for $b \in B_0$. Let A be the class of non-negative multiples of characteristic functions of intervals of type $[a,b)$, and let $J(h) := {}_0\int^1 h(x)dx$. For the system (B_0,I_0,A,J) the condition A holds because of the result of Example 1.2.5 (strictly speaking the functions must be extended to the entire real line by putting them zero outside S). The norm $\|\cdot\|_0$ generated is just the Lebesgue norm, and the integrable functions are those that are constant a.e. in the Lebesgue sense. It follows that if $\varphi \in \Phi$, then

$$\|\varphi\|_1 = \inf\{a \in \mathbb{R}^* \mid |\varphi| \leq a \text{ a.e. in the Lebesgue sense}\} .$$

Hence there are many functions φ with $\|\varphi\|_0 \neq \|\varphi\|_1$; for instance, the characteristic function of $[0,\frac{1}{2})$ is one.

1.4.6. The phenomenon of the preceding example is not very common, though. And it cannot occur if $\|\cdot\|_0$ can be described by means of integrable functions, that is, if the condition $A5$ of Subsection 1.2.6 is satisfied.

Proposition. Assume $A(B_0,I_0,A_0,J_0)$, and assume that $A5$ is satisfied. Follow the route

$$A(B_0,I_0,A_0,J_0) \to N(B_0,I_0,\|\cdot\|_0) \to L(L_0,I_0) \to D \to A(B_1,I_1,A_1,J_1) \to N(B_1,I_1,\|\cdot\|_1) .$$

Then $\|\varphi\|_0 = \|\varphi\|_1$ for every $\varphi \in \Phi$.

Proof. Let $\varphi \in \Phi$. In view of Lemma 1.4.3 we need only show that $\|\varphi\|_1 \leq \|\varphi\|_0$. We may assume $\|\varphi\|_0 < \infty$. By Proposition 1.2.7,

$$\|\varphi\|_0 = \inf\left\{\sum_{n=1}^{\infty} I_0(f_n) \mid f_n \in L_0^+ \ (n \in \mathbb{N}),\ |\varphi| \leq \sum_{n=1}^{\infty} f_n\right\} .$$

Since $B_1 = L_0$, the last infimum is just the definition of $\|\varphi\|_1$. □

1.4.7. We summarize the results obtained in this section. If one starts from anywhere except L, then both the class of integrable functions and the integral are definitely determined at the first passage through L. If one starts from L, D, D', or A with $A5$ satisfied, then the norm is definitely determined at the first passage through N. However, the norm may change if the starting-point is A or N.

1.4.8. We shall see later that $A5$ is satisfied in the case $R \to A$ (see the end of 3.2.10), or $T \to D' \to A$, or $T \to D' \to D \to A$ (see 4.1.7). Hence, if one starts from R or T (that is, in the cases commonly of interest), the class of integrable functions, the integral, and the norm are all fixed at the first confrontation with L or N.

Exercises Section 1.4

1. Follow the line $N(B_0,I_0,\|\cdot\|_0) \to L \to \mathcal{D} \to A \to N(B_1,I_1,\|\cdot\|_1)$. Show that $\|\varphi\|_0 = 0$ if and only if $\|\varphi\|_1 = 0$ for $\varphi \in \Phi$.

2. Follow the line $L(L_0,I_0) \to \mathcal{D} \to A \to N \to L(L_1,I_1)$. Give an example where L_0 is a proper subset of L_1.

3. Let $S = \{1,\ldots,n\} \times \{1,\ldots,n\}$ where $n \in \mathbb{N}$. Let B be the set of all functions $f : S \to \mathbb{R}$ which depend only on the first variable, and put $I(b) := \sum_{i=1}^n \sum_{j=1}^n b(i,j)$ for $b \in B$, $\|\varphi\|_0 := \sum_{i=1}^n \sum_{j=1}^n |\varphi(i,j)|$ for $\varphi \in \Phi$. What happens to the norm when we follow the line $N \to L \to \mathcal{D} \to A \to N$?

4. Start from $N(B_0,I_0,\|\cdot\|_0)$ and assume that the norm $\|\cdot\|_1$ obtained in the process $N \to L \to \mathcal{D} \to A \to N$ is the same as $\|\cdot\|_0$. Show that $\|\cdot\|_0$ satisfies 1.2.6 (*) (with L_0^+ instead of L^+.)

CHAPTER TWO

FURTHER DEVELOPMENT OF THE THEORY OF INTEGRATION

To increase its usefulness, the theory of the integral will now be developed into several directions. In particular, manipulation of integrals will be made easier.

The first section of this chapter is about measurability. In Lebesgue's original theory this concept was fundamental: first the measurable sets were described, then the measurable functions, and finally the integral was defined for certain measurable functions. In our set-up things go the other way round: we already know what the integrable functions are, and now define measurability by means of these, starting with L. Measurable functions behave much nicer than integrable ones. For instance, a pointwise limit of a sequence of measurable functions is always measurable, whether the convergence is dominated or not. The importance of the new concept is shown by Theorem 2.1.17: measurable functions that are not too big are integrable. Hence it is interesting to know which functions are measurable. In this connection we discuss measurability of certain composite functions. Here, and in some other places as well, we need the condition M (Stone's condition), which requires the underlying set S to be measurable. Fortunately, this condition is satisfied in most cases.

In the next section the starting point is station N. As shown in Section 1.1, the transition $N \to L$ furnishes integrable functions and an integral. Although the integral and the associated convergence theorems thus obtained are sufficient for many purposes, a further extension is desirable to facilitate handling of integrals. For instance, when talking about ${}_0\int^1 x^{-\frac{1}{2}}dx$, we don't like to be bothered by the fact that the integrand is not defined at the single point 0. The extension we shall give is based on the observation made at the end of 1.1.15: functions that are equal almost everywhere exhibit identical behavior with respect to integration. Thus we are able to integrate functions that are defined (possibly with infinite values) almost everywhere only, if only they are almost everywhere equal to a function that is integrable in the sense of Chapter 1.

This new extension, however, destroys the linear structure of the class of integrable functions. One of the difficulties is that expressions like $\infty - \infty$ cannot be handled consistently, and another that now functions are defined almost everywhere only. To cope with these difficulties we introduce equivalence classes of functions equal almost everywhere, and this set of equivalence classes is then provided with a natural linear structure. There are still problems left, though, for in many situations we start with classes, while in the course of the discussion we want to shift to functions. This predicament seems unavoidable, but we can alleviate it by not being too strict in our terminology.

The third section is about L^p-spaces, which occur in many parts of mathematics. These spaces come with a natural linear structure and a norm. If one takes as the distance between two elements of L^p the norm of their difference, then the metric space thus obtained is complete. This completeness property is very important in applications, because it enables us to assert for certain sequences in L^p that they are convergent even if we have no idea of what their limits might be.

The final section of this chapter treats the local norm, which in some cases gives a further possibility to extend the integral. Since the local norm is used just once in the sequel, and then for a minor point only, this section may be omitted at a first reading.

2.1. Measurable functions and measurable sets

In this section measurability is defined. This concept establishes a link with Lebesgue's approach, and it enables us to treat questions of more practical interest. For instance, if f and g are integrable functions, then $f \cdot g$, $\exp(f)$, g^2, and functions like these are not a priori integrable. But they are always measurable, and to ensure their integrability, it is (under M) sufficient that they have finite norms. Intimately connected with measurability are measures, which here are derived from integrals. (In Section 3.2 integrals will be constructed from measures.)

2.1.1. Our point of departure is station L of Section 1.1. Let L be a Riesz function space of real-valued functions defined on the non-empty set S, let I be a positive linear functional on L, and assume that $L1$ of 1.1.1 is satisfied.

The following notation will be convenient in the sequel. If f and g are $\mathbb{R}^*$-valued functions on S, and if $f \geq 0$, then $(g)_f$ is the function whose values are given by

$$(g)_f(s) := \begin{cases} f(s) & \text{if } g(s) > f(s) , \\ g(s) & \text{if } -f(s) \leq g(s) \leq f(s) , \\ -f(s) & \text{if } g(s) < -f(s) . \end{cases}$$

Roughly: the graph of $(g)_f$ is obtained from the graph of g by cutting away everything outside the area enclosed by the graphs of f and -f. In terms of sup and inf we have $(g)_f = \sup(\inf(g,f),-f)$.

2.1.2. Definition. Assume $L(L,I)$. An $\mathbb{R}^*$-valued function g on S is called *measurable* if $(g)_f \in L$ for every $f \in L^+$. A subset E of S is called *measurable* if its characteristic function χ_E is measurable.

The class of measurable functions is denoted by M, the subclass of finite-valued functions in M by M_F, and the class of measurable sets by Σ.

2.1.3. The following proposition shows first that measurable functions do exist, and then lists some simple properties of M which will be used time and again.

Proposition. Assume $L(L,I)$. Then the following assertions hold.

(i) $L \subset M$.

(ii) If $g \in M$, $\alpha \in \mathbb{R}$, then $\alpha g \in M$.

(iii) If $g \in M$, $h \in M$, then $\sup(g,h) \in M$, $\inf(g,h) \in M$.

(iv) If $(g_n)_{n\in\mathbb{N}}$ is a sequence in M, and if $g_n \to g$ pointwise, then $g \in M$.

(v) If $(g_n)_{n\in\mathbb{N}}$ is a sequence in M, then $\sup\limits_{n\in\mathbb{N}} g_n \in M$, $\inf\limits_{n\in\mathbb{N}} g_n \in M$, $\limsup\limits_{n\to\infty} g_n \in M$, $\liminf\limits_{n\to\infty} g_n \in M$.

Proof.

(i) This is obvious from the expression for $(g)_f$ at the end of 2.1.1 and the fact that L is a Riesz function space.

(ii) Let $g \in M$. First let $\alpha > 0$. If $f \in L^+$, then $(\alpha g)_f = \alpha(g)_{(1/\alpha)f} \in L$. Hence $\alpha g \in M$ in this case. Next, if $\alpha = 0$, the assertion is trivial. Finally, if $\alpha < 0$, use that $(-g)_f = -(g)_f$ together with the result for positive scalar multiples.

(iii) Write $k := \sup(g,h)$, and let $f \in L^+$. Then $(k)_f = \sup((g)_f,(h)_f)) \in L$. Hence $k \in M$. This result, however, together with (ii) and the observation $\inf(g,h) = -\sup(-g,-h)$, also gives $\inf(g,h) \in M$.

(iv) Let $f \in L^+$. It is clear that $(g_n)_f \to (g)_f$ pointwise. Since $(g_n)_f \in L$ and $|(g_n)_f| \le f$ $(n \in \mathbb{N})$, application of Lebesgue's theorem 1.1.4 gives $(g)_f \in L$. Hence $g \in M$.

(v) Apply (iii) repeatedly to get $\sup_{1\le n\le m} g_n \in M$ for all $m \in \mathbb{N}$, note that $\sup_{n\in\mathbb{N}} g_n = \lim_{m\to\infty} \sup_{1\le n\le m} g_n$, and use (iv). The remaining proofs are similar. □

M is not a Riesz function space, of course, since it is not closed under pointwise addition. However, if we restrict ourselves to the finite-valued members of M, then there are no difficulties with addition.

2.1.4. Theorem. M_F is a Riesz function space.

Proof. It is sufficient to show that M_F is closed under addition. Let $g \in M_F$, $h \in M_F$. Let $f \in L^+$, and define $\varphi := \lim_{n\to\infty} nf$. Then φ vanishes when f does, and takes the value ∞ elsewhere. Now $(g + h)_\varphi = (g)_\varphi + (h)_\varphi = \lim_{n\to\infty}\{(g)_{nf} + (h)_{nf}\}$, and 2.1.3(iv) shows that $(g + h)_\varphi$ is measurable. Since $(g + h)_f = ((g + h)_\varphi)_f \in L$, it follows that $g + h \in M$. It is obvious that $g + h$ is finite-valued. Therefore $g + h \in M_F$. □

2.1.5. Assertions (i) and (iv) of Proposition 2.1.3 show that the pointwise limit of a sequence of integrable functions is measurable. If the system is σ-finite (see 1.1.1), there is a converse, which provides us with a constructive characterization of the measurable functions.

Theorem. Let the system (L,I) be σ-finite, and let $g \in M$. Then there exists a sequence $(g_n)_{n\in\mathbb{N}}$ in L such that $g = \lim_{n\to\infty} g_n$.

Proof. Let $(f_n)_{n\in\mathbb{N}}$ be a sequence in L^+ such that $\sum_{n=1}^\infty f_n > 0$. For each $n \in \mathbb{N}$, put $h_n := n \sum_{k=1}^n f_k$ and $g_n := (g)_{h_n}$. Then $h_n \in L^+$, so $g_n \in L$ for each n. Moreover, $h_n \to \infty$ as $n \to \infty$, so $g_n \to g$ as $n \to \infty$. □

2.1.6. Now we turn to the measurable sets. Since these are characterized as sets whose characteristic functions belong to M_F, the algebraic properties of M_F as described in 2.1.3 and 2.1.4 are partly reflected in the behavior of the measurable sets under set theoretic operations such as unions and intersections. The appropriate notions are those of a σ-ring and a σ-algebra.

Definition. A non-empty class C of subsets of S is called a *σ-ring* if

(i) $A\setminus B \in C$ whenever $A \in C$, $B \in C$, and

(ii) $\bigcup_{n=1}^{\infty} A_n \in C$ whenever $A_n \in C$ $(n \in \mathbb{N})$.

A σ-ring C of subsets of S is called a *σ-algebra* if $S \in C$.

2.1.7. Proposition. Let C be a σ-ring of subsets of S. Then the following assertions hold.

(i) $\emptyset \in C$.

(ii) If $A \in C$, $B \in C$, then $A \cup B \in C$, and also $A \div B := (A\setminus B) \cup (B\setminus A) \in C$ (symmetric difference).

(iii) If $(A_n)_{n\in\mathbb{N}}$ is a sequence in C, then $\bigcap_{n=1}^\infty A_n \in C$.

(iv) If C is a σ-algebra, then $S\setminus A \in C$ whenever $A \in C$.

Proof. (i), (ii) and (iv) are trivial; (iii) follows from

$$\bigcap_{n=1}^{\infty} A_n = A_1 \setminus \bigcup_{n=1}^{\infty} (A_1 \setminus A_n) . \qquad \square$$

Thus, a σ-ring is closed under the formation of unions and intersections of countably many of its members, and also under symmetric differences. A σ-algebra is closed under complementation.

2.1.8. Recall that Σ denotes the class of measurable subsets of S.

Theorem. Assume $L(L,I)$. Then Σ is a σ-ring.

Proof. (i) Let $A,B \in \Sigma$. Then $\chi_{A\setminus B} = \chi_A - \inf(\chi_A, \chi_B) \in M_F$ by Theorem 2.1.4. Hence $A\setminus B \in \Sigma$.

(ii) Let $(A_n)_{n\in\mathbb{N}}$ be a sequence in Σ. The characteristic function of $\bigcup_{n=1}^{\infty} A_n$ equals $\sup_{n\geq 1} \chi_{A_n}$, and is therefore measurable by 2.1.3(v). $\square$

2.1.9. The integral I gives rise to a function μ on Σ in the following way: $\mu(A) := I(\chi_A)$ if $\chi_A \in L$, $\mu(A) := \infty$ if $\chi_A \in M\setminus L$. Obviously, μ is a much simpler object than I. Nevertheless, it is often possible to reconstruct I from μ (see Section 3.4). Here we point out some properties of μ that will turn out to be crucial in this reconstruction. Together they show that μ is a measure (the formal definition of this concept is in Section 3.1).

2.1.10. Theorem. Assume $L(L,I)$. Introduce μ as in 2.1.9. The following assertions hold.

(i) $\mu(\emptyset) = 0$.

(ii) Let $(A_n)_{n\in\mathbb{N}}$ be a sequence of pairwise disjoint measurable sets. Then

$$\mu\left(\bigcup_{n=1}^{\infty} A_n\right) = \sum_{n=1}^{\infty} \mu(A_n) .$$

(iii) Let $(A_n)_{n\in\mathbb{N}}$ be a decreasing sequence of measurable sets, that is, $A_{n+1} \subset A_n$ $(n \in \mathbb{N})$, and let $\mu(A_1) < \infty$. Then

$$\lim_{n\to\infty} \mu(A_n) = \mu\left(\bigcap_{n=1}^{\infty} A_n\right) .$$

Proof. (i) Trivial.

(ii) Let χ_0 be the characteristic function of $\bigcup_{n=1}^{\infty} A_n$, and χ_n that of A_n $(n \in \mathbb{N})$. Then $\chi_0 = \sum_{n=1}^{\infty} \chi_n$. First, if $\sum_{n=1}^{\infty} \mu(A_n) < \infty$, then $L1$ gives

$$\sum_{n=1}^{\infty} \mu(A_n) = \sum_{n=1}^{\infty} I(\chi_n) = I(\chi_0) = \mu\left(\bigcup_{n=1}^{\infty} A_n\right) .$$

Second, if $\sum_{n=1}^{\infty} \mu(A_n) = \infty$, then $\mu(\bigcup_{n=1}^{\infty} A_n)$ cannot be finite. To show this, assume $\mu(\bigcup_{n=1}^{\infty} A_n) < \infty$. Then $\chi_0 \in L$, and therefore $\chi_n = (\chi_n)_{\chi_0} \in L$ for each $n \in \mathbb{N}$. Since $0 \leq \sum_{n=1}^{N} \chi_n \leq \chi_0$, and I is linear and positive,

$$\sum_{n=1}^{N} \mu(A_n) = \sum_{n=1}^{N} I(\chi_n) = I\left(\sum_{n=1}^{N} \chi_n\right) \leq I(\chi_0) = \mu\left(\sum_{n=1}^{\infty} A_n\right)$$

for each N. This shows $\sum_{n=1}^{\infty} \mu(A_n) < \infty$, which is impossible. Hence $\sum_{n=1}^{\infty} \mu(A_n) = \infty = \mu(\bigcup_{n=1}^{\infty} A_n)$ in this case.

(iii) Let χ_0 be the characteristic function of $\bigcap_{n=1}^{\infty} A_n$. Then $\chi_{A_n} \in L$, $|\chi_{A_n}| \leq \chi_{A_1}$ $(n \in \mathbb{N})$, and $\chi_{A_n} \to \chi_0$, so the assertion follows from Lebesgue's theorem 1.1.4. □

2.1.11. In Lebesgue's approach the measurable functions are defined by means of measurable sets. In a situation $\mathcal{L}(L,I)$ on a set S we can obtain a characterization similar to Lebesgue's definition if we add an additional condition $\mathcal{M}$ (due to Stone), which reads as follows.

$\mathcal{M}$. S is a measurable set.

An obvious reformulation of $\mathcal{M}$ is: Σ is a σ-algebra, or also, the constant functions on S are measurable. We note that $\mathcal{M}$ and the σ-finiteness condition $\mathcal{L}2$ are unrelated, in the sense that neither implies the other; see Exercises 2.1.10 and 2.1.11. The condition $\mathcal{M}$ is vital in the discussion of composite functions later in this section, and also in the reconstruction of I from μ in Section 3.4.

2.1.12. Theorem. Assume $\mathcal{L}(L,I)$ with $\mathcal{M}$. Let g be an $\mathbb{R}^*$-valued function defined on S. Then g is measurable if and only if for every $a \in \mathbb{R}$ the set $E_a := \{s \in S \mid g(s) < a\}$ is measurable.

Proof. First, let g be measurable, and $a \in \mathbb{R}$. Assume for the moment that g does not take the value ∞. For each $n \in \mathbb{N}$, put

$$h_n := n\{\sup(g, a\chi_S) - \sup(g, (a - \tfrac{1}{n})\chi_S)\} .$$

By $\mathcal{M}$, Proposition 2.1.3 and Theorem 2.1.4, each h_n is measurable. Also, $h_n(s) = 0$ if $g(s) \geq a$, $h_n(s) = 1$ if $g(s) < a - 1/n$, so the sequence $(h_n)_{n\in\mathbb{N}}$ tends pointwise to the characteristic functions of E_a, which is therefore measurable by Proposition 2.1.3(iv). Now drop the restriction on g. Let $h := \inf(g, a\chi_S)$. Then h is measurable by $\mathcal{M}$ and Proposition 2.1.3(iii), so E_a, which is equal to $\{s \in S \mid h(s) < a\}$, is measurable by the special case applied to h.

Conversely, assume that each E_a is measurable. It is easy to see that $F := \{s \in S \mid g(s) = -\infty\}$ and $G := \{s \in S \mid g(s) = \infty\}$ are measurable. For $n \in \mathbb{N}$ and $-2^{2n} \le k \le 2^{2n}$, define χ_{kn} to be the characteristic function of $E_{(k+1)2^{-n}} \setminus E_{k2^{-n}}$, and put

$$g_n := \sum_{k=-2^{2n}}^{2^{2n}} k \cdot 2^{-n} \chi_{kn} - 2^n \chi_F + 2^n \chi_G .$$

The g_n are measurable, and $g_n \to g$. So g is measurable by 2.1.3(iv). □

2.1.13. The second part of the preceding proof contains a result worth to be noted separately. To describe it neatly, let us agree to call a real-valued function on S *simple*, if it takes only finitely many values. It follows easily from Theorem 2.1.12 that under condition M a measurable simple function is just a linear combination of finitely many characteristic functions of measurable sets. The result meant is: if M is satisfied, then every measurable function is a pointwise limit of a sequence of measurable simple functions. So if M holds we have here a further constructive characterization of the measurable functions besides the one contained in Theorem 2.1.5 for the σ-finite case.

2.1.14. Often we would like to know whether a certain composite function is measurable. This occurs, for instance, if we want to apply Theorem 2.1.17 below. If M is satisfied, there is a neat answer to this question. Since M is not very restrictive (it is satisfied in the cases we shall consider in Chapters 3 and 4), this answer suffices for most cases of interest. To formulate it, we need the notion of a Baire function on $\mathbb{R}$. (In Section 4.4 we shall meet Baire functions in a more general context.)

The class of *Baire functions* on $\mathbb{R}$ consists of $\mathbb{R}^*$-valued functions. It is obtained in a somewhat peculiar way. Consider the family $\mathcal{X}$ of classes X, where each class X consists of $\mathbb{R}^*$-valued functions on $\mathbb{R}$ and has the following two properties:

(i) X contains all continuous real-valued functions on $\mathbb{R}$,

(ii) X is closed under pointwise sequential limits, that is, if $(\varphi_n)_{n \in \mathbb{N}}$ is a sequence in X that tends pointwise on $\mathbb{R}$ to an $\mathbb{R}^*$-valued function φ, then $\varphi \in X$.

Obviously, $\mathcal{X}$ is non-empty, for the class of all $\mathbb{R}^*$-valued functions on $\mathbb{R}$ trivially belongs to it. The *class of Baire functions* on $\mathbb{R}$ (the *Baire class*) is the intersection $\bigcap_{X \in \mathcal{X}} X$.

It is clear that the class of Baire functions itself belongs to $\mathcal{X}$, so we can also describe the Baire class as the smallest class of real-valued functions on $\mathbb{R}$ that has properties (i) and (ii) above. ("Smallest" means that it is contained in every member of $\mathcal{X}$.)

From (i) it follows that any continuous real-valued function on $\mathbb{R}$ is a Baire function, and so is any pointwise limit of a sequence of continuous functions, by (ii). The procedure of taking pointwise limits can be iterated, but we never leave the class of Baire functions in this way. This implies that it cannot be easy to write down a non-Baire function, and the example we shall meet later in 3.3.10 will be described in a very non-constructive way.

2.1.15. The usage of the definition of the Baire functions is well illustrated in the proof of the following useful theorem. (We shall meet more examples of the same kind of reasoning later.)

Theorem. Assume $\mathcal{L}(L,I)$ and $\mathcal{M}$. Let $f \in M_F$, and let φ be a Baire function on $\mathbb{R}$. Then the composite function $\varphi \circ f$ is measurable.

Proof. Let X be the class of $\mathbb{R}^*$-valued functions ψ on $\mathbb{R}$ such that $\psi \circ f$ is measurable. We are going to show that $X \in \mathcal{X}$, the family introduced in the second paragraph of 2.1.14. Then X will contain all Baire functions, which will show that $\varphi \circ f$ is measurable.

First, let ψ be a continuous real-valued function on $\mathbb{R}$, and let $a \in \mathbb{R}$. Since ψ is continuous, the set $U := \{u \in \mathbb{R} \mid \psi(u) < a\}$ is open. Hence U is the union of countably many open intervals, $U = \bigcup_{n=1}^{\infty} U_n$, say (see Exercise 0.2-7). Now note that

$$\{s \in S \mid (\psi \circ f)(s) < a\} = \bigcup_{n=1}^{\infty} \{s \in S \mid f(s) \in U_n\} .$$

It follows easily from Theorem 2.1.12 that each of the sets in the union is measurable. Therefore the union is itself measurable, by Theorem 2.1.8. Another application of Theorem 2.1.12 (but now in the reverse direction) shows that $\psi \circ f$ is measurable, that is, $\psi \in X$. So X satisfies (i) of 2.1.14.

Next, let $(\psi_n)_{n\in\mathbb{N}}$ be a sequence in X that converges pointwise to an $\mathbb{R}^*$-valued function ψ on $\mathbb{R}$. By assumption, each $\psi_n \circ f$ is measurable. Since $\psi_n \circ f \to \psi \circ f$ pointwise, $\psi \circ f$ is measurable by 2.1.3(iv). Hence $\psi \in X$, and therefore X also satisfies (ii) of 2.1.14.

So $X \in \mathcal{X}$, as we set out to prove. □

2.1.16. Examples. Assume $\mathcal{L}(L,I)$ along with $\mathcal{M}$.

(i) If $p > 0$, $f \in M_F$, then $|f|^p \in M_F$. This follows from Theorem 2.1.15 if we take $\varphi := \curlyvee_{x\in\mathbb{R}} |x|^p$.

(ii) If $f \in M_F$, $g \in M_F$, then the product function $f\cdot g \in M_F$. In fact,

$$f\cdot g = \tfrac{1}{4}(f + g)^2 - \tfrac{1}{4}(f - g)^2 ,$$

and the result follows from (i) with $p = 2$, and Theorem 2.1.4.

2.1.17. The importance of the notion of measurability may be seen from the final result of this section, which shows that often measurability is not very far from integrability.

Theorem. Assume $N(B,I,\|\cdot\|)$, and follow the route $N \to L(L,I)$. Let $f \in M_F$, $\|f\| < \infty$, and assume that there exists a sequence $(f_n)_{n\in\mathbb{N}}$ in L^+ such that $|f| \le \sum_{n=1}^{\infty} f_n$. Then $f \in L$.

Proof. For $k \in \mathbb{N}$, put $g_k := \sum_{n=1}^{k} f_n$. Then $(g_k)_{k\in\mathbb{N}}$ is a pointwise increasing sequence in L^+, with the property that $(f)_{g_k} \to f$, while $|(f)_{g_k}| \le |f|$ for all k. Since f is measurable, $(f)_{g_k} \in L$ for all k. Now Theorem 1.1.16 shows that $f \in L$. □

In case we arrived at N from a station A where $A5$ was satisfied, the theorem takes a much simpler form. In fact, if $\|f\| < \infty$, then $A5$ ensures the existence of a sequence $(f_n)_{n\in\mathbb{N}}$ in L^+ such that $|f| \le \sum_{n=1}^{\infty} f_n$. The theorem then reads: a finite measurable function with finite norm is integrable. If the system $N(B,I,\|\cdot\|)$ is σ-finite, the situation is similar: now for every real-valued f on S there is a sequence $(b_n)_{n\in\mathbb{N}}$ in B^+ with $|f| \le \sum_{n=1}^{\infty} b_n$.

Exercises Section 2.1

1. Start from $N(B,I,\|\cdot\|)$ and follow $N \to L$. Show that $f \in \Phi$ is measurable if and only if $(f)_b \in L$ for every $b \in B^+$.

2. Start from $N(B,I,\|\cdot\|)$ and follow $N \to L$. Show that if $f \in \Phi$ and $(f_n)_{n\in\mathbb{N}}$ is a sequence in M with $f_n \to f$ (a.e.) then f is measurable.

3. Start from L and assume that M is satisfied. Show that $E \subset S$ is measurable if and only if $E \cap F$ is measurable for every measurable set F of finite measure.

4. Start from $L(L,I)$. Let $f \in L$ and let $E \subset S$ be measurable. Show that $f\cdot\chi_E \in L$.

5. Start from $A + A5$ and follow the route $A \to N \to L$. Let $(E_n)_{n\in\mathbb{N}}$ be a sequence of measurable subsets of S such that $E_n \cap E_m$ is a null set if $n \ne m$. Let $\varphi \in \Phi$ and put $E := \bigcup_{n=1}^{\infty} E_n$. Show that

$$\|\varphi\cdot\chi_E\| = \sum_{n=1}^{\infty} \|\varphi\cdot\chi_{E_N}\| .$$

(Hint. See Exercise 1.2-16 and use Exercise 4.)

6. Show by an example that the condition $A5$ in Exercise 5 cannot be omitted. (Hint. Take $S = \{0,1\}$, B and I as simple as possible and A and J such that the resulting norm $\|\cdot\|$ satisfies $\|\varphi\| = \max(|\varphi(0)|,|\varphi(1)|)$ for $\varphi \in \Phi$.)

7. Start from $A + A5$, follow the route $A \to N \to L$ and assume that M is satisfied. Show that $E \subset S$ is measurable if and only if

$$\|\chi_A\| = \|\chi_{A\cap E}\| + \|\chi_{A\cap E^*}\|$$

for every measurable set A of finite measure. (Hint. For the "if" part one can assume, by Exercise 3, that $\|\chi_E\| < \infty$. Now take $f \in L^+$ with $f \geq \chi_E$ (a.e.), $I(f) = \|\chi_E\|$ and consider $F := \{s \in S \mid f(s) \geq 1\}$.)

8. Start from $A + A5$, follow the route $A \to N \to L$ and assume that M is satisfied. Show that $\varphi \in \Phi^+$ is measurable if and only if for every $f \in L^+$

$$\|f\| = \|f\cdot\chi_E\| + \|f\cdot\chi_{E^*}\|$$

where $E := \{s \in S \mid \varphi(s) \leq f(s)\}$. (Hint. Use 2.1.12 and Exercise 7.)

9. Let C be a class of subsets of S such that

(i) if $A,B \in C$ then $A\setminus B \in C$,

(ii) if $(A_n)_{n\in\mathbb{N}}$ is a sequence of pairwise disjoint elements of C then $\bigcup_{n=1}^{\infty} A_n \in C$.

Show that C is a σ-ring.

10. Let $S := [0,1]$, $L := \{\Upsilon_{s\in S}\, \alpha s \mid \alpha \in \mathbb{R}\}$, $I(f) := f(1)$. Show that

(i) (L,I) satisfies the conditions $L1$ and $L2$ of 1.1.1,

(ii) $\emptyset$ is the only measurable set.

11. Give an example of a non σ-finite situation $L(L,I)$ where M is satisfied.

12. Give an example of a station $N(B,I,\|\cdot\|)$ where M is satisfied and where there is an $f \in M_F$ with $\|f\| < \infty$, $f \notin L$. Try to find an example with $B \neq \{0\}$.

13. Start from $L(L_0,I_0)$ and follow the route $L(L_0,I_0) \to D \to A \to N(B_1,I_1,\|\cdot\|_1) \to$ $\to L(L_1,I_1)$. Denote the respective classes of measurable functions by M_0 and M_1. Show that $M_0 \subset M_1$ and give an example where the inclusion is proper. Show that if $f \in M_1$ is real-valued and $\|f\|_1 < \infty$ then $f \in L_1$. Assume in addition that $L(L_0,I_0)$ is derived from $N(B_0,I_0,\|\cdot\|_0)$. Show that if $f \in M_0\setminus L_0$ is real-valued and $\|f\|_0 < \infty$, then $\|f\|_1 = \infty$.

14. Start from $L(L,I)$ and assume that M is satisfied. Let $(f_n)_{n\in\mathbb{N}}$ be a sequence in M. Show that $V := \{s \in S \mid \lim_{n\to\infty} f_n(s)$ exists and is finite$\}$ is measurable. (Hint. Consider

$$\bigcap_{\ell=1}^{\infty} \bigcup_{N=1}^{\infty} \bigcap_{n=N}^{\infty} \bigcap_{m=N}^{\infty} \{s \in S \mid |f_n(s) - f_m(s)| < \ell^{-1}\} .)$$

15. Introduce the Baire class of $\mathbb{R}^*$-valued functions of several real variables, formulate and prove Theorem 2.1.15 for this case and derive a new proof of 2.1.16, Example (ii).

16. Start from $L(L,I)$ and assume that M is satisfied. Let $f \in M$ and let φ be a Baire function for which $\lim_{x\to\infty} \varphi(x)$ and $\lim_{x\to-\infty} \varphi(x)$ exist. Define $\varphi(\pm\infty) := \lim_{x\to\pm\infty} \varphi(x)$. Show that $\varphi \circ f$ is measurable.

17. Start from $L(L,I)$ and assume that M is satisfied. Let f and g be integrable. Show that $f\cdot g$, $\sin f$, f^2 are integrable if they have finite norm.

2.2. Further extension of the integral

Integration is a practical tool, so the more functions we can integrate the better. We know already that if we start from N, and then follow $N \to L$, two integrable functions that are equal almost everywhere have the same integral. This is the clue to our final extension of the integral: if f is integrable, and $g = f$ almost everywhere, then g gets the same integral as f.

Simple as the definition looks, some complications arise from dealing with functions that are defined almost everywhere only: the algebraic operations must be defined carefully. The final solution of these difficulties is the introduction of equivalence classes of functions, to be treated in the next section.

The mathematical contents of the present section are rather meagre; once the underlying idea is grasped, everything runs along predictable lines.

2.2.1. Let S be a non-empty set. We start from station N. By an $\mathbb{R}^*$-valued function f defined almost everywhere on S we mean the following: there exists a null set $N \subset S$ such that f is defined on $S\backslash N$ with values in $\mathbb{R}^*$. What happens on N is immaterial: f need not be defined there or f may take values in some strange set. The class of $\mathbb{R}^*$-valued functions defined a.e. is denoted by Φ^*, and the subclass of functions that are finite a.e. is denoted by Φ_F^*. Note that $\Phi \subset \Phi^*$ and $\Phi_F \subset \Phi_F^*$.

2.2.2. Relations between members of Φ^* must be interpreted in the almost everywhere sense. For instance, if $f,g \in \Phi^*$, then $f \geq g$ (a.e.) means that there exists a null set N such that both f and g are defined on $S \setminus N$, while $f(s) \geq g(s)$ for $s \in S \setminus N$.

2.2.3. Let $f,g \in \Phi^*$. Since the union of two null sets is a null set, there exists a null set $N \subset S$ such that both f and g are defined on $S \setminus N$. Now $\sup(f,g)$ denotes any $h \in \Phi^*$ such that h is defined on $S \setminus N$ with values in $\mathbb{R}^*$, while $h(s) = \sup(f(s),g(s))$ if $s \in S \setminus N$. The meanings of $\inf(f,g)$ and $|f|$ are defined similarly. Finally, if, in addition, $f \geq 0$ (a.e.), then it may also be assumed that $f(s) \geq 0$ if $s \in S \setminus N$, and $(g)_f$ is defined to be any member of Φ^* that takes at $s \in S \setminus N$ the value

$$\begin{cases} g(s) & \text{if } |g(s)| \leq f(s) , \\ f(s) & \text{if } g(s) > f(s) , \\ -f(s) & \text{if } g(s) < -f(s) . \end{cases}$$

The reader will note that we did not define $\sup(f,g)$, $\inf(f,g)$, etc. as unique members of Φ^*. This lack of uniqueness is not serious, because the candidates are equivalent anyway, and we shall soon get accustomed to it.

2.2.4. Let $(f_n)_{n \in \mathbb{N}}$ be a sequence in Φ^*. By $\sup_{n \in \mathbb{N}} f_n$, $\inf_{n \in \mathbb{N}} g_n$, $\limsup_{n \to \infty} f_n$, $\liminf_{n \to \infty} f_n$ we denote functions defined on $S \setminus N$, where N is a null set, taking the values $\sup_{n \in \mathbb{N}} f_n(s)$, $\inf_{n \in \mathbb{N}} f_n(s)$, $\limsup_{n \to \infty} f_n(s)$, $\liminf_{n \to \infty} f_n(s)$, respectively, at $s \in S \setminus N$. (Note that there is such an N, since a countable union of null sets is still a null set.) If $f \in \Phi^*$, $f = \limsup_{n \to \infty} f_n$ (a.e.) and $f = \liminf_{n \to \infty} f_n$(a.e.), then we write $f = \lim_{n \to \infty} f_n$ (a.e.), or also $f_n \to f$ (a.e.).

2.2.5. Scalar multiplication in Φ^* is defined as follows. Let $f \in \Phi^*$ and $\alpha \in \mathbb{R}$. Let $N \subset S$ be a null set outside of which f is defined with values in $\mathbb{R}^*$. Then αf is any member h of Φ^* with $h(s) = \alpha f(s)$ if $s \in S \setminus N$. (Recall that the expression $\alpha f(s)$ makes sense even if $f(s) = \pm \infty$.) If, in addition, $f \in \Phi_F^*$, then $\alpha f \in \Phi_F^*$.

If $f,g \in \Phi^*$, then the product function $f \cdot g$ can be defined similarly. This function $f \cdot g$ is again an element of Φ^*.

Things are a little different with addition. Here we have to restrict ourselves to those members of Φ^* that are finite almost everywhere. Let $f,g \in \Phi_F^*$. There exists a null set N outside of which both f and g are defined and have values in $\mathbb{R}$. Then $f + g$ is any function $k \in \Phi^*$ such that $k(s) = f(s) + g(s)$ if $s \in S \setminus N$. Obviously, such a k still belongs to Φ_F^*. Infinite sums can now be defined in Φ_F^* by combining the sum definition with the limit definition in the preceding subsection. Note that if $f_n \in \Phi_F^*$ $(n \in \mathbb{N})$, and if $\sum_{n=1}^{\infty} f_n$ is defined, the result need not be in Φ_F^*.

Obviously, Φ_F^* with addition and scalar multiplication as defined here is not in general a linear space. For instance, every null function (that is, every function that is zero almost everywhere) acts as a neutral element for the operation of addition, and there may be many null functions. This undesirable state of affairs will be remedied in the next section by the formation of equivalence classes of functions. For the time being, however, it is more convenient to talk about functions and function values.

2.2.6. The members of Φ^* do not differ substantially from those of Φ, and it is easy to extend the norm to Φ^*.

Definition. Let $f \in \Phi^*$, and let $f_0 \in \Phi$ be such that $f = f_0$ (a.e.). The norm of f is defined by $\|f\| := \|f_0\|$.

The value of $\|f\|$ obtained does not depend on the particular element $f_0 \in \Phi$ chosen in defining it (see 1.1.15(iii)), and on Φ the new norm coincides with the old one, which justifies use of the old notation. The norm on Φ^* satisfies the conditions $N1$ through $N4$ of 1.1.5, and assertions (i) through (iv) of Proposition 1.1.15 also hold, if one extends the notion of null function in the obvious way. (Proofs are left to the reader.)

2.2.7. At this point we might introduce integrability in Φ^* as in Definition 1.1.8, and then repeat most of the proofs of Chapter 1. We prefer a different approach, however, which enables us to use the relevant results directly, perhaps by throwing in an occasional "almost everywhere".

Definition. Assume $N(B,I,\|\cdot\|)$, and follow $N \to L(L,I)$. If $f \in \Phi^*$ and there exists $f_0 \in L$ such that $f = f_0$ (a.e.), then f is called *integrable*, and the *integral* of f is $I(f) := I(f_0)$. The class of integrable functions in Φ^* is denoted by L^*.

Again, it must be checked that the value of I(f) does not depend on the particular element $f_0 \in L$ chosen in the definition, and also that we do not get into conflicts with the old definition on L. This is easily done.

The extended integral is still positive and linear. Indeed, if $f \in L^*$ and $f \geq 0$, then for every $f_0 \in L$ with $f_0 = f$ (a.e.) we have $f_0 \geq 0$ (a.e.), hence $I(f) = I(f_0) \geq 0$. Also, if $f,g \in L^*$, then f and g are in Φ_F^*. So if $\alpha,\beta \in \mathbb{R}$, then $\alpha f + \beta g \in \Phi_F^*$, and taking auxiliary elements in L we see that $I(\alpha f + \beta g) = \alpha I(f) + \beta I(g)$. Moreover, we still have the estimate $|I(f)| \leq \|f\|$ if $f \in L^*$, with equality if $f \geq 0$ (a.e.).

The relation of our definition of L^* to the old one for L is expressed by the following result: $f \in \Phi^*$ is integrable if and only if for every $\varepsilon > 0$ there exists $b \in B$ such that $\|b - f\| < \varepsilon$. Proof of this fact is left to the reader.

2.2.8. To show how to adapt the results of Chapter 1 to the present situation, we treat Lebesgue's theorem 1.1.16.

Theorem (Dominated convergence theorem). Assume $N(B,I,\|\cdot\|)$ and $L(L,I)$ derived from it. Let (f_n) be a sequence in L^*, let $h \in \Phi^*$ with $\|h\| < \infty$ and $h \geq 0$ (a.e.). Assume that $|f_n| \leq h$ (a.e.) for $n \in \mathbb{N}$, and that the sequence $(f_n)_{n\in\mathbb{N}}$ converges a.e. to $f \in \Phi^*$. Then $f \in L^*$, and $I(f) = \lim_{n\to\infty} I(f_n)$.

Proof. For each $n \in \mathbb{N}$, take $f'_n \in L$ with $f'_n = f_n$ (a.e.), and take $h' \in \Phi$ such that $h' = h$ (a.e.). Since $|f| \leq |h|$ (a.e.), we have $\|f\| < \infty$, and therefore there exists $f' \in \Phi_F$ such that $f' = f$ (a.e.). Now the original form of Lebesgue's theorem 1.1.16 shows that $f' \in L$, and $I(f'_n) \to I(f')$. So by Definition 2.2.7, $f \in L^*$ and $I(f_n) \to I(f)$. □

2.2.9. To define measurability in Φ^* we proceed as with integrability.

Definition. Assume $N(B,I,\|\cdot\|)$, follow the route $N \to L(L,I)$, and denote by M the class of measurable functions as defined in 2.1.2. A function $g \in \Phi^*$ is called *measurable* if there exists $g_0 \in M$ such that $g = g_0$ (a.e.). The set of measurable members of Φ^* is denoted by M^*, and the subset of measurable functions that are finite a.e. is denoted by M^*_F. So $M^*_F = M^* \cap \Phi^*_F$.

In this case too we might choose to copy the original definition. It is a fortunate fact, though, that this leads to the same result. That is: $g \in \Phi^*$ belongs to M^* if and only if $(g)_f \in L^*$ for every $f \in L^+$. (L^+ may also be replaced by $\{f \in L^* \mid f \geq 0 \text{ (a.e.)}\}$.)

As in 2.1.3 we have $L^* \subset M^*$; M^* is closed with respect to the operations sup, inf, lim sup, lim inf if countably many members of M^* are involved; an almost everywhere limit of a sequence in M^* is still in M^*; M^*_F is closed with respect to addition and scalar multiplication; and so on. The proofs consist in redefining the functions on suitable null sets so that they become measurable in the familiar sense of Section 2.1, employing the old results, and going back by means of the definition; see the proof of Theorem 2.2.8 for an example.

2.2.10. Finally we extend the integral to all non-negative measurable functions.

Definition. Assume $N(B,I,\|\cdot\|)$, and follow $N \to L(L,I)$. If $f \in M^*$, $f \notin L^*$, $f \geq 0$ (a.e.), then $I(f) := \infty$.

It is easy to see that this extended I is still positive, that $I(\alpha f) = \alpha I(f)$ if $f \in M^*$, $f \geq 0$ (a.e.), $\alpha \geq 0$, and that $I(f + g) = I(f) + I(g)$ if $f \in M^*$, $g \in M^*$, $f \geq 0$ (a.e.), $g \geq 0$ (a.e.). Moreover, the extended I is related to the set function μ

introduced in 2.1.9: μ is just the restriction of the extended I to the measurable characteristic functions in Φ^*. (We ignore here the difference between a set and its characteristic function!)

2.2.11. The last definition does not constitute a very significant extension, of course, but nevertheless it is rather convenient. For instance, the monotone convergence theorem 1.1.18 takes the following pleasant form now.

Theorem (Monotone convergence theorem). Assume $N(B,I,\|\cdot\|)$ and follow $N \to L(L,I)$. Let $(f_n)_{n\in\mathbb{N}}$ be a sequence in M^*, and assume $0 \le f_n \le f_{n+1}$ (a.e.) for all $n \in \mathbb{N}$. Let $f \in \Phi^*$ satisfy $f = \lim_{n\to\infty} f_n$ (a.e.). Then $f \in M^*$, $f \ge 0$ (a.e.) and $I(f_n) \to I(f)$.

Proof. We know from the remarks in 2.2.9 that $f \in M^*$. Moreover, it is obvious that $f \ge 0$ (a.e.). Hence $I(f)$ is well-defined.

Since $0 \le f_n \le f_{n+1} \le f$ (a.e.) for $n \in \mathbb{N}$, it follows from the remarks in 2.2.10 that $I(f_n) \le I(f_{n+1}) \le I(f)$ for all n. This shows that $\lim_{n\to\infty} I(f_n)$ exists, and also that $\lim_{n\to\infty} I(f_n) \le I(f)$. In the proof of the reverse inequality we may assume $\lim_{n\to\infty} I(f_n) < \infty$. For each $n \in \mathbb{N}$, Definition 2.2.10 shows that $f_n \in L^*$, and hence there exists $f'_n \in L$ such that $f'_n = f_n$ (a.e.). Now $0 \le f'_n \le f'_{n+1}$ (a.e.) for all n, and $\lim_{n\to\infty} I(f'_n)$ is finite, so the original monotone convergence theorem 1.1.18 implies the existence of $f' \in L^+$ such that $f' = \lim_{n\to\infty} f'_n$ (a.e.), while $I(f'_n) \to I(f')$. Since obviously $f = f'$ (a.e.), we have $f \in L^*$, and $I(f_n) \to I(f)$. □

2.2.12. The final result of this section is another example of the convenience of the definition in 2.2.10. We label the result as a theorem, but it is always referred to as Fatou's lemma. Occasionally it comes in handy in proving that a particular function is integrable (two typical examples occur in the proof of Theorem 2.3.11).

Theorem (Fatou's lemma). Assume $N(B,I,\|\cdot\|)$, and follow $N \to L(L,I)$. Let $(f_n)_{n\in\mathbb{N}}$ be a sequence in M^* with $f_n \ge 0$ (a.e.) for $n \in \mathbb{N}$, and let $f := \liminf_{n\to\infty} f_n$. Then $f \in M^*$, $f \ge 0$ (a.e.) and $I(f) \le \liminf_{n\to\infty} I(f_n)$. In particular, $f \in L^*$ whenever $\liminf_{n\to\infty} I(f_n)$ is finite.

Proof. For $n \in \mathbb{N}$, put $g_n := \inf_{m\ge n} f_m$. Then $g_n \in M^*$, $g_n \ge 0$ (a.e.) for each n, and $g_n \uparrow f$ (a.e.). Hence $f \ge 0$ (a.e.), $f \in M^*$, and $I(f) = \lim_{n\to\infty} I(g_n)$ by monotone convergence (2.2.11). Since $g_n \le f_m$ (a.e.) if $n,m \in \mathbb{N}$, $m \ge n$, we have $I(g_n) \le \inf_{m\ge n} I(f_m)$. This implies $\lim_{n\to\infty} I(g_n) \le \liminf_{n\to\infty} I(f_n)$. □

We note that it may well occur that $I(f) < \liminf_{n\to\infty} I(f_n)$ in the above theorem. (See exercise 2.2-6.)

Exercises Section 2.2

1. Show that the extension of the integral in 2.2.10 is positive and that $I(\sum_{n=1}^{\infty} f_n) = \sum_{n=1}^{\infty} I(f_n)$ for a sequence $(f_n)_{n\in\mathbb{N}}$ in M^* with $f_n \geq 0$ (a.e.) for all n.

2. If A is a measurable subset of S, then $\mu(A) = I(\chi_A)$. Here μ is the set function defined in 2.1.9 and I is the extension of the integral according to 2.2.10.

3. Show that $I(f) = \|f\|$ for $f \in M^*$, $f \geq 0$ (a.e.) if the system is σ-finite. Give an example of a system $(B, I, \|\cdot\|)$ and an $f \in M^*$, $f \geq 0$ (a.e.) such that $I(f) \neq \|f\|$.

4. Let $f \in M^*$, $f \geq 0$ (a.e.) and assume that there is a sequence $(f_n)_{n\in\mathbb{N}}$ in L^+ with $f_n \geq 0$ (a.e.) such that $\sum_{n=1}^{\infty} f_n \geq f$ (a.e.). Show that $I(f) = \|f\|$.

5. Assume that the station N is derived from A + $A5$. Show that $I(f) = \|f\|$ for $f \in M^*$, $f \geq 0$ (a.e.).

6. Give an example of a system $(B, I, \|\cdot\|)$ and a sequence $(f_n)_{n\in\mathbb{N}}$ in M^* with $f_n \geq 0$ (a.e.) such that $I(\liminf_{n\to\infty} f_n) < \liminf_{n\to\infty} I(f_n)$.

7. Generalize Theorem 2.2.12 as follows. Assume that the norm $\|\cdot\|$ has the Fatou property of 1.2.9. Let $\varphi_n \in \Phi$, $\varphi_n \geq 0$ (a.e.). Show that $\|\liminf_{n\to\infty} \varphi_n\| \leq \liminf_{n\to\infty} \|\varphi_n\|$.

2.3. L^p-spaces

In the preceding section we agreed on what to understand by sup(f,g), f + g, and so on, if the functions f and g are defined almost everywhere. We worked with these objects without being much troubled by the fact that they were not really defined in an unambiguous way. Yet, things look untidy, and we shall clarify this point in the present section. Roughly, the solution is that we no longer distinguish between functions that differ by a null function only.

This is also a natural place to introduce the L^p-spaces, which occur in many applications. Fundamental in the L^p-theory are the Hölder and Minkowski inequalities. We turn to these first.

2.3.1. Start with a situation $N(B, I, \|\cdot\|)$ for a non-empty set S. Recall that Φ_F^* is the class of $\mathbb{R}^*$-valued functions that are defined and finite almost everywhere on S.

Let $f \in \Phi_F^*$. For $1 \le p \le \infty$ the *p-norm* of f is defined by

$$\|f\|_p := \||f|^p\|^{1/p} \quad \text{if } 1 \le p < \infty ,$$

$$\|f\|_p := \inf\{a \in \mathbb{R} \mid |f| \le a \text{ (a.e.)}\} \quad \text{if } p = \infty .$$

In the first of these expressions, both ∞^p and $\infty^{1/p}$ must be read as ∞. Obviously, $\|f\|_1$ is just $\|f\|$, and the subscript is often omitted in this case. The number $\|f\|_\infty$ is often called the *essential supremum* of $|f|$, where the adjective essential refers to our disregarding what happens on null sets.

2.3.2. In the context of L^p-spaces, each p with $1 \le p \le \infty$ has a q associated with it. This q, called the *conjugate exponent* of p, is defined by

$$q := \begin{cases} 1 & \text{if } p = \infty , \\ \infty & \text{if } p = 1 , \\ \dfrac{p}{p-1} & \text{if } 1 < p < \infty . \end{cases}$$

So if we interpret $1/\infty$ as 0, then $1/p + 1/q = 1$ in each case. Note that q and p are conjugate exponents of each other.

2.3.3. Our first result is called *Hölder's inequality*. (The special case $p = q = 2$ is called the *Cauchy-Schwarz inequality*.)

Theorem. Assume $N(B, I, \|\cdot\|)$. Let $1 \le p \le \infty$, and let q be the conjugate exponent of p. Let $f, g \in \Phi_F^*$. Then

$$(*) \qquad \|f \cdot g\| \le \|f\|_p \cdot \|g\|_q .$$

Proof. First we consider the case that $\|f\|_p = 0$. Then f is a null function, and so is $f \cdot g$. This gives $(*)$. Next, if $\|f\|_p = \infty$, $(*)$ is trivial. So from now on we assume $0 < \|f\|_p < \infty$ and $0 < \|g\|_q < \infty$.

We start with the case $1 < p < \infty$. Let N be a null set in S outside of which both f and g are defined and finite. The inequality

$$uv \le \frac{u^p}{p} + \frac{v^q}{q} \qquad (u \ge 0,\ v \ge 0)$$

implies

$$|f(s)g(s)| \le \frac{\alpha^p}{p} |f(s)|^p + \frac{1}{\alpha^q q} |g(s)|^q \qquad (s \in S \setminus N)$$

for every $\alpha > 0$. Now choose $\alpha := \|f\|_p^{-1/q} \|g\|_q^{1/p}$ and note that N is a null set to obtain

$$\|f\cdot g\| \le \frac{\alpha^p}{p}\|f\|_p^p + \frac{1}{\alpha^q q}\|g\|_q^q = \|f\|_p\cdot\|g\|_q .$$

If $p = 1$, then $q = \infty$, and it is obvious that $|f\cdot g| \le \|g\|_\infty\cdot|f|$ (a.e.). Taking norms in the last inequality, we obtain (*) for this case. □

2.3.4. Next we prove *Minkowski's inequality*.

Theorem. Assume $N(B,I,\|\cdot\|)$. Let $1 \le p \le \infty$, and $f,g \in \Phi_F$. Then

(**) $$\|f + g\|_p \le \|f\|_p + \|g\|_p .$$

Proof. For $p = 1$, (**) is a consequence of $N3$ extended to Φ_F^* (see 2.2.6). For $p = \infty$, (**) is trivial. So we assume $1 < p < \infty$ in the remainder of the proof.

Let $h := f + g$. Then $h \in \Phi_F^*$ (see 2.2.5). Let N be a null set in S outside of which f and g are defined and finite. Obviously,

$$|h(s)|^p \le |h(s)|^{p-1}\cdot|f(s)| + |h(s)|^{p-1}\cdot|g(s)| \qquad (s \in S\backslash N) .$$

Hence, by the properties of the norm and Hölder's inequality 2.3.3,

$$\begin{aligned}\|h\|_p^p = \||h|^p\| &\le \||h|^{p-1}\cdot f\| + \||h|^{p-1}\cdot g\| \\ &\le \||h|^{p-1}\|_q\cdot\|f\|_p + \||h|^{p-1}\|_q\cdot\|g\|_p ,\end{aligned}$$

where q is the conjugate exponent of p. Since $\||h|^{p-1}\|_q = \|h\|_p^{p-1}$, division by $\|h\|_p^{p-1}$ leads to (**). (If $\|h\|_p^{p-1} = 0$, (**) is trivial.) □

2.3.5. Before introducing the L^p-spaces, we consider the question of an algebraic structure for Φ_F^*.

First we define an equivalence relation in Φ_F^*: two elements f and g of Φ_F^* are called *equivalent* (written $f \sim g$) if and only if $f = g$ (a.e.). This is really an equivalence relation. That is, if f, g and h are in Φ_F^*, then

(i) $f \sim f$,

(ii) $f \sim g$ implies $g \sim f$,

(iii) $f \sim g$ together with $g \sim h$ implies $f \sim h$.

The proofs consist in the observation that $f \sim g$ is merely another way of writing $f = g$ (a.e.).

Next we form classes of equivalent functions in the usual way: f and g are in the same class if and only if $f \sim g$. The class that f belongs to is temporarily denoted by $[f]$, and f is called a *representative* of its class $[f]$. In general, each class has many representatives. There are simple examples, though, where each class has just one representative; this occurs exactly if the empty set is the only null set in S. The set of equivalence classes is denoted by E.

It is easy to see that if a member of E contains a measurable function, all the elements of the class are measurable. Thus measurability is a property of the classes. Similarly, if $f \in \Phi_F^*$, then for all $g \in [f]$ we have $\|f\|_p = \|g\|_p$ $(1 \le p \le \infty)$. This means that the p-norms are numbers which can be attached to the classes as well: we write $\|[f]\|_p := \|f\|_p$, and omit the subscript if $p = 1$.

2.3.6. Now we introduce algebraic operations in E. Let f and g be representatives of classes [f] and [g], respectively, and let $\alpha \in \mathbb{R}$. We use the representatives to define sum and scalar multiple:

$$[f] + [g] := [f + g] ,$$

$$\alpha[f] := [\alpha f] .$$

It is easy to see that neither the sum nor the scalar multiple depends on the particular representatives chosen in defining them, that is, the operations are well-defined. It is also easy, though rather tedious, to check that E is a linear space with these operations. The class [0], which consists precisely of the null functions, acts as the zero element of the linear space.

It is also possible to define a multiplication of classes, reflecting the point-wise multiplication of functions: for $[f],[g] \in E$ put

$$[f]\cdot[g] := [f\cdot g] .$$

Again, it is easy to check that multiplication is well-defined and that it is commutative and distributive with respect to the other operations in the linear space E.

Finally, an order relation may be defined in E: $[f] \ge [g]$ whenever $f \ge g$ (a.e.).

2.3.7. At last we are ready to define the L^p-spaces. Assume that we have started from $N(B,I,\|\cdot\|)$ and run through the process $N \to L(L,I)$, so that we can talk about measurability. Assume in addition that M is satisfied, so that the constant functions are measurable. (See the proofs of the next two theorems to find why we cannot do without the last condition.) For $1 \le p < \infty$ we put

$$L^p := \{[f] \in E \mid f \in M_F^* \text{ and } |f|^p \in L^*\} ,$$

and

$$L^\infty := \{[f] \in E \mid f \in M_F^* \text{ and } \|f\|_\infty < \infty\} .$$

2.3.8. In the language of L^p-spaces, Hölder's inequality takes the following form. (We simplify the notation for classes by dropping the square brackets; after 2.3.12 we shall simplify even further.)

<u>Theorem</u>. Assume $N(B,I,\|\cdot\|)$, follow $N \to L(L,I)$, and assume M. Let $1 \le p \le \infty$, and let q be the conjugate exponent of p. Let $f \in L^p$, $g \in L^q$. Then $f\cdot g \in L^1$ and $\|f\cdot g\| \le$ $\le \|f\|_p \cdot \|g\|_q$.

<u>Proof</u>. Let f' be a representative of the class f, and g' one of the class g. We may as well assume that f' and g' are defined and finite on all of S. Now $f'\cdot g'$ is a representative of $f\cdot g$, and it is measurable by Example 2.1.16(ii). Since Theorem 2.3.3 implies that $\|f'\cdot g'\| < \infty$, we need only show that $f'\cdot g'$ satisfies the covering condition of Theorem 2.1.17. This is easy. For if $1 < p < \infty$, then $|f'\cdot g'| \le \sum_{n=1}^{\infty} f_n$, where $f_n := |f'|^p$ if n is even, $f_n := |g'|^q$ if n is odd. And if $p = 1$, then $|f'\cdot g'| \le \sum_{n=1}^{\infty} f_n$ where $f_n := |f'|$, which also takes care of the case $p = \infty$ if we interchange the roles of f and g. □

2.3.9. Minkowski's inequality gives the following result.

<u>Theorem</u>. Assume $N(B,I,\|\cdot\|)$, follow $N \to L(L,I)$, and assume M. Let $1 \le p \le \infty$, $f \in L^p$, $g \in L^p$. Then $f + g \in L^p$ and $\|f + g\|_p \le \|f\|_p + \|g\|_p$.

<u>Proof</u>. The case $p = \infty$ is very simple, so we assume $1 \le p < \infty$. Again, let f' be a representative of the class f, g' one of the class g. Then $f' + g'$ is a representative of $f + g$, belongs to M_F by 2.2.9, and satisfies $\||f' + g'|^p\| \le (\|f'\|_p + \|g'\|_p)^p$ by Theorem 2.3.4. Since $|f' + g'|^p$ is measurable by Example 2.1.16(i), it is sufficient to show that $f' + g'$ can be "covered" by countably many elements of L^*. This can be done as in the preceding proof. □

2.3.10. Let $1 \le p \le \infty$. Since L^p is obviously closed under multiplication by real scalars, and the preceding theorem shows that it is closed under addition, it follows that L^p is a linear subspace of E. Furthermore, Minkowski's inequality in 2.3.9 is exactly the statement that $\|\cdot\|_p$ satisfies the triangle inequality for the norm on a linear space (see 0.3.1). It is also clear that $\|\alpha f\|_p = |\alpha|\cdot\|f\|_p$ for all $\alpha \in \mathbb{R}$, $f \in L^p$. It follows that the p-norm makes L^p into a normed linear space, if only we can show that $\|f\|_p = 0$ implies $f = 0$. This is just what our equivalence classes are meant for. For, let $f \in L^p$, $\|f\|_p = 0$, and let f' be a representative of the class f. Then $\||f'|^p\| = 0$, so $|f'|^p$ is a null function. Hence f' is a null function, and $f = 0$. Summing up, we have the following important result.

<u>Theorem</u>. Assume $N(B,I,\|\cdot\|)$, follow $N \to L(L,I)$, and assume M. Let $1 \le p \le \infty$. With the p-norm, L^p is a normed linear space.

2.3.11. Still more can be said. We know that the norm in a normed linear space leads to a natural metric if one takes as the distance between two elements the norm of their difference (see 0.3.1). With this metric, L^p becomes a complete metric space as we shall prove now.

Theorem. Assume $N(B,I,\|\cdot\|)$, follow $N \to L(L,I)$, and assume M. Let $1 \le p \le \infty$. With the p-norm, L^p is a Banach space.

Proof. Assume $1 \le p < \infty$. Let $(f_n)_{n\in\mathbb{N}}$ be a fundamental sequence in L^p, and suppose in addition $\|f_{n+1} - f_n\|_p < 2^{-n}$ for all n. Take representatives $f'_1, f'_2, \ldots$ in the classes $f_1, f_2, \ldots$. Our first aim is to show that $(f'_n)_{n\in\mathbb{N}}$ converges to a function in an L^p-class.

Define the sequence $(g_n)_{n\in\mathbb{N}}$ by

$$g_n := |f'_1| + \sum_{k=1}^{n-1} |f'_{k+1} - f'_k| \qquad (n \in \mathbb{N}).$$

Minkowski's inequality 2.3.4, applied repeatedly, gives

$$\|g_n\|_p \le \|f'_1\|_p + \sum_{k=1}^{n-1} \|f'_{k+1} - f'_k\|_p \le \|f'_1\|_p + 1 .$$

Hence, by Fatou's lemma 2.2.12,

$$I(\liminf_{n\to\infty} g_n^p) \le (\|f'_1\|_p + 1)^p .$$

The sequence $(g_n)_{n\in\mathbb{N}}$ is a.e. pointwise increasing and non-negative, so it converges a.e. to a function $g \in \Phi^*$ which is also non-negative a.e. The inequality that we just derived shows that g^p is integrable, and therefore finite a.e. It follows that the partial sums of $f'_1 + \sum_{k=1}^{\infty}(f'_{k+1} - f'_k)$ converge a.e. to a measurable function $f' \in \Phi^*_F$. Let f be the class in E that f' belongs to. Now f' is measurable, $|f'|^p \le g^p$ and g^p is integrable, so $|f'|^p$ is integrable by 2.1.16(i) and 2.1.17. Hence $f \in L^p$.

Next, we show that f is the L^p-limit of the sequence $(f_n)_{n\in\mathbb{N}}$. Again applying Fatou's lemma and Minkowski's inequality, we obtain

$$\begin{aligned} I(|f'_n - f'|^p) &= I(\liminf_{m\to\infty} |f'_n - f'_m|^p) \le \liminf_{m\to\infty} I(|f'_n - f'_m|^p) = \\ &= \liminf_{m\to\infty} \|f'_n - f'_m\|_p^p \le \liminf_{m\to\infty} (\sum_{k=n}^{m} \|f'_{k+1} - f'_k\|_p)^p \le \\ &\le 2^{(1-n)p} \end{aligned}$$

for all n. Hence $\|f'_n - f'\|_p \to 0$ as $n \to \infty$, that is, $\|f_n - f\|_p \to 0$ as $n \to \infty$.

The general case is now easily deduced. Let $(f_n)_{n\in\mathbb{N}}$ be a fundamental sequence in L^p. Pick a subsequence $(f_{n_k})_{k\in\mathbb{N}}$ such that $\|f_{n_{k+1}} - f_{n_k}\|_p < 2^{-k}$ $(k \in \mathbb{N})$. The special case applied to this subsequence furnishes an $f \in L^p$ such that $\|f_{n_k} - f\|_p \to 0$ as $k \to \infty$, and it is not difficult to show that $\|f_n - f\|_p \to 0$ as $n \to \infty$.

The case $p = \infty$ is much easier. Now the problem is essentially one about uniform convergence. Details are left to the reader; one does not even need Fatou's lemma. □

2.3.12. The preceding theorems show that the L^p-spaces are well-behaved objects from the algebraic point of view. Working with function classes instead of with functions, however, means that we cannot talk about function values. This is sometimes very awkward, and there are even cases where it is completely undesirable to work with function classes (for an example from the theory of probability, see Section 6.5).

Now we have seen some situations where we essentially identified the elements of L^p with their representatives. For instance, analyzing the proofs of 2.3.8, 2.3.9 and 2.3.11 we see that they all follow the same pattern. First representatives are chosen, then the assertion to be proved is translated into one about the representatives, which is then proved and finally translated back to the classes. This is a hint about the right direction to take.

The solution we choose is the following one. If we write $f \in L^p$ in the sequel, it will depend on the context whether we mean a function, or a class of functions; and such an f will be called an L^p-function even if it is really a class of functions. Often we shall not state explicitly which interpretation is meant. Experience shows that there need be no fear for misunderstandings.

2.3.13. The final result of this section is an approximation theorem. It shows that L^p-functions with $1 \le p < \infty$ can be approximated by basic functions, in the sense appropriate to L^p. The case $p = 1$ is just the definition of integrability. Note that the b of the theorem is also in L^p.

Theorem. Assume $N(B,I,\|\cdot\|)$, follow $N \to L(L,I)$ and assume M. Let $1 \le p < \infty$, $f \in L^p$ and $\varepsilon > 0$. Then there exists $b \in B$ such that $\|b - f\|_p < \varepsilon$.

Proof. For each $n \in \mathbb{N}$, let $V_n := \{s \in S \mid n^{-1} \le |f(s)| \le n\}$. Since M is satisfied and f is measurable, each V_n is measurable by 2.1.12 and 2.1.8. Hence, if $f_n := f \cdot \chi_{V_n}$, then f_n is measurable by Example 2.1.16(ii). Now the sequence $(f_n)_{n \in \mathbb{N}}$ is dominated by $|f|$. Hence each $|f_n|^p$ is integrable by 2.1.17, and therefore $f_n \in L^p$. By Lebesgue's theorem 2.2.8, $\|f_n - f\|_p \to 0$ as $n \to \infty$.

Choose n so large that $\|f_n - f\|_p < \varepsilon/2$. We shall approximate f_n. The inequality $|f_n| \le n^{p-1}|f|^p$, together with the measurability of f_n, shows that $f_n \in L^*$. Hence there exists $b \in B$ such that $\|f_n - b\| < (\varepsilon/4n)^p$. Replacing b by the truncated function $(b)_{n\chi_S}$, we see that we may assume $|b| \le n$. It follows that $|f_n - b|^p \le (2n)^{p-1}|f_n - b|$, and therefore $\|f_n - b\|_p \le (2n)^{(p-1)/p}\|f_n - b\|^{1/p} < \varepsilon/2$. Hence $\|f - b\|_p < \varepsilon$. □

Exercises Section 2.3

1. Assume that $\|\chi_S\| < \infty$ and let $f \in \Phi_F^*$. Show that $\lim_{p\to\infty}\|f\|_p = \|f\|_\infty$. Show by an example that the condition $\|\chi_S\| < \infty$ cannot be dropped.

2. Assume that $\|\chi_S\| < \infty$ and let $1 \le q < p < \infty$. Show that $\|f\|_q < \infty$ if $f \in \Phi_F^*$ and $\|f\|_p < \infty$. Give an example of a situation $N(B,I,\|\cdot\|)$ and an $f \in \Phi_F^*$ with $\|f\|_p = \infty$, $\|f\|_q < \infty$.

3. When are there equality signs in 2.3.3 and 2.3.4?

4. Let $1 \le p,q,r \le \infty$ and assume that $1/p + 1/q = 1/r$. Show that $\|f\cdot g\|_r \le \|f\|_p \cdot \|g\|_q$ for $f \in \Phi_F^*$, $g \in \Phi_F^*$.

5. Show that $\|\chi_S\| < \infty$ follows from $f \in \Phi_F^*$, $\|f\|_p < \infty \Rightarrow \|f\|_q < \infty$ if

(a) M is satisfied,

(b) the system is σ-finite.

Here $1 \le q < p < \infty$. Show by examples that omission of (a) or (b) makes this conclusion invalid. (Hint. To show that the conclusion is valid if (a) and (b) are satisfied, assume that $\|\chi_S\| = \infty$ and split S into disjoint measurable parts A_n with $1 \le \|\chi_{A_n}\| < \infty$. Consider f's of the form $\sum_{n=1}^\infty \alpha_n\chi_{A_n}$. To show that (a) cannot be omitted, consider Exercise 2.1-10 with norm $\|\cdot\|$ given by

$$\|f\| = \begin{cases} \sup\{|f(x)| \mid x \in S\} & \text{if } f(x) \to 0 \ (x \to 0)\ , \\ \infty & \text{otherwise.} \end{cases}$$

Show that $\|\cdot\|$ is a norm compatible with I, and that the system is σ-finite. As to (b), there is an example where S consists of 2 points.)

6. Take $S := \mathbb{N}$, $B := \{b \in \mathbb{R}^{\mathbb{N}} \mid \sum_{n=1}^\infty |b(n)| < \infty\}$, $I(b) := \sum_{n=1}^\infty b(n)$ $(b \in B)$, $\|\varphi\| := \sum_{n=1}^\infty |\varphi(n)|$ $(\varphi \in \Phi)$. Show that $(B,I,\|\cdot\|)$ satisfies N. Let $1 \le p \le \infty$, and define the spaces ℓ^p as the set of all $f \in \Phi_F^*$ with $\sum_{n=1}^\infty |f(n)|^p < \infty$ if $p < \infty$, and with $\sup\{|f(n)| \mid n \in \mathbb{N}\} < \infty$ if $p = \infty$.

(i) Show that L^p can be identified with ℓ^p.

(ii) Show that $\ell^q \subset \ell^p$ if $1 \le q \le p \le \infty$.

7. Assume M, and let $1 \le p \le r \le q \le \infty$. Show that $L^p \cap L^q \subset L^r$.

8. Take S, B, I, $\|\cdot\|$ as in Example 1.2.4, and let $f \in L$. Show that

$$\Upsilon_x\, f(x)(1+|x|)^{-\frac{1}{2}} \in L\ ,\quad \Upsilon_x\, f(x)(\sin x)/x \in L\ .$$

9. Assume M and $\|\chi_S\| < \infty$. Show that $|f|(1+|f|)^{-1} \in L^*$ for $f \in M_F^*$. Define

$$d(f_1,f_2) = I(|f_1 - f_2| \,/\, (1 + |f_1 - f_2|))$$

for $f_1, f_2 \in M_F^*$, and show that d is a metric for M_F^* if we identify functions that are equal a.e. Also show that M_F is complete with this metric.

10. Assume that the system $N(B,I,\|\cdot\|)$ is σ-finite and that M is satisfied. Let $1 \le p \le \infty$, q its conjugate exponent, and let $g \in M$ be such that $f\cdot g \in L^1$ for every $f \in L^p$. Show that $g \in L^q$ as follows.

(i) Treat the case $p = \infty$ separately.

Assume in what follows that $1 \le p < \infty$.

(ii) Show that there is an $M_0 > 0$ such that $\|f\cdot g\|_1 \le M_0\|f\|_p$ for all $f \in L^p$. (Hint. If such an M_0 does not exist we can find a sequence $(f_n)_{n\in\mathbb{N}}$ in L^p such that $f_n\cdot g \ge 0$, $\|f_n\|_p \le 2^{-n}$, $\|f_n\cdot g\|_1 \ge 1$.)

Assume in what follows that M_0 is as in (ii).

(iii) Show that there is an $f_0 \in L^p$ such that $f_0 > 0$ everywhere on S. (Hint. Use the result of Exercise 1.1-5.)

Assume in what follows that f_0 is as in (iii).

(iv) Put $q_n := 1 + 1/p + \ldots + 1/p^n$ for $n \in \mathbb{N}$. Show by induction that $|g|^{q_n}\cdot f_0^{1/p^n} \in L^1$, $\|\,|g|^{q_n}\cdot f_0^{1/p^n}\| \le M_0^{q_n}\|f_0\|_p^{1/p^n}$ for $n \in \mathbb{N}$.

(v) Show that $g \in L^q$, $\|g\|_q \le M_0$. (Hint. Use (iv) and Fatou's lemma; treat the cases $p = 1$ and $p > 1$ separately.)

11. Show that the conclusion of Exercise 10 holds if the σ-finiteness condition is replaced by: there is a sequence $(f_n)_{n\in\mathbb{N}}$ in L^+ such that $|g| \le \sum_{n=1}^{\infty} f_n$. Show that this condition may not be deleted.

12. Let $(a_n)_{n\in\mathbb{N}}$ be a sequence in $\mathbb{R}$ such that $\lim_{N\to\infty} \sum_{n=1}^{N} a_n b_n$ exists for every sequence $(b_n)_{n\in\mathbb{N}}$ in $\mathbb{R}$ with $\sum_{n=1}^{\infty} b_n^2 < \infty$. Show that $\sum_{n=1}^{\infty} a_n^2 < \infty$.

13. Let $1 \le p \le \infty$, $\varphi \in \Phi$, $\varphi_n \in \Phi$ $(n \in \mathbb{N})$, $|\varphi| \le \sum_{n=1}^{\infty} |\varphi_n|$. Show that $\|\varphi\|_p \le \sum_{n=1}^{\infty} \|\varphi_n\|_p$. (Hint. Assume $p < \infty$ and $\sum_{n=1}^{\infty} \|\varphi_n\|_p = 1$. If $F : [0,\infty) \to [0,\infty)$ is a convex function and $a_n \ge 0$, $b_n \ge 0$, $\sum_{n=1}^{\infty} a_n = 1$, $\sum_{n=1}^{\infty} a_n b_n < \infty$, then $F(\sum_{n=1}^{\infty} a_n b_n) \le \sum_{n=1}^{\infty} a_n F(b_n)$. Use this with $F(x) = \curlyvee_x x^p$, $a_n = \|\varphi_n\|_p$, $b_n = \varphi_n(s) / \|\varphi_n\|_p$.)

14. Show that L^2 is an inner product space if we define $(f,g) := I(f \cdot g)$ for $f \in L^2$, $g \in L^2$.

15. Does the approximation theorem 2.3.13 hold for $p = \infty$?

16. Let S, B, I and $\|\cdot\|$ be as in Example 1.2.4. Show that $\|f_h - f\|_p \to 0$ $(h \to 0)$ if $f \in L^p$ and $1 \le p < \infty$. Here $f_h := \curlyvee_x f(x + h)$ for $f \in \Phi_F^*$, $h \in \mathbb{R}$. (Hint. Use Theorem 2.3.13 and Exercise 1.2-4.) What about $p = \infty$?

17. Let R be the set of all measurable functions defined on $\mathbb{R}$ and real-valued a.e. that are periodic with period 1. Put $d(f) := I(|f|/(1 + |f|))$ for $f \in R$ where I means Lebesgue integration over $[0,1]$. Show that $d(f_h) = d(f)$ for $f \in R$, $h \in \mathbb{R}$, and that $\lim_{h \to 0} d(f_h - f) = 0$ for $f \in R$.

18. If $[a,b]$ is a finite interval, then $\int_{-\infty}^{\infty} \chi_{[a,b]}(t) \sin(xt)\,dt \to 0$ if $x \to \infty$. Show that $\int_{-\infty}^{\infty} f(t) \sin xt\,dt \to 0$ if $x \to \infty$ and $f \in L^1$.

2.4. The local norm

In Section 1.1 we introduced the fundamental construction $N(B,I,\|\cdot\|) \to L(L,I)$, which starts with a norm on Φ and leads to integrable functions together with an integral. From the system $L(L,I)$ we can derive a new norm $\|\cdot\|_1$ by means of $L \to D \to A \to$ $\to N(L,I,\|\cdot\|_1)$, and then again follow $N(L,I,\|\cdot\|_1) \to L(L_1,I_1)$. But the result is not very imposing: although $\|\cdot\|$ and $\|\cdot\|_1$ may be different, the class of integrable functions and the integral do not change anymore (see 1.4.7).

In this section we present another way to derive a norm from $L(L,I)$ for the case that the original norm has the Fatou property. (See 1.2.9; norms derived from R or T have this property, so this is not a serious restriction.) The new norm is called the local norm; it leads to locally integrable functions, and so on. The extension obtained by introducing the local norm is only slight, and in the σ-finite case even nil.

The local norm will be used at only one place in the main text (Theorem 4.1.9, this point may be skipped by then), and seldom in the exercises. So the reader may pass directly to Chapter 3, without loosing the general line of the development.

2.4.1. The setting is a system $N(B,I,\|\cdot\|)$ for a non-empty set S. Assume that the norm has the Fatou property introduced in 1.2.9, that is, if $(\varphi_n)_{n\in\mathbb{N}}$ is a non-decreasing sequence in Φ^+, then $\|\lim_{n\to\infty}\varphi_n\| = \lim_{n\to\infty}\|\varphi_n\|$. Follow the route $N \to L(L,I)$ and put

$$\|\varphi\|_{loc} := \sup\{\|(\varphi)_f\| \mid f \in L^+\} \qquad (\varphi \in \Phi) .$$

This gives a norm on Φ, called the *local norm*. A function $\varphi \in \Phi$ is a *local null function* if $\|\varphi\|_{loc} = 0$, and a *local null set* in S is a subset of S whose characteristic function is a local null function.

We stated in the introduction to this section that the assumption that the norms have the Fatou property is not very restrictive. Indeed, 1.2.9 says that if the norm was derived from A with $A5$, it automatically has the Fatou property. And 1.4.8 says that if A was derived from R or T, then $A5$ holds. So in most of the cases of interest to us we are not hampered by this restriction.

2.4.2. Example. Let S be a non-empty set, α a non-negative $\mathbb{R}^*$-valued function defined on S. Let B_α be the class of $\mathbb{R}$-valued functions b defined on S that have $\alpha\cdot b$ summable over S in the sense of 0.4.1, and put

$$I_\alpha(b) := \sum_{s\in S} \alpha(s)b(s) \qquad (b \in B_\alpha) .$$

Note that if $b \in B_\alpha$, then b must vanish at the points where α is infinite. Put

$$\|\varphi\|_\alpha := \inf\{I_\alpha(b) \mid b \in B^+, |\varphi| \le b\} \qquad (\varphi \in \Phi) ,$$

where $\inf \emptyset = \infty$. Then the triple $(B_\alpha, I_\alpha, \|\cdot\|_\alpha)$ satisfies N. (If α is finite-valued, this is just Example 1.1.6.) It is not hard to show that $\|\cdot\|_\alpha$ has the Fatou property (see the proof of Theorem 1.2.9). If $V := \{s \in S \mid \alpha(s) = \infty\}$, then V is a local null set, and $\varphi \in \Phi$ is a local null function if and only if φ vanishes on $\{s \in S \mid 0 < \alpha(s) < \infty\}$.

2.4.3. Proposition. Assume $N(B,I,\|\cdot\|)$ and the Fatou property for the norm. Then $\|\cdot\|_{loc}$ is a norm on Φ compatible with I in the sense of 1.1.5. Moreover, if $\varphi \in \Phi$, then $\|\varphi\|_{loc} \le \|\varphi\|$, and there is equality if there exists a sequence $(f_n)_{n\in\mathbb{N}}$ in L^+, such that $|\varphi| \le \sum_{n=1}^\infty f_n$.

Proof. The verification that $\|\cdot\|_{loc}$ is a norm dominated by the original norm is left to the reader. (In this part of the proof we can do without the Fatou property.)

Now let $\varphi \in \Phi$, and assume that there exists a sequence $(f_n)_{n\in\mathbb{N}}$ in L^+ such that $|\varphi| \le \sum_{n=1}^{\infty} f_n$. Let $h_k := \sum_{n=1}^{k} f_n$ $(k \in \mathbb{N})$, then $h_k \in L^+$, and $\|(\varphi)_{h_k}\| \to \|\varphi\|$ by the Fatou property. Since $\|\varphi\|_{loc} \ge \|(\varphi)_{h_k}\|$ for each k, we have $\|\varphi\|_{loc} \ge \|\varphi\|$. Hence $\|\varphi\|_{loc} = \|\varphi\|$ in this case. □

2.4.4. It is easy to find examples where the norm and the local norm are different (Exercise 2.4-2), but one has to go beyond the σ-finite spaces. For if the original system is σ-finite, then there exists a sequence $(f_n)_{n\in\mathbb{N}}$ in L^+ (or even in B^+) such that $\sum_{n=1}^{\infty} f_n = \infty$, so norm and local norm coincide in this case.

2.4.5. Example. Let $S := (0,1]$, $B := \{\alpha \overset{\cup}{|}_x x \mid \alpha \in \mathbb{R}\}$, $I(b) := b(1)$ for $b \in B$. For $\varphi \in \Phi$, put

$$\|\varphi\| := \begin{cases} \sup\limits_{x\in(0,1]} |\varphi(x)| & \text{if } \varphi(x) \to 0 \text{ as } x \to 0 \,, \\ \infty & \text{otherwise.} \end{cases}$$

It is not difficult to show that $(B,I,\|\cdot\|)$ satisfies N. Follow $N \to L(L,I)$. Then it is easy to see that $L = B$.

Now if $\varphi := \chi_S$, then $\|\varphi\| = \infty$. If the local norm is defined as in 2.4.1, then $\|\varphi\|_{loc} = 1$. Nevertheless the system is σ-finite. The discrepancy with Proposition 2.4.3 is due to the missing Fatou property for the norm.

2.4.6. Proposition. Assume A and $A5$. Follow $A \to N \to L$. Let $\varphi \in \Phi$, $\|\varphi\| < \infty$. Then $\|\varphi\|_{loc} = \|\varphi\|$.

Proof. As shown in the proof of Proposition 1.2.7, there exists a sequence $(f_n)_{n\in\mathbb{N}}$ in L^+ such that $|\varphi| \le \sum_{n=1}^{\infty} f_n$. Apply Proposition 2.4.3. □

2.4.7. We can now introduce the notions of integrability, integral and measurability for the local norm. The integrable functions obtained by following the route $N(B,I,\|\cdot\|_{loc}) \to L$ are called *locally integrable*, and the resulting integral is the *local integral*. By 1.1.20, each integrable function is locally integrable, and its integral is equal to its local integral. In the reverse direction we know that a locally integrable function is equal to an integrable function, except for a local null function (again by 1.1.20). Example 2.4.2 shows that there may be local null functions which are not integrable, that is, the class of locally integrable functions may be strictly larger than the class of integrable functions. However, the class of locally measurable functions coincides with the class of measurable functions (Exercise 2.4-8).

Exercises Section 2.4

1. Prove the first part of Proposition 2.4.3 without using the Fatou property.

2. Give an example of a station $N(B,I,\|\cdot\|)$ as in 2.4.1 and a $\varphi \in \Phi$ with $\|\varphi\| < \infty$, $\|\varphi\|_{loc} \neq \|\varphi\|$.

3. Show that $\|\varphi\|_{loc} = \sup\{\|(\varphi)_b\| \mid b \in B^+\}$.

4. In Example 2.4.5 condition M is not satisfied. Modify Example 2.4.5 in such a way that M is satisfied and nevertheless $\|\cdot\|_{loc} \neq \|\cdot\|$.

5. Let $\varphi \in \Phi$ and assume that $\|\varphi\|_{loc} = \|\varphi\| < \infty$. Show that there is a sequence $(f_n)_{n\in\mathbb{N}}$ in L^+ such that $\sum_{n=1}^{\infty} f_n \geq |\varphi|$.

6. Assume that $(B,I,\|\cdot\|)$ is derived from $A + A5$. Show that $\|\varphi\|_{loc} = 0$ if and only if $\|\varphi\| = 0$ or $\|\varphi\| = \infty$.

7. With the notation of 2.4.7 we have: f is locally integrable if and only if $f = f_0 + g$ where $f_0 \in L$, $\|g\|_{loc} = 0$.

8. Prove the remark about measurable functions at the end of 2.4.7.

9. Let (B,I) satisfy the conditions D' of 1.3.5 and follow the routes $D' \to D \to A \to N(B_1,I_1,\|\cdot\|_1)$, $D' \to A \to N(B_2,I_2,\|\cdot\|_2)$. Give an example where $\|\varphi\|_{1,loc} > \|\varphi\|_2$ for some $\varphi \in \Phi$. (Hint. Take $S = [0,1]$, $B := \{f : S \to \mathbb{R} \mid \exists_{\alpha\in\mathbb{R}} [f(x) = \alpha \ (x \in (\frac{1}{2},1]), f(x) \neq \alpha$ for at most countably many $x \in [0,\frac{1}{2}]]\}$, $I(f) = f(1)$. Show that $\|\chi_V\|_{1,loc} = 1$, $\|\chi_V\|_2 = 0$ for every $V \subset [0,\frac{1}{2}]$ with both V and $[0,\frac{1}{2}]\backslash V$ uncountable.)

10. For every $b \in B$ let $N_b := \{s \in S \mid b(s) \neq 0\}$. Show that $A \subset S$ is a local null set if and only if $A \cap N_b$ is a null set for every $b \in B$.

11. Let $S_\alpha := \mathbb{R}$ for $\alpha \in [0,1]$. Let S be the disjoint union of the S_α's, i.e. S consists of uncountably many copies of $\mathbb{R}$ (S can be pictured as $[0,1] \times \mathbb{R}$). Let B, I, A, J and $\|\cdot\|$ as in Example 1.2.4. Let A_1 be the class of functions $h : S \to \mathbb{R}^+$ defined as follows: there exists an $\alpha \in [0,1]$ such that $h(s) = 0$ if $s \notin S_\alpha$ while $h|_{S_\alpha} \in A$; for such an h we put $J_1(h) = J(h|_{S_\alpha})$. Let B_1 be the class of functions $b : S \to \mathbb{R}$ defined as follows: there exists $n \in \mathbb{N}$, $\alpha_1 \in [0,1],\ldots,\alpha_n \in [0,1]$ such

that $b(s) = 0$ if $s \notin \bigcup_{k=1}^{n} S_{\alpha_k}$ while $b|S_{\alpha_k} \in B$ $(k = 1,\dots,n)$; for such a b we put $I_1(b) = \sum_{k=1}^{n} I(b|S_{\alpha_k})$.

(i) Show that (B_1,I_1,A_1,J_1) satisfies $A + A5$.

Follow the route $A \to N$, and denote the resulting norm by $\|\cdot\|_1$.

(ii) Let $\varphi \in \Phi$, $\|\varphi\|_1 < \infty$. Show that there is a countable set $E \subset [0,1]$ such that $\varphi|S_\alpha = 0$ if $\alpha \notin E$.

(iii) For each $\alpha \in [0,1]$ let $N_\alpha \subset \mathbb{R} = S_\alpha$ be a non-empty null set, and let $N := \bigcup_{\alpha\in[0,1]} N_\alpha$. Show that N is a local null set but not a null set.

12. Show that $\|\cdot\|_{loc}$ has the Fatou property. Let $\|\cdot\|_1$ be the local norm obtained by taking $\|\cdot\| = \|\cdot\|_{loc}$ in Definition 2.4.1. Show that $\|\cdot\|_1 = \|\cdot\|_{loc}$.

13. Assume $A + A5$ and follow the route $A \to N \to L$. Define

$$\|\varphi\|_{ess} := \inf\{\|\varphi(1 - \chi_M)\| \mid M \text{ local null set}\}$$

for $\varphi \in \Phi$. Let $\varphi \in \Phi$. This exercise is meant to show that $\|\varphi\|_{ess} = \|\varphi\|_{loc}$ and that there exists a local null set M such that $\|\varphi\|_{ess} = \|\varphi(1 - \chi_M)\|$. We may assume that $\varphi \geq 0$.

(i) Show that $\|\varphi\|_{loc} \leq \|\varphi\|_{ess}$.

In what follows we assume that $\|\varphi\|_{loc} < \infty$.

(ii) Show that there exists a sequence $(f_n)_{n\in\mathbb{N}}$ in L^+ such that $\|(\varphi)_{f_n}\| \uparrow \|\varphi\|_{loc}$ while for $s \in S$ either $f_n(s) = 0$ $(n \in \mathbb{N})$ or $f_n(s) \uparrow \infty$ $(n \to \infty)$.

(iii) Let $\psi := \lim_{n\to\infty} (\varphi)_{f_n}$, where f_n is as in (ii). Show that $\psi(s) = \varphi(s)$ on the set $E := \{s \in S \mid f_n(s) \uparrow \infty \text{ if } n \to \infty\}$ and that $\|\psi\| = \|\varphi\|_{loc}$.

(iv) Suppose that $\|\varphi\cdot\chi_{E^*}\|_{loc} > 0$. Show that there exists an $f \in L^+$ with $f\cdot\chi_E = 0$ and $\|(\varphi)_f\| > 0$. (Hint. If $g \in L$ then $g\cdot\chi_{E^*} \in L^+$ and $(\varphi\cdot\chi_{E^*})_g = (\varphi)_{g\cdot\chi_{E^*}}$.) Use Exercise 1.2-16 to show that $\|(\varphi)_{f_n+f}\| = \|(\varphi)_{f_n}\| + \|(\varphi)_f\|$ for all n and derive a contradiction.

(v) Show that $M := \{s \in E^* \mid \varphi(s) > 0\}$ is a local null set and that

$$\|\varphi(1 - \chi_M)\| = \|\varphi\cdot\chi_E\| = \|\psi\| = \|\varphi\|_{loc} .$$

CHAPTER THREE

INTEGRATION ON MEASURE SPACES

About fifteen years after the introduction of Lebesgue's integral, and after development of the new tool into several directions, Carathéodory began the formulation of measure theory in axiomatic form. In this approach the starting point is usually a ring Γ of subsets of a set S (a ring is stable under the formation of finite unions and differences) together with a measure μ defined on Γ (that is, $\mu : \Gamma \to \mathbb{R}_+^*$ is σ-additive: $\mu(\bigcup_{n=1}^{\infty} A_n) = \sum_{n=1}^{\infty} \mu(A_n)$ whenever the $A_n \in \Gamma$ are pairwise disjoint and $\bigcup_{n=1}^{\infty} A_n \in \Gamma$). The aim is to extend the initial measure μ on Γ to one defined on a σ-ring containing Γ, in such a way that the extension is still σ-additive on the σ-ring. Once this aim is attained, the construction of an integral is not difficult. The terminology of measure theory was used by Kolmogorov in the 1930's to give an axiomatization of probability theory which has no serious rival as yet. This is one of the reasons why the Carathéodory approach is very popular.

In this chapter we present our version of measure theory. In the first section semirings, measures on semirings, and measure spaces are introduced. (Although a semiring is almost a ring, we prefer the semiring as the initial object because it better suits the discussion of Lebesgue-Stieltjes measures and product measures.) In the next section the starting point is a measure space (Γ,μ), where Γ is a semiring of subsets of S and μ is the measure on Γ. From Γ we construct a class B of basic functions together with a positive linear functional I on B such that the pair (B,I) satisfies condition $\mathcal{D}$ of Section 1.3. The extension theory of Chapter 1 then furnishes an integral I on a class L of integrable functions. The pair (L,I) in turn leads to a class M_μ of measurable functions, a class Σ_μ of measurable sets, and a measure $\tilde{\mu}$ on Σ_μ, as described in Section 2.1. This new measure space $(\Sigma_\mu,\tilde{\mu})$ is an extension of (Γ,μ), so there is no need to distinguish between μ and $\tilde{\mu}$, after all. Thus we obtain an extension of the original measure on the semiring Γ to a measure on a σ-ring containing Γ, as aimed at in the Carathéodory set-up.

The concept of measurability mentioned in the preceding paragraph depends on an integral. In general measure theory there is a related notion, we call it Σ-measurability, that depends only on a σ-algebra Σ of subsets of S; it is particularly useful when one considers several measures at the same time. The two kinds of measurability are compared in Section 3.3.

In Section 2.1 a measure space (Σ,μ) was derived from a situation $\mathcal{L}(L,I)$. If one takes such a pair (Σ,μ) as a starting point for the procedure described in the first two sections of the present chapter, one obtains a new integral $\bar{I}$. In the final section we show that $\bar{I}$ is an extension of I if the original pair (L,I) satisfies Stone's condition $\mathcal{M}$. Thus, certain abstract integrals turn out to be integrals with respect to a measure.

3.1. Semirings and measures

We shall now discuss semirings and measures on semirings. Examples of measures have occurred before already. The first one was in 1.2.5, where A3 was so hard to prove - we shall soon recognize that A3 is closely connected with the fact that "length" on $\mathbb{R}$ is a measure. The second example was in 2.1.9; the set function μ introduced there is also a measure. To illustrate the new concepts, additional examples will be treated below. The most important of these is 3.1.12, which deals with Lebesgue-Stieltjes measures on $\mathbb{R}$.

3.1.1. Let S be a non-empty set, Γ a class of subsets of S. By Ω, or $\Omega(\Gamma)$ if we want to be more specific, we denote the class of sets $\bigcup_{n=1}^{N} A_n$, where $N \in \mathbb{N}$ is arbitrary, and the A_n are pairwise disjoint elements of Γ. In words: Ω is the collection of all finite disjoint unions of sets in Γ. The class Γ is called a *semiring* if it has the following two properties:

R1. $\emptyset \in \Gamma$.

R2. If $A \in \Gamma$, $B \in \Gamma$, then $A \cap B \in \Omega$ and $A \setminus B \in \Omega$.

If Γ is a semiring and $\Gamma = \Omega$, then Γ is called a *ring*. A ring Γ such that $S \in \Gamma$ is called an *algebra*.

Thus a ring is a class of subsets of S which is closed with respect to taking differences, finite unions, and finite intersections of its members. An algebra has the additional property that it is closed with respect to complementation. In 2.1.6 we met σ-rings and σ-algebras; Proposition 2.1.7 shows that any σ-ring is a ring and any σ-algebra is an algebra in the sense just defined. The additional σ denotes that unions and intersections of countably many members of the σ-ring or σ-algebra can be formed without leaving the σ-ring or σ-algebra.

3.1.2. First we show that each class of subsets has a particular ring, algebra and so on associated with it. The qualification "smallest" in the proposition means that the object whose existence is asserted, is contained in every ring (or algebra, etc.) that contains Γ.

Proposition. Let Γ be a class of subsets of S. There exists a unique smallest ring (algebra, σ-ring, σ-algebra) of subsets of S that contains Γ.

Proof. Consider the ring case. Let $\mathcal{F}$ be the family of all rings of subsets of S that contain Γ. It is obvious that P(S), the class of all subsets of S, belongs to $\mathcal{F}$. Let Λ be the intersection of the members of $\mathcal{F}$. It is clear that $\Gamma \subset \Lambda$, and it is easy to show that Λ is still a ring. Hence $\Lambda \in \mathcal{F}$, and it is evidently the smallest member of $\mathcal{F}$. Thus Λ is the object we were looking for.

The remaining proofs run similarly. □

The ring Λ in the proof is called the *ring generated by* Γ. In the same way we talk about the *algebra*, *σ-ring* or the *σ-algebra generated by* Γ. This last σ-algebra will be important later; it is denoted by $\sigma(\Gamma)$.

3.1.3. The next proposition gives some simple properties of semirings; the proofs are left to the reader.

Proposition. Let Γ be a semiring of subsets of S, and $\Omega := \Omega(\Gamma)$. Then the following assertions hold.

(i) $\Gamma \subset \Omega$.

(ii) If $A_n \in \Gamma$ $(1 \le n \le N)$, then $\bigcup_{n=1}^{N} A_n \in \Omega$.

(iii) If $A \in \Omega$, $B \in \Omega$, then $A \backslash B \in \Omega$ and $A \cap B \in \Omega$.

(iv) If $A_n \in \Omega$ $(1 \le n \le N)$, then $\bigcup_{n=1}^{N} A_n \in \Omega$ and $\bigcap_{n=1}^{N} A_n \in \Omega$.

Note that the last two assertions of the proposition imply that if Γ is a semiring, then $\Omega(\Gamma)$ is just the ring generated by Γ.

3.1.4. Example. Take $S := \mathbb{R}$. Let

$$\Gamma_1 := \{(a,b] \mid a \in \mathbb{R},\ b \in \mathbb{R},\ a < b\} \cup \{\emptyset\} ,$$

$$\Gamma_2 := \{(k,\ell] \mid k \in \mathbb{Z},\ \ell \in \mathbb{Z},\ k < \ell\} \cup \{\emptyset\} ,$$

$$\Gamma_3 := \{(n,n+1] \mid n \in \mathbb{Z}\} \cup \{\emptyset\} .$$

Then each of these classes is a semiring. The elements of Γ_1 are called *nails*. Note that Γ_2 and Γ_3 generate the same Ω.

3.1.5. Example. Take $S := \mathbb{R}^n$ for some $n \in \mathbb{N}$. Let Γ consist of $\emptyset$, and all sets of the form

$$\{(x_1,x_2,\ldots,x_n) \mid a_i < x_i \le b_i \ (1 \le i \le n)\} ,$$

where $a_i, b_i \in \mathbb{R}$ and $a_i < b_i$ $(1 \le i \le n)$. This Γ is a semiring of subsets of $\mathbb{R}^n$; it is the n-dimensional equivalent of Γ_1 in the preceding example. The elements of Γ are called *cells*.

3.1.6. Example. Let S be a non-empty set. Let Γ_1 consist of all subsets of S, and let Γ_2 consist of all subsets of S containing at most one point. Then Γ_1 is a σ-ring, and even a σ-algebra, while Γ_2 is only a semiring in general. The ring generated by

Γ_2 is the class of finite subsets of S. We have $\Gamma_1 = \sigma(\Gamma_2)$ if and only if S is countable (Exercise 3.1-3).

3.1.7. Let S be a non-empty set, Γ a semiring on S. A mapping μ of Γ into $\mathbb{R}_+^*$ is called a *measure* if it has the following two properties.

$\mathcal{R}3$. $\mu(\emptyset) = 0$.

$\mathcal{R}4$. If A_n $(n \in \mathbb{N})$ are pairwise disjoint elements of Γ with $\bigcup_{n=1}^{\infty} A_n \in \Gamma$, then

$$\mu\left(\bigcup_{n=1}^{\infty} A_n\right) = \sum_{n=1}^{\infty} \mu(A_n) .$$

The property expressed by $\mathcal{R}4$ is usually called the *σ-additivity* of μ. Because of $\mathcal{R}3$, a measure μ is also (*finitely*) *additive*: if $A_1,\ldots,A_N$ are pairwise disjoint elements of Γ with $\bigcup_{n=1}^{N} A_n \in \Gamma$, then $\mu(\bigcup_{n=1}^{N} A_n) = \sum_{n=1}^{N} \mu(A_n)$ (Exercise 3.1-4).

Let Γ be a collection of subsets of S, and let μ be an $\mathbb{R}_+^*$-valued function on Γ. The triple (S,Γ,μ) is called a *measure space* if Γ is a semiring of subsets of S and μ is a measure on Γ, that is, if the pair (Γ,μ) satisfies $\mathcal{R}1$ through $\mathcal{R}4$. This situation is denoted by $\mathcal{R}(\Gamma,\mu)$, or just $\mathcal{R}$.

Occasionally we consider the following additional condition.

$\mathcal{R}5$. S is the union of countably many elements of Γ, each of finite measure.

If the measure space (S,Γ,μ) also satisfies $\mathcal{R}5$, then it is called *σ-finite*, we write $\mathcal{R}_\sigma(\Gamma,\mu)$, or $\mathcal{R}_\sigma$, and μ is called a *σ-finite measure*.

3.1.8. There is a consequence of $\mathcal{R}4$ that will be used often. (See Exercise 3.1-7 for a partial converse.)

Proposition. Let (S,Γ,μ) be a measure space. Let $(A_n)_{n\in\mathbb{N}}$ be a decreasing sequence in Γ (that is, $A_{n+1} \subset A_n$ $(n \in \mathbb{N})$) such that $\bigcap_{n=1}^{\infty} A_n = \emptyset$. Assume $\mu(A_1) < \infty$. Then $\mu(A_n) \to 0$ as $n \to \infty$.

Proof. For each $n \in \mathbb{N}$ we have $A_n \setminus A_{n+1} \in \Omega$, so there exists disjoint members of Γ, B_{kn} $(1 \le k \le m_n)$, say, such that $A_n \setminus A_{n+1} = \bigcup_{k=1}^{m_n} B_{kn}$. Obviously,

$$A_1 = \bigcup_{n=1}^{\infty} (A_n \setminus A_{n+1}) = \bigcup_{n=1}^{\infty} \bigcup_{k=1}^{m_n} B_{kn} ,$$

where the B_{kn} are pairwise disjoint. Hence, by $\mathcal{R}4$, $\mu(A_1) = \sum_{n=1}^{\infty} \sum_{k=1}^{m_n} \mu(B_{kn})$. Similarly, $\mu(A_N) = \sum_{n=N}^{\infty} \sum_{k=1}^{m_n} \mu(B_{kn})$. Since $\mu(A_1) < \infty$, we conclude that $\mu(A_N) \to 0$. □

3.1.9. In the preceding proof we have $A_N \subset A_1$ and $\mu(A_N) \leq \mu(A_1)$, that is, the larger set has the larger measure. This is generally so: we say that measures are *monotone* mappings.

Proposition. Let (S,Γ,μ) be a measure space. Let $A,B \in \Gamma$, $A \subset B$. Then $\mu(A) \leq \mu(B)$.

Proof. We have $B\backslash A = \bigcup_{n=1}^N A_n$, where the A_n are pairwise disjoint elements of Γ. Hence $B = A \cup \bigcup_{n=1}^N A_n$, and therefore $\mu(B) = \mu(A) + \sum_{n=1}^N \mu(A_n) \geq \mu(A)$ by the additivity of μ (see 3.1.7). □

3.1.10. Example. Let S be a non-empty set, let L be a Riesz function space of real-valued functions on S, and let I be a positive linear functional defined on L. Assume that the condition *L*1 of 1.1.1 is satisfied. According to 2.1.8 the class Σ of measurable sets as defined in 2.1.2 is a σ-ring, and 2.1.10 says that the set function μ defined on Σ has all the properties of a measure. Hence (S,Γ,μ) is a measure space.

3.1.11. Example. Let S and Γ_1 be as in Example 3.1.6. For $A \in \Gamma_1$ put

$$\mu_1(A) := \begin{cases} \text{the number of elements of } A & \text{if } A \text{ is finite,} \\ \infty & \text{otherwise.} \end{cases}$$

Then (S,Γ_1,μ_1) is a measure space, which is σ-finite if and only if S is countable. Often μ_1 is called the *counting measure*.

Now take S with Γ_2 of Example 3.1.6. Let $\mu_2 : \Gamma_2 \to \mathbb{R}_+^*$ satisfy $\mu_2(\emptyset) = 0$. Then (S,Γ_2,μ_2) is a measure space. This measure space is σ-finite if and only if S is countable and μ_2 does not take the value ∞.

3.1.12. Example. Let $S := \mathbb{R}$. Let Γ be the collection of nails together with $\emptyset$ (so Γ is the semiring Γ_1 of Example 3.1.4). Let g be a real-valued non-decreasing function on $\mathbb{R}$. Put $\mu_g(\emptyset) := 0$, and for $a \in \mathbb{R}$, $b \in \mathbb{R}$, $a < b$, put $\mu_g((a,b]) := g(b + 0) + - g(a + 0)$. (If $c \in \mathbb{R}$, then $g(c + 0)$ denotes the right-hand limit of g at the point c, that is, $g(c + 0) := \lim_{x \downarrow c} g(x)$.) We shall prove that μ_g is a measure on Γ; it is called the *Lebesgue-Stieltjes measure generated by* g. The particular case $g := \curlyvee_{x \in \mathbb{R}} x$ leads to what is called (*ordinary*) *Lebesgue measure* where the measure of a nail $(a,b]$ is just its length $b - a$. (This is not strictly true: the names belong to the extensions of these measures to much larger classes of subsets of $\mathbb{R}$, which will be constructed in the next section. For the moment, we ignore this distinction.)

It is convenient to assume that the function g is continuous from the right on $\mathbb{R}$. This is no restriction. For if $g^* := \curlyvee_{c \in \mathbb{R}} g(c + 0)$, then g^* is still non-decreasing, g^* is continuous from the right, and it is obvious that g and g^* generate the same Lebesgue-Stieltjes measure. With this continuity assumption on g the definition of μ_g simplifies to $\mu_g((a,b]) := g(b) - g(a)$.

Let us abbreviate μ_g to μ. We have to show that μ is a measure. We know already that Γ is a semiring, and $\mu(\emptyset) = 0$ by definition, so we need only show that μ is σ-additive, that is, we must show that μ satisfies $\mathcal{R}4$. The proof depends on the fact that closed bounded subsets of $\mathbb{R}$ are compact. Since the elements of Γ (except $\emptyset$) are neither open nor closed, the compactness argument must be applied cautiously.

In the first place, note that μ is additive: if A_n $(1 \le n \le N)$ are pairwise disjoint elements of Γ, and if their union A also belongs to Γ, then $\mu(A) = \sum_{n=1}^{N} \mu(A_n)$. To prove this, it is no restriction to assume that none of the A_n is empty. Write $A_n = (a_n, b_n]$, $A = (a,b]$, and assume $a_1 < a_2 < \ldots < a_N$ (reorder if necessary). Then obviously $b_1 = a_2, \ldots, b_{N-1} = a_N$, $a = a_1$, $b_N = b$. Hence

$$\mu(A) = g(b) - g(a) = \sum_{n=1}^{N} \{g(b_n) - g(a_n)\} = \sum_{n=1}^{N} \mu(A_n) .$$

Now we turn to σ-additivity. Let A_n $(n \in \mathbb{N})$ be pairwise disjoint elements of Γ, and assume $A := \bigcup_{n=1}^{\infty} A_n \in \Gamma$. Again we may assume that none of the A_n is empty. Write $A_n = (a_n, b_n]$, $A = (a,b]$. Fix $N \in \mathbb{N}$. It is easy to see that A may be written as the disjoint union of $A_1, A_2, \ldots, A_N$ and certain other members of Γ, $B_1, B_2, \ldots, B_M$, say. Since μ is additive,

$$\mu(A) = \sum_{n=1}^{N} \mu(A_n) + \sum_{m=1}^{M} \mu(B_m) \ge \sum_{n=1}^{N} \mu(A_n) .$$

Hence $\sum_{n=1}^{\infty} \mu(A_n) \le \mu(A)$.

To prove the reverse inequality, let $\varepsilon > 0$. We shall show that $\sum_{n=1}^{\infty} \mu(A_n) > \mu(A) - \varepsilon$. Take positive numbers δ, δ_n $(n \in \mathbb{N})$ such that $g(a + \delta) < \frac{1}{2}\varepsilon + g(a)$, $g(b_n + \delta_n) < 2^{-n-1}\varepsilon + g(b_n)$, and define $A' := [a+\delta, b]$, $A'_n := (a_n, b_n + \delta_n)$, where we assume $a + \delta < b$. The compact set A' is covered by the collection $\{A'_n \mid n \in \mathbb{N}\}$ which consists of open sets. Hence (see 0.2.7), there exists $N \in \mathbb{N}$ such that $A' \subset \bigcup_{n=1}^{N} A'_n$. Put $A'' := (a+\delta, b]$, $A''_n := (a_n, b_n + \delta_n]$ $(1 \le n \le N)$. Then these sets are in Γ, and $\mu(A'') > \mu(A) - \frac{1}{2}\varepsilon$, $\mu(A''_n) < \mu(A_n) + 2^{-n-1}\varepsilon$ $(1 \le n \le N)$. Therefore

$$\mu(A) - \sum_{n=1}^{N} \mu(A_n) < \mu(A'') + \varepsilon - \sum_{n=1}^{N} \mu(A''_n) .$$

Since we can find pairwise disjoint C_n's in Γ such that $C_n \subset A''_n$ $(1 \le n \le N)$ and $A'' = \bigcup_{n=1}^{N} C_n$, it follows that

$$\mu(A'') = \sum_{n=1}^{N} \mu(C_n) \le \sum_{n=1}^{N} \mu(A''_n) .$$

This implies $\sum_{n=1}^{N} \mu(A_n) > \mu(A) - \varepsilon$. Since ε is arbitrary, we conclude that $\sum_{n=1}^{\infty} \mu(A_n) \ge \mu(A)$. Hence μ is σ-additive, and therefore a measure.

3.1.13. Remark. It is interesting to discuss the consequence of a minor change in the preceding example. Replace $\mathbb{R}$ in the role of S by $\mathbb{Q}$. The resulting μ need no longer be a measure. This is already illustrated by the ordinary Lebesgue case. Indeed, take $g := \bigcup_{x\in\mathbb{R}} x$, and proceed as in 3.1.12 in defining Γ and μ, but now on $\mathbb{Q}$ only. That is, Γ consists of the empty set together with nails $(a,b] := \{r \in \mathbb{Q} \mid a < r \le b\}$ with $a \in \mathbb{Q}$, $b \in \mathbb{Q}$, $a < b$, and $\mu((a,b]) := b - a$.

Let $0 < \varepsilon < 1$. Let $(r_n)_{n\in\mathbb{N}}$ be an enumeration of $\mathbb{Q} \cap [0,1]$. Put $A_0 := \mathbb{Q}\backslash(0,1]$, and proceed by induction:

$$A_n := \begin{cases} \emptyset & \text{if } r_n \in \bigcup_{k=1}^{n-1} A_k , \\ (s_n,r_n] & \text{otherwise,} \end{cases}$$

where $s_n < r_n$ is chosen in such a way that $s_n - r_n < \varepsilon\cdot 2^{-n}$ and $(s_n,r_n] \cap A_k = \emptyset$ for $k = 0,1,\ldots,n-1$. The sets A_n $(n \in \mathbb{N})$ are pairwise disjoint by construction, and their union is $(0,1]$, while

$$\sum_{n=1}^{\infty} \mu(A_n) \le \sum_{n=1}^{\infty} \varepsilon\cdot 2^{-n} = \varepsilon < 1 = \mu\left(\bigcup_{n=1}^{\infty} A_n\right) .$$

Hence, μ is not σ-additive, and therefore not a measure.

This example further stresses the importance of the compactness argument in 3.1.12.

Exercises Section 3.1

1. Give the proof of Proposition 3.1.3.

2. Elaborate the Examples 3.1.4 and 3.1.5.

3. Let Γ_1 and Γ_2 be as in Example 3.1.6. Show that the collection Γ_3 of all countable subsets with their complements equals $\sigma(\Gamma_2)$. Show that $\Gamma_1 = \sigma(\Gamma_2)$ if and only if S is countable.

4. Let (S,Γ,μ) be a measure space and let $A_1,\ldots,A_N$ be pairwise disjoint elements of Γ with $A := \bigcup_{n=1}^N A_n \in \Gamma$. Show that $\mu(A) = \sum_{n=1}^N \mu(A_n)$.

5. Elaborate Example 3.1.11.

6. Give an example of a measure space (S,Γ,μ) and a sequence $(A_n)_{n\in\mathbb{N}}$ in Γ with $A_{n+1} \subset A_n$ $(n \in \mathbb{N})$, $\bigcap_{n=1}^{\infty} A_n = \emptyset$ such that $\lim_{n\to\infty} \mu(A_n) > 0$.

7. Let Γ be a ring of subsets and let $\mu : \Gamma \to \mathbb{R}_+^*$ satisfy $\mu(\emptyset) = 0$. Consider the following conditions.

(i) $\mu(A) < \infty$ $(A \in \Gamma)$.

(ii) μ is additive (cf. 3.1.7).

(iii) $\mu(A_n) \to 0$ if $A_n \in \Gamma$, $A_{n+1} \subset A_n$ $(n \in \mathbb{N})$, $\bigcap_{n=1}^{\infty} A_n = \emptyset$.

Show that (S,Γ,μ) is a σ-finite measure space if (i), (ii) and (iii) are satisfied. Show that if μ satisfies (i) and (ii), or (ii) and (iii), then μ need not be a measure.

8. Let $S = \mathbb{Z}$. We call $A \subset S$ periodic with period $p \in \mathbb{N}$ if $A + p := \{n + p \mid n \in A\} = A$. Let Γ consist of all periodic subsets of S, and put $\mu(A) := 1/p \sum_{n=1}^{p} \chi_A(n)$ where $A \in \Gamma$ and p is such that A is periodic with period p. Show that Γ is a semiring and that μ is well-defined and additive on Γ. Is μ a measure?

9. Let Γ be the set of nails in $\mathbb{R}$ (cf. 3.1.4). Let μ be a measure on Γ with $\mu(A) < \infty$ $(A \in \Gamma)$. Show that there is exactly one real-valued function g defined on $\mathbb{R}$, non-decreasing, continuous from the right, with $g(0) = 0$ and such that g generates μ as in Example 3.1.12.

10. Let Γ be a semiring (or ring, algebra, σ-ring, σ-algebra) of subsets of S, and let $S_0 \subset S$. Show that $\Gamma_0 := \{E \cap S_0 \mid E \in \Gamma\}$ is a semiring (or ring, algebra, σ-ring, σ-algebra) of subsets of S_0. Let μ be a measure on Γ and assume that $E \cap S_0 \in \Gamma$ if $E \in \Gamma$. Show that $\mu_0 := \curlyvee_{E\in\Gamma_0} \mu(E)$ is a measure on Γ_0 and that μ_0 is σ-finite if μ is σ-finite. This Γ_0 is called the *induced semiring* (*ring, algebra, σ-ring, σ-algebra*) and μ_0 is called the *induced measure*.

11. Let $f : S_1 \to S_2$ where S_1 and S_2 are non-empty sets. Let Γ be a class of subsets of S_2 and let $\overset{\leftarrow}{f}(\Gamma) := \{\overset{\leftarrow}{f}(A) \mid A \in \Gamma\}$. Show that if Γ is a semiring (ring, algebra, σ-ring, σ-algebra) then so is $\overset{\leftarrow}{f}(\Gamma)$.

Let (S_1,Γ_1,μ) be a measure space, let $S_2 \neq \emptyset$, and let Γ_2 be a semiring of subsets of S_2. Assume that $f : S_1 \to S_2$ satisfies $\overset{\leftarrow}{f}(\Gamma_2) \subset \Gamma_1$. Show that $\curlyvee_{A\in\Gamma_2} \mu(\overset{\leftarrow}{f}(A))$ is a measure on Γ_2.

13. Assume that (B,I) satisfies $\mathcal{D}$ (cf. 1.3.1). Let $S_0 := S \times [0,\infty)$ and define

$$F_b := \{(s,t) \mid s \in S,\ 0 \le t \le b(s)\} \quad \text{for } b \in B^+ .$$

Also, let

$$\Gamma := \{F_b \backslash F_c \mid b,c \in B^+,\ c \le b\} ,$$

and put $\mu(A) := I(b) - I(c)$ if $A = F_b \backslash F_c$ (with $b,c \in B^+$, $c \le b$). Show that (S_0,Γ,μ) is a measure space. When is the system σ-finite?

3.2. Measure spaces and integration

From a measure space (S,Γ,μ) as defined in the preceding section we shall now construct an integral and a space of integrable functions. The construction is in two steps. The first step consists in defining a suitable Riesz function space B together with a positive linear functional I on B that satisfies the condition $\mathcal{D}$ of Section 1.3. There are no real difficulties, but we have to do some work. The second step is still easier, for here we invoke the extension process $\mathcal{D} \to \mathcal{A} \to \mathcal{N} \to \mathcal{L}$, described in Chapter 1, to obtain the desired integral.

3.2.1. Let S be a non-empty set, let Γ be a semiring of subsets of S, and let μ be a measure defined on Γ. Let B denote the collection of all real-valued functions b defined on S that can be written as $b = \sum_{n=1}^{N} b_n \chi_{B_n}$, where $b_n \in \mathbb{R}$, $B_n \in \Gamma$, $\mu(B_n) < \infty$ $(1 \le n \le N)$, and the number N may depend on b. So B consists of the linear combinations of characteristic functions of those members of Γ that have finite measure.

3.2.2. Before showing that B is a Riesz function space, we rephrase the definition of B. As a preparation we extend the measure μ to the ring $\Omega := \Omega(\Gamma)$ generated by Γ (see 3.1.3). First we derive an auxiliary result.

<u>Lemma</u>. Let both $(A_n)_{n\in\mathbb{N}}$ and $(B_n)_{n\in\mathbb{N}}$ be sequences of pairwise disjoint elements of Γ, and assume $\bigcup_{n=1}^{\infty} A_n = \bigcup_{n=1}^{\infty} B_n$. Then

$$\sum_{n=1}^{\infty} \mu(A_n) = \sum_{n=1}^{\infty} \mu(B_n) .$$

<u>Proof</u>. If $(n,m) \in \mathbb{N} \times \mathbb{N}$, then $A_n \cap B_m \in \Omega$, so there exist pairwise disjoint $C_{nmk} \in \Gamma$ $(k \in \mathbb{N})$ such that $A_n \cap B_m = \bigcup_{k=1}^{\infty} C_{nmk}$, while $C_{nmk} = \emptyset$ for all large k. (We write infinite unions to make the notation simpler.)

If $n \in \mathbb{N}$, $m \in \mathbb{N}$, then by $R4$,

$$\mu(A_n) = \sum_{m=1}^{\infty} \sum_{k=1}^{\infty} \mu(C_{nmk}) , \quad \mu(B_m) = \sum_{n=1}^{\infty} \sum_{k=1}^{\infty} \mu(C_{nmk}) .$$

Hence

$$\sum_{m=1}^{\infty} \mu(B_m) = \sum_{m=1}^{\infty} \sum_{n=1}^{\infty} \sum_{k=1}^{\infty} \mu(C_{nmk}) = \sum_{n=1}^{\infty} \sum_{m=1}^{\infty} \sum_{k=1}^{\infty} \mu(C_{nmk}) = \sum_{n=1}^{\infty} \mu(A_n) ,$$

where the rearrangement of the triple sum is justified by the fact that the terms are non-negative (see 0.4.7). ∎

3.2.3. The extension of μ to Ω is very easy now. Let $P \in \Omega$. Then by definition $P = \bigcup_{n=1}^{N} A_n$, where the A_n are pairwise disjoint elements of Γ, and N may depend on P. We put $\tilde{\mu}(P) := \bigcup_{n=1}^{N} \mu(A_n)$. Σ

Lemma 3.2.2 shows that $\tilde{\mu}$ is well-defined, that is, the value of $\tilde{\mu}(P)$ does not depend on the representation of P as a disjoint union of elements of Γ. The lemma also shows that $\mu(A) = \tilde{\mu}(A)$ if $A \in \Gamma$. Hence $\tilde{\mu}$ is an extension of Γ. Therefore there is no danger in writing μ instead of $\tilde{\mu}$, and we shall do this from now on.

3.2.4. The extended μ is a measure on Ω. To prove this, it is enough to show that μ is σ-additive on Ω.

Proposition. Let $(P_n)_{n\in\mathbb{N}}$ be a sequence of pairwise disjoint elements of Ω, and assume that $P := \bigcup_{n=1}^{\infty} P_n \in \Omega$. Then $\mu(P) = \sum_{n=1}^{\infty} \mu(P_n)$.

Proof. For each $n \in \mathbb{N}$, let $(A_{nm})_{m\in\mathbb{N}}$ be a sequence of pairwise disjoint elements of Γ such that $P_n = \bigcup_{m=1}^{\infty} A_{nm}$, with $A_{nm} = \emptyset$ for large m. Then $\mu(P_n) = \sum_{m=1}^{\infty} \mu(A_{nm})$ by definition. Clearly, $P = \bigcup_{n=1}^{\infty} \bigcup_{m=1}^{\infty} A_{nm}$, and the A_{nm} are pairwise disjoint. Hence

$$\mu(P) = \sum_{n=1}^{\infty} \sum_{m=1}^{\infty} \mu(A_{nm}) = \sum_{n=1}^{\infty} \mu(P_n)$$

by Lemma 3.2.2. ∎

3.2.5. Now B can be described in terms of Ω. For it is easy to see that Γ may be replaced by Ω in 3.2.1 without any change in the resulting B. So from now on B consists of the real-valued functions on S that have the form $\sum_{n=1}^{N} b_n \chi_{B_n}$, where $b_n \in \mathbb{R}$, $B_n \in \Omega$, $\mu(B_n) < \infty$ $(1 \le n \le N)$, and N may depend on b. It is also easy to see that if $b \in B$ and $b = \sum_{n=1}^{N} b_n \chi_{B_n}$ in the way just described, then we may even arrange the B_n to be pairwise disjoint; such a representation will be called a *standard representation* of b. In general, there are many of these standard representations for any $b \in B$. Before being able to use them in the definition of the functional I, we need the following result.

3.2.6. <u>Lemma</u>. Let $b \in B$ have standard representations $\sum_{n=1}^{N} b_n \chi_{B_n}$ and $\sum_{m=1}^{M} c_m \chi_{C_m}$. Then

$$\sum_{n=1}^{N} b_n \mu(B_n) = \sum_{m=1}^{M} c_m \mu(C_m) .$$

<u>Proof</u>. We may assume that $b_n \neq 0 \neq c_m$ for $1 \leq n \leq N$, $1 \leq m \leq M$, so if $P := \bigcup_{n=1}^{N} B_n$, then $P = \bigcup_{m=1}^{M} C_m$. Since μ is a measure on the ring Ω,

$$\sum_{n=1}^{N} b_n \mu(B_n) = \sum_{n=1}^{N} b_n \mu(B_n \cap P) = \sum_{n=1}^{N} \sum_{m=1}^{M} b_n \mu(B_n \cap C_m) .$$

Since $B_n \cap C_m \neq \emptyset$ implies $b_n = c_m$, while $\mu(\emptyset) = 0$, it follows that $b_n \mu(B_n \cap C_m) = c_m \mu(B_n \cap C_m)$ in any case. Tracing back, we see that the double sum equals

$$\sum_{m=1}^{M} \sum_{n=1}^{N} c_m \mu(B_n \cap C_m) = \sum_{m=1}^{M} c_m \mu(C_m) . \qquad \square$$

3.2.7. Finally we are in a position to define I. Let $b \in B$, and let $\sum_{n=1}^{N} b_n \chi_{B_n}$ be a standard representation of b. Put $I(b) := \sum_{n=1}^{N} b_n \mu(B_n)$.

Lemma 3.2.6 ensures that I is well-defined on B. It also shows how μ and I are connected on Ω: if $A \in \Omega$ and $\mu(A) < \infty$, then $\chi_A \in B$ and $I(\chi_A) = \mu(A)$. (The last relation will return in more general form in 3.2.10.)

3.2.8. <u>Theorem</u>. B is a Riesz function space, and I is a positive linear functional on B.

<u>Proof</u>. That B is a Riesz function space is obvious from the rephrased definition in 3.2.5, so only the assertion about I needs a proof.

Let $b \in B$, $c \in B$ have standard representations $\sum_{n=1}^{N} b_n \chi_{B_n}$ and $\sum_{m=1}^{M} c_m \chi_{C_m}$, respectively. Let $S_1 := \bigcup_{n=1}^{N} B_n$, $S_2 := \bigcup_{m=1}^{M} C_m$. Then

$$b + c = \sum_{n=1}^{N} \sum_{m=1}^{M} (b_n + c_m) \chi_{B_n \cap C_m} + \sum_{n=1}^{N} b_n \chi_{B_n \setminus S_2} + \sum_{m=1}^{M} c_m \chi_{C_m \setminus S_1} ,$$

and this is easily seen to be a standard representation of b + c. Hence, by the definition of I,

$$I(b + c) = \sum_{n=1}^{N} \sum_{m=1}^{M} (b_n + c_m) \mu(B_n \cap C_m) + \sum_{n=1}^{N} b_n \mu(B_n \setminus S_2) + \sum_{m=1}^{M} c_m \mu(C_m \setminus S_1) .$$

The double sum can be rewritten as

$$\sum_{n=1}^{N} b_n \sum_{m=1}^{M} \mu(B_n \cap C_m) + \sum_{m=1}^{M} c_m \sum_{n=1}^{N} \mu(B_n \cap C_m) =$$

$$= \sum_{n=1}^{N} b_n \mu(B_n \cap S_2) + \sum_{m=1}^{M} c_m \mu(C_m \cap S_1) ,$$

and therefore

$$I(b + c) = \sum_{n=1}^{N} b_n\{\mu(B_n \cap S_2) + \mu(B_n \backslash S_2)\} + \sum_{m=1}^{M} c_m\{\mu(C_m \cap S_1) + \mu(C_m \backslash S_1)\} =$$

$$= \sum_{n=1}^{N} b_n \mu(B_n) + \sum_{m=1}^{M} c_m \mu(C_m) = I(b) + I(c) .$$

Hence I is additive.

Let $b \in B$ and $\alpha \in \mathbb{R}$. If b has standard representation $\sum_{n=1}^{N} b_n \chi_{B_n}$, then $\sum_{n=1}^{N} \alpha b_n \chi_{B_n}$ is a standard representation for αb. Hence $I(\alpha b) = \alpha I(b)$.

Finally, if $b \in B^+$ has standard representation $\sum_{n=1}^{N} b_n \chi_{B_n}$, then all coefficients b_n can be assumed to be non-negative. Hence $I(b) \geq 0$. □

3.2.9. We shall now show that the system (B,I) satisfies the condition $\mathcal{D}1$ of 1.3.1.

<u>Theorem</u>. Let $(b_n)_{n \in \mathbb{N}}$ be a sequence in B^+. Assume that $b_{n+1} \leq b_n$ $(n \in \mathbb{N})$ and $b_n \to 0$. Then $I(b_n) \to 0$.

<u>Proof</u>. Let $\sum_{m=1}^{M} c_m \chi_{C_m}$ be a standard representation of b_1. Put $C := \bigcup_{m=1}^{M} C_m$, $c := \max\{|c_m| \mid 1 \leq m \leq M\}$. Let $\varepsilon > 0$. Define

$$A_n := \{s \in S \mid b_n(s) > \varepsilon\{1 + 2\mu(C)\}^{-1}\}$$

for $n \in \mathbb{N}$. It is easy to see that $A_n \in \Omega$, $A_{n+1} \subset A_n$ $(n \in \mathbb{N})$, $\bigcap_{n=1}^{\infty} A_n = \emptyset$, and $\mu(A_1) < \infty$. Since (S,Ω,μ) is a measure space by 3.2.4, Proposition 3.1.8 shows that $\mu(A_n) \to 0$ as $n \to \infty$.

If $n \in \mathbb{N}$, then

$$0 \leq b_n \leq c\chi_{A_n} + \frac{\varepsilon}{1 + 2\mu(C)} \chi_{C \backslash A_n} ,$$

and therefore

$$0 \leq I(b_n) \leq c\mu(A_n) + \frac{\varepsilon}{2} .$$

Since $\mu(A_n) \to 0$, it follows that $0 \leq I(b_n) < \varepsilon$ for all large n. That is, $I(b_n) \to 0$ as $n \to \infty$. □

3.2.10. The construction of B and I just completed leads from a situation $\mathcal{R}$ to a situation $\mathcal{D}$; the process is denoted by $\mathcal{R}(\Gamma,\mu) \to \mathcal{D}(B,I)$, or $\mathcal{R} \to \mathcal{D}$. It is clear that if the $\mathcal{R}$-system is σ-finite, then so is the $\mathcal{D}$-system, so there is also a transition $\mathcal{R}_\sigma \to \mathcal{D}_\sigma$.

We can now follow the route $\mathcal{R} \to \mathcal{D} \to \mathcal{A} \to \mathcal{N} \to \mathcal{L}$ to introduce the class of integrable functions together with the integral. This newly introduced class will be denoted by L_μ, the corresponding class of measurable functions by M_μ, and the σ-ring of measurable subsets by Σ_μ. The subscript μ may be deleted if there is no danger of confusion. As usual, the integral on L_μ is still denoted by I. (In the next subsection we shall introduce additional notation for this special situation.) The members of L_μ will occasionally be called μ-*integrable functions*; the corresponding term for members of M_μ and Σ_μ is μ-*measurable*.

In the manner of 2.1.9 we now introduce a set function $\tilde{\mu}$ on Σ_μ: if $E \in \Sigma_\mu$, then $\tilde{\mu}(E) := I(\chi_E)$ if $\chi_E \in L$, and $\tilde{\mu}(E) := \infty$ otherwise. According to Example 3.1.10 this results in a measure space $(S,\Sigma_\mu,\tilde{\mu})$. It is an important fact that $\Gamma \subset \Sigma_\mu$, and $\tilde{\mu}(A) = \mu(A)$ if $A \in \Gamma$ (Exercise 3.2-4). So $\tilde{\mu}$ is an extension of μ. What happens if we start the process anew from $(S,\Sigma_\mu,\tilde{\mu})$? Nothing, as we shall see in 3.4.6.

Another point of interest is that condition $\mathcal{M}$ of 2.1.11 is satisfied, that is, Σ_μ is a σ-algebra. The proof is simple: note that $\inf(b,\chi_S) \in B^+$ whenever $b \in B^+$, and apply the result of Exercise 2.1-1. It follows that measurability of functions can be described in terms of Σ_μ (see Theorem 2.1.12).

Finally we point out that in the present situation $\mathcal{A}5$ of 1.2.6 is satisfied, because the line $\mathcal{D} \to \mathcal{A}$ is part of the route from $\mathcal{R}$ to $\mathcal{L}$ (see 1.3.9). This means that the norm is definitely determined at the first passage through $\mathcal{N}$. Also, the set of integrable functions together with the integral is definitely determined at the first passage through $\mathcal{L}$ (see 1.4.7). Thus further trips in the circle line give no change at all.

3.2.11. Here is some more notation. Let (S,Γ,μ), L_μ, M_μ and Σ_μ be as in 3.2.10. Let L^*_μ and M^*_μ be derived from L_μ and M_μ in the manner of Section 2.2. Let $E \in \Sigma_\mu$. For a real-valued f defined on S such that $f \cdot \chi_E \in L_\mu$ it is customary to write

$$\int_E f\,d\mu := I(f\chi_E) ,$$

and f is said to be *integrable over* E. An obvious extension of this agreement will be admitted: f need not be real-valued, nor need it be defined on all of S. To be precise, let $E \in \Sigma_\mu$ and let $E_0 \subset E$ be a μ-null set. Let f be a function defined on $E \backslash E_0$ with values in $\mathbb{R}^*$. Define the extension $\tilde{f}$ by

$$\tilde{f}(s) := \begin{cases} f(s) & \text{if } s \in E\backslash E_0 , \\ 0 & \text{if } s \in S\backslash E . \end{cases}$$

Suppose $I(\tilde{f})$ is defined according to Section 2.2, that is, suppose that $\tilde{f} \in L^*_\mu$, or $\tilde{f} \in M^*_\mu$ and $\tilde{f} \geq 0$ (a.e.). Then the *integral of* f *over* E is defined as

$$\int_E f\,d\mu := I(\tilde{f}) .$$

If $I(\tilde{f})$ is finite, then f is said to be *integrable over* E. In the special case where E = S, the name of the set if often suppressed in the notation; so $\int f d\mu$ is the same as $_S\int f\,d\mu$.

For Lebesgue and Lebesgue-Stieltjes measures on the real line there is a special notation. Consider Example 3.1.12, where a measure μ_g defined on the class Γ of nails in $\mathbb{R}$ is generated by a non-decreasing $g : \mathbb{R} \to \mathbb{R}$. In this case one often writes $_{-\infty}\int^{\infty} f(x)dg(x)$ instead of $\int f d\mu_g$. If $g = \Upsilon_{x \in \mathbb{R}}\, x$, that is, in the Lebesgue case, this is further simplified to $_{-\infty}\int^{\infty} f(x)dx$. For Lebesgue measure countable sets are null sets, so $_E\int f\,d\mu_g$ has the same value for each of the intervals [a,b), [a,b], (a,b), (a,b] in the role of E; this common value is denoted by $_a\int^b f(x)dx$. (For Lebesgue-Stieltjes measures the corresponding notation is ambiguous, however (see Exercise 3.2-1).) We note that the Lebesgue integral as introduced here coincides with the Lebesgue integral as introduced in Examples 1.2.4 and 1.2.5 (Exercise 3.2-2).

Exercises Section 3.2

1. Let $g := \chi_{[0,\infty)}$. Assuming the notation of 3.2.11 show that $_{[0,\infty)}\int d\mu_g = 1$, $_{(0,\infty)}\int d\mu_g = 0$. (The function behind the integral sign, which seems to be absent, is identically 1.)

2. Show that integrable functions and integrals of 1.2.4 and 1.2.5 and those of 3.2.11 for $\mu = \mu_g$ with $g(x) = x$ are the same.

3. Can a Lebesgue null set have a non-empty interior?

4. Let (S,Γ,μ) be a measure space. Show that $\Gamma \subset \Sigma_\mu$. Assume in addition that the system is σ-finite and that Γ is a σ-algebra containing all $A \in \Sigma_\mu$ with $\mu(A) = 0$. Show that $\Gamma = \Sigma_\mu$. (Hint. Let $A \in \Sigma_\mu$, $\mu(A) < \infty$. By Theorem 1.1.19(i) there is a sequence $(b_n)_n$ in B such that $b_n \to \chi_A$ (a.e.).)

5. Let (S,Γ,μ) be a σ-finite measure space. Assume that $\mathcal{E} \subset \Sigma_\mu$ satisfies

(i) $\Gamma \subset \mathcal{E}$,

(ii) $A \in \mathcal{E}$, $B \in \mathcal{E}$, $B \subset A \Rightarrow A \backslash B \in \mathcal{E}$,

(iii) $A_n \in E$, $A_n \cap A_m = \emptyset$ $(n,m \in \mathbb{N},\ n \neq m) \Rightarrow \bigcup_{n=1}^{\infty} A_n \in E$,

(iv) $A \in \Sigma_\mu$, $\mu(A) = 0 \Rightarrow A \in E$.

Show that $E = \Sigma_\mu$. Is this also true in the non σ-finite case?

6. Let $S := \mathbb{R}$, and consider S with Lebesgue measure μ. Let $S_0 := [0,1]$. Each $x \in S_0$ has at least one triadic representation $x = \sum_{n=1}^{\infty} a_n(x)3^{-n}$ with $a_n(x) \in \{0,1,2\}$ for all n. (Some have even two; for instance 1/3 has besides the obvious one also the representation $\sum_{n=2}^{\infty} 2\cdot 3^{-n}$.) The *Cantor set* C consists of all $x \in S_0$ that have a representation $\sum_{n=1}^{\infty} a_n(x)3^{-n}$ with $a_n(x) \neq 1$ for all n. Show that C is uncountable and that $\mu(C) = 0$. The Cantor set can also be described as follows. From S_0 we remove the middle open third $I_{11} := (\frac{1}{3},\frac{2}{3})$. There remain two closed intervals. From each of these intervals we remove the middle open thirds $I_{21} := (\frac{1}{9},\frac{2}{9})$, $I_{23} := (\frac{7}{9},\frac{8}{9})$. In general we put $I_{kj} := \left(\frac{3j-2}{3^k},\frac{3j-1}{3^k}\right)$ for $j = 1,\ldots,3^{k-1}$, $k \in \mathbb{N}$, and we remove these intervals successively from S_0. What remains is the Cantor set.

7. Let $f : [0,1] \to \mathbb{R}$ be bounded. Show that f is Riemann integrable over $[0,1]$ if and only if the set of discontinuities of f has measure zero. (Hint. If I is an open interval, let $\Delta(f,I) := \sup\{f(x) - f(y) \mid x \in I,\ y \in I\}$. The set of all discontinuities of f equals

$$\bigcup_{n=1}^{\infty} \bigcap_{m=1}^{\infty} \{x \mid \Delta(f,(x-\tfrac{1}{m},x+\tfrac{1}{m})) > \tfrac{1}{n}\} .$$

Assume that f is Riemann integrable. Then by Exercise 0.5-3(ii) we can find for any $n \in \mathbb{N}$, $\varepsilon > 0$ an $m \in \mathbb{N}$ such that $\{x \mid \Delta(f,(x-\frac{1}{m},x+\frac{1}{m})) > \frac{1}{n}\}$ is contained in a set whose measure is less than ε.)

8. Show that χ_C, where C is the Cantor set, is Riemann integrable over $[0,1]$ with integral 0.

9. Let (S,Γ,μ) be a measure space. If f is integrable and $\int_A f\,d\mu = 0$ for all $A \in \Gamma$, then $f = 0$ (a.e.).

10. Do Exercise 9 with the condition "f is integrable" replaced by "there is a sequence $(A_n)_{n\in\mathbb{N}}$ in Γ with $\mu(A_n) < \infty$ $(n \in \mathbb{N})$ such that the support of f is contained in $\bigcup_{n=1}^{\infty} A_n$."

11. Let (S,Γ,μ) be a measure space, where Γ is a σ-algebra of subsets of S. Let $1 \leq p \leq q \leq \infty$ and assume that $L^p \subset L^q$. Show that there is a $\delta > 0$ such that if $A \in \Gamma$

and $\mu(A) > 0$, then $\mu(A) > \delta$. Use the following steps. Suppose that such a $\delta > 0$ does not exist.

(i) Construct sets $A_k^{(n)} \in \Gamma$ ($n \in \mathbb{N}$, $k = 1,\dots,n$) successively as follows. Let $A_1^{(1)} \in \Gamma$ be such that $0 < \mu(A_1^{(1)}) < 1$. If $n \in \mathbb{N}$, and $A_k^{(n)} \in \Gamma$ ($k = 1,\dots,n$) have been found already, let $B \in \Gamma$ be such that $0 < \mu(B) < 4^{-n}\mu(A_k^{(n)})$ ($k = 1,\dots,n$) and put $A_k^{(n+1)} := A_k^{(n)} \setminus B$, $A_{n+1}^{(n+1)} := B$. Show that $A_k^{(n)}$ ($k = 1,\dots,n$) are pairwise disjoint elements of Γ and that $0 < \frac{2}{3}\mu(A_k^{(1)}) \le \mu(A_k^{(n)}) < 4^{-k}$ for all $n \in \mathbb{N}$ and all $k = 1,\dots,n$.

(ii) Define $A_k := \bigcap_{n=1}^{\infty} A_k^{(n)}$ and show that $A_k \in \Gamma$, $0 < \mu(A_k) < 4^{-k}$, $A_k \cap A_\ell = \emptyset$ ($k \in \mathbb{N}$, $\ell \in \mathbb{N}$, $k \neq \ell$).

(iii) Derive a contradiction by considering certain functions of the form $\sum_{k=1}^{\infty} \alpha_k \chi_{A_k}$ with $\alpha_k > 0$.

12. Let (S,Γ,μ) be a measure space, where Γ is a σ-algebra of subsets of S, and assume that $\mu(S) < \infty$. Let $(f_n)_{n\in\mathbb{N}}$ be a sequence of measurable functions with $f_n \to 0$. Show that for every $\varepsilon > 0$ there is an $E \in \Gamma$, $\mu(E) < \varepsilon$ such that $f_n \to 0$ uniformly outside E. This result is usually called *Egorov's theorem*. (Hint. Let $\varepsilon > 0$, $m \in \mathbb{N}$. Since $S = \bigcup_{N=1}^{\infty} \bigcap_{n=N}^{\infty} \{s \mid |f_n(s)| < m^{-1}\}$, there is an $N(m)$ such that

$$\mu\Big(S \setminus \bigcap_{n=N(m)}^{\infty} \{s \mid |f_n(s)| < m^{-1}\}\Big) < \varepsilon \cdot 2^{-m-1} .$$

Let

$$F := \bigcap_{m=1}^{\infty} \bigcap_{n=N(m)}^{\infty} \{s \mid |f_n(s)| < m^{-1}\} ,$$

and show that $\mu(S\setminus F) < \varepsilon$, $f_n \to 0$ uniformly on F.) Show by an example that the condition "$\mu(S) < \infty$" cannot be omitted.

13. Let (S,Γ,μ) be a measure space, where Γ is a σ-algebra of subsets of S, and assume $\mu(S) < \infty$. Let $p > 1$. Let $(f_n)_{n\in\mathbb{N}}$ be a sequence in L^p such that $f_n \to 0$, $\|f_n\|_p \le 1$ ($n \in \mathbb{N}$). Show that $\|f_n\|_r \to 0$ for every r with $1 \le r < p$. (Hint. Use the previous exercise.) Give an example of a measure space (S,Γ,μ) (with Γ a σ-algebra and $\mu(S) < \infty$) and a sequence $(f_n)_{n\in\mathbb{N}}$ in L^2 such that $f_n \to 0$, $\|f_n\|_2 \le 1$, $\|f_n\|_2 \not\to 0$.

14. Take $S = [0,1]$ with Lebesgue measure μ. Let $f \in L_\mu$ and show that for every $\varepsilon > 0$ there exists a set $C \in \Sigma_\mu$ such that the restriction of f to C is continuous and $\mu([0,1]\setminus C) < \varepsilon$. (Hint. Take a sequence of continuous functions $(b_n)_{n\in\mathbb{N}}$ with $b_n \to f$ (a.e.); use Exercise 12.) This result is usually called *Lusin's theorem*; we present a more general version of this theorem in Section 4.2.

15. The following approach to measure theory starts from a slightly more general kind of semiring than in 3.1.1. Let $\Gamma \subset P(S)$ and let Ω_0 be the set of countable unions of pairwise disjoint elements of Γ. Assume that $\emptyset \in \Gamma$ and $A \in \Gamma$, $B \in \Gamma \Rightarrow A \cup B \in \Omega_0$. $A \backslash B \in \Omega_0$. Let $\mu : \Gamma \to \mathbb{R}_+$ satisfy $\mu(\emptyset) = 0$, $\mu(A) = \sum_{n=1}^{\infty} \mu(A_n)$ if $A \in \Gamma$, $A_n \in \Gamma$ pairwise disjoint, $A = \bigcup_{n=1}^{\infty} A_n$. Finally assume that there are pairwise disjoint $A_n \in \Gamma$ such that $S = \bigcup_{n=1}^{\infty} A_n$ and $\mu(A_n) < \infty$ for all n. We define B as the set of functions b of the form $\Sigma_n \alpha_n \chi_{A_n}$, where $\alpha_n \in \mathbb{R}$, $A_n \in \Gamma$, $\Sigma_n |\alpha_n| \mu(A_n) < \infty$ and S is the disjoint union of the A_n's, and we put $I(b) = \Sigma_n \alpha_n \mu(A_n)$. Show that B is a Riesz function space, that I is well-defined on B, that I is a positive linear functional on B, and that $\chi_A \in B$, $I(\chi_A) = \mu(A)$ if $A \in \Gamma$, $\mu(A) < \infty$. To show that (B,I) satisfies $\mathcal{D}$, let $b_n \in B^+$, $b \in B^+$, $b = \Sigma_n b_n$. Show that $I(b) = \Sigma_n I(b_n)$ as follows.

(i) Show that $I(b) \geq \Sigma_n I(b_n)$.

(ii) First treat the case $b_n \cdot b_m = 0$ $(n \neq m)$, and conclude that we may assume that $b = \chi_A$ with $A \in \Gamma$, $\mu(A) < \infty$.

(iii) Let $0 < \gamma < 1$, and define c_n for $n \in \mathbb{N}$ by $c_n(s) = 1$ if n is the first index such that $\sum_{k=1}^{n} b_k(s) \geq \gamma$, and $c_n(s) = 0$ otherwise. Show that $c_n \in B^+$, $c_n \cdot c_m = 0$ $(n \neq m)$, $\Sigma_n c_n = \chi_A$, $I(\sum_{k=1}^{n} c_k) \leq \gamma^{-1} I(\sum_{k=1}^{n} b_k)$, and conclude that $\sum_{k=1}^{\infty} I(b_k) \geq \gamma I(\chi_A)$. Finish the proof.

3.3. Measurability with respect to a σ-algebra; approximation properties

Sometimes one has to consider simultaneously several measures on the same set. We shall meet this situation in Chapter 5 only, but in functional analysis and probability theory it occurs quite often. There is no problem in having different measures on the same set, of course. However, the sets and functions that we talk about in such a situation must be well-behaved with respect to the measures, that is, they must be measurable. But measurable in what sense? The definition of measurability in Section 2.1 is stated in terms of integrable functions, and since each measure comes with its own class of integrable functions, all measures under consideration must be taken into account. We eliminate these difficulties by introducing a new concept, called measurability with respect to a σ-algebra, where measures do not enter into the definition.

The new concept must be compared with the old one, of course. In this connection we meet the following question: If one starts from a measure space (S,Γ,μ), introduces the integral and so on, how are the μ-measurable sets in Σ_μ related to those in $\sigma(\Gamma)$, the σ-ring generated by Γ? It will turn out that an $E \in \Sigma_\mu$ with $\mu(E) < \infty$ is "almost equal" to a member of $\sigma(\Gamma)$. This approximation will be further refined for the Lebesgue case.

This section is perhaps also a good place to present the construction of a non Lebesgue measurable subset of $\mathbb{R}$.

3.3.1. The result of Theorem 2.1.12 leads us to the following definition.

<u>Definition</u>. Let S be a non-empty set, Σ a σ-algebra of subsets of S, f an $\mathbb{R}^*$-valued function defined on S. The function f is called *Σ-measurable* if $\{s \in S \mid f(s) < a\} \in \Sigma$ for every $a \in \mathbb{R}$.

We may also say that if f is *measurable with respect to* Σ, or just *measurable* if it is clear which σ-algebra is meant.

3.3.2. <u>Example</u>. Consider the situation of Example 3.1.10, and assume that M is satisfied, so that Σ is a σ-algebra. Now Theorem 2.1.12 shows that $f : S \to \mathbb{R}^*$ is Σ-measurable if and only if f is measurable in the sense of Section 2.1.

3.3.3. Let Σ be a σ-algebra consisting of subsets of the non-empty set S. Let M denote the class of $\mathbb{R}^*$-valued functions that are measurable with respect to Σ. The following result parallels 2.1.3 and 2.1.4.

<u>Theorem</u>. (i) Let $f,g \in M$. Then $-f \in M$, $\sup(f,g) \in M$, $\inf(f,g) \in M$, and $|f| \in M$.

(ii) Let $f,g \in M$ be finite-valued, and $\gamma \in \mathbb{R}$. Then $f + \gamma g \in M$.

(iii) Let $(f_n)_{n\in\mathbb{N}}$ be a sequence in M, and assume $f_n \to f$. Then $f \in M$.

(iv) Let $(f_n)_{n\in\mathbb{N}}$ be a sequence in M. Then $\sup_{n\in\mathbb{N}} f_n \in M$, $\inf_{n\in\mathbb{N}} f_n \in M$, $\limsup_{n\to\infty} f_n \in M$, $\liminf_{n\to\infty} f_n \in M$.

<u>Proof</u>. (i) We only show that $h := \sup(f,g)$ belongs to M. Indeed, if $\alpha \in \mathbb{R}$, then $\{s \in S \mid h(s) < \alpha\} = \{s \in S \mid f(s) < \alpha\} \cap \{s \in S \mid g(s) < \alpha\} \in \Sigma$.

(ii) We may assume $\gamma > 0$ (otherwise replace g by $-g$). Let $\alpha \in \mathbb{R}$. For each $q \in \mathbb{Q}$ let $A_q := \{s \in S \mid f(s) < q\} \cap \{s \in S \mid g(s) < (\alpha - q)\gamma^{-1}\}$. Obviously, each A_q belongs to Σ. Now $\{s \in S \mid (f + \gamma g)(s) < \alpha\}$, is just $\bigcup_{q\in\mathbb{Q}} A_q$, and therefore belongs to Σ. Hence $f + \gamma g \in M$.

(iii) Let $\alpha \in \mathbb{R}$, and let $(q_n)_{n\in\mathbb{N}}$ be an enumeration of the rationals less than α. It is easy to check that

$$\{s \in S \mid f(s) < \alpha\} = \bigcup_{n\in\mathbb{N}} \bigcup_{p\in\mathbb{N}} \bigcap_{m>p} \{s \in S \mid f_m(s) < q_n\},$$

and the latter set belongs to Σ, since Σ is closed under countable unions and intersections. This gives $f \in M$.

(iv) This follows easily from (i) and (iii). □

3.3.4. For a measure space (S,Γ,μ), let Σ_μ, L_μ and M_μ be as defined in 3.2.10, and let M denote the class of $\mathbb{R}^*$-valued functions on S that are measurable with respect to $\sigma(\Gamma)$, the σ-algebra generated by Γ.

<u>Theorem</u>. Let (S,Γ,μ) be a measure space. Then

(i) $\sigma(\Gamma) \subset \Sigma_\mu$, and

(ii) if $f : S \to \mathbb{R}^*$ is $\sigma(\Gamma)$-measurable, then $f \in M_\mu$.

<u>Proof</u>. (i) It is easy to see that $\Gamma \subset \Sigma_\mu$, for if $A \in \Gamma$, $b \in B^+$, then $(\chi_A)_b \in B^+$. (Here B is as in 3.2.1.) So Σ_μ is a σ-algebra of subsets of S which contains Γ, and the definition of $\sigma(\Gamma)$ as the smallest such σ-algebra shows that $\sigma(\Gamma) \subset \Sigma_\mu$.
(ii) Let $f : S \to \mathbb{R}^*$ be $\sigma(\Gamma)$-measurable. The system (L_μ, I) satisfies $\mathcal{M}$ (see 3.2.10), so by Theorem 2.1.12 it is sufficient to prove that $\{s \in S \mid f(s) < a\} \in \Sigma_\mu$ for every $a \in \mathbb{R}$. Since this follows at once from the definition of $\sigma(\Gamma)$-measurability together with (i), we are done. □

3.3.5. We shall show next that $\sigma(\Gamma)$ and Σ_μ are not very far apart, at least for sets whose measure is not "too infinite". The hard work is concentrated in the following lemma.

<u>Lemma</u>. Let (S,Γ,μ) be a measure space. Let $A \in \Sigma_\mu$, $\mu(A) < \infty$. For every $\varepsilon > 0$ there exists a sequence $(E_n)_{n\in\mathbb{N}}$ of pairwise disjoint elements of Γ such that $A \subset \bigcup_{n=1}^\infty E_n$ and $\sum_{n=1}^\infty \mu(E_n) \le \mu(A) + \varepsilon$.

<u>Proof</u>. Let $\varepsilon > 0$. According to the definition of the norm obtained via the route $\mathcal{R} \to \mathcal{D} \to \mathcal{A} \to \mathcal{N}$, there is a sequence $(b_n)_{n\in\mathbb{N}}$ of positive real numbers and a sequence $(B_n)_{n\in\mathbb{N}}$ of elements of Γ such that

$$\chi_A \le \sum_{n=1}^\infty b_n \chi_{B_n} ,$$

$$\sum_{n=1}^\infty b_n \mu(B_n) \le \|\chi_A\| + \frac{\varepsilon}{2} = \mu(A) + \frac{\varepsilon}{2} .$$

The problem is to show that the B_n's may be chosen pairwise disjoint.

For $n \in \mathbb{N}$, let

$$V_n := \left\{ s \in S \mid \sum_{k=1}^{n-1} b_k \chi_{B_k}(s) < \gamma \le \sum_{k=1}^{n} b_k \chi_{B_k}(s) \right\} ,$$

where $\gamma := (\mu(A) + \varepsilon/2)(\mu(A) + \varepsilon)^{-1}$ and a sum $\sum_{k=1}^{0}$ is zero. Then $A \subset \bigcup_{n=1}^{\infty} V_n$, the V_n's are pairwise disjoint, and

$$\sum_{n=1}^{\infty} \chi_{V_n} \le \gamma^{-1} \sum_{n=1}^{\infty} b_n \chi_{B_n} .$$

Moreover, each V_n belongs to $\Omega(\Gamma)$. Hence, there exists a sequence $(E_n)_{n\in\mathbb{N}}$ of pairwise disjoint elements of Γ such that $\bigcup_{n=1}^{\infty} V_n = \bigcup_{n=1}^{\infty} E_n$, and

$$\sum_{n=1}^{\infty} \mu(E_n) = \sum_{n=1}^{\infty} \mu(V_n) \le \gamma^{-1} \sum_{n=1}^{\infty} b_n \mu(B_n) \le \mu(A) + \varepsilon . \qquad \square$$

3.3.6. <u>Theorem</u>. Let (S,Γ,μ) be a measure space. Let $A \in \Sigma_\mu$, $\mu(A) < \infty$. Then there exist $E,F \in \sigma(\Gamma)$ such that $E \subset A \subset F$ and $\mu(E) = \mu(A) = \mu(F)$.

<u>Proof</u>. By Lemma 3.3.5, there is for every $k \in \mathbb{N}$ an $F_k \in \sigma(\Gamma)$ such that $A \subset F_k$, $\mu(F_k) \le \mu(A) + 2^{-k}$. Put $F := \bigcap_{k=1}^{\infty} F_k$. Then $F \in \sigma(\Gamma)$, $A \subset F$, and $\mu(A) = \mu(F)$.

Next consider $B := F\setminus A$. Since $B \in \Sigma_\mu$ and $\mu(B) = 0$, the first part of the proof shows that there exists $G \in \sigma(\Gamma)$ such that $B \subset G$, $\mu(G) = 0$. Now $E := F\setminus G$ satisfies the requirements. $\square$

3.3.7. The preceding approximation result can be extended to certain sets of infinite measure, namely to those that have the covering property mentioned in the theorem below. (Such sets are often said to have *σ-finite measure*.)

<u>Theorem</u>. Let (S,Γ,μ) be a measure space. Let $A \in \Sigma_\mu$, and assume that there exists a sequence $(A_n)_{n\in\mathbb{N}}$ in Σ_μ such that $A \subset \bigcup_{n=1}^{\infty} A_n$ and $\mu(A_n) < \infty$ $(n \in \mathbb{N})$. Then there exist $E,F \in \sigma(\Gamma)$ such that $E \subset A \subset F$ and $\mu(F\setminus E) = 0$.

<u>Proof</u>. Left to the reader. $\square$

3.3.8. Each subset of S corresponds to a function on S, namely to its characteristic function. So the approximation results that we derived may also be couched in terms of functions. If one does this, they appear as special cases of the next theorem, which asserts that certain μ-measurable functions are almost $\sigma(\Gamma)$-measurable functions. Of course, something like the σ-finiteness of the set A in 3.3.7 must be provided in this case, too (Exercise 3.3-2). Recall that M is the class of $\sigma(\Gamma)$-measurable $\mathbb{R}^*$-valued functions on S.

<u>Theorem</u>. Let (S,Γ,μ) be a measure space. Let $f \in M_\mu$, and assume that there exists a set $A \in \Sigma_\mu$ of σ-finite measure outside of which f is constant. Then there exists $g \in M$ such that $f = g$ (a.e.).

Proof. We shall assume the space S itself to have σ-finite measure, because the general case is not very different.

By 2.1.13 there exists a sequence $(t_n)_{n\in\mathbb{N}}$ of μ-measurable simple functions such that $t_n \to f$. Moreover, 3.3.7 implies that for each $n \in \mathbb{N}$ there exists $t_n' \in M$ such that $t_n' = t_n$ (a.e.). Obviously, $t_n' \to f$ (a.e.). Hence, if $A = \{s \in S \mid t_n'(s) \not\to f(s)\}$, then $A \in \Sigma_\mu$ and $\mu(A) = 0$. By Theorem 3.3.6 there exists $F \in \sigma(\Gamma)$ such that $A \subset F$, $\mu(F) = 0$. Now for each $n \in \mathbb{N}$, $s \in S$ define $t_n''(s) := t_n'(s)$ if $s \in S\backslash F$ and $t_n'(s) := 0$ if $s \in F$. Then the sequence $(t_n'')_{n\in\mathbb{N}}$ is in M, and it converges pointwise to a function g, which is in M by 3.3.3(iii). Clearly $f = g$ (a.e.). □

3.3.9. Lebesgue measure on the real line deserves somewhat more attention. Since the result of this subsection will return in much more general form in Section 4.2, it is not formally labeled as a theorem.

Let $S := \mathbb{R}$, let Γ be the semiring of nails, and let μ be Lebesgue measure (see 3.1.12). The σ-algebra $\sigma(\Gamma)$ generated by Γ occurs quite often in the theory; it is called the class of *Borel sets of* $\mathbb{R}$, and denoted by $\mathcal{B}$. The reader should verify that $\mathcal{B}$ can also be described as the smallest σ-ring (or the smallest σ-algebra) containing the open sets (or the closed sets, or the compact sets) in $\mathbb{R}$. Since the measure space $(\mathbb{R},\Gamma,\mu)$ itself is σ-finite, Theorem 3.3.7 gives approximations of the Lebesgue measurable sets by Borel sets. However, there also exist approximations by open sets and compact sets, which are nicer from a topological point of view. We turn to this now.

Let $E \subset \mathbb{R}$ be a Lebesgue measurable set of finite measure, and let $\varepsilon > 0$. We are going to show that there exist a compact $C \subset E$ and an open set $O \supset E$ such that $\mu(O\backslash E) < \varepsilon$ and $\mu(E\backslash C) < \varepsilon$.

The fundamental Lemma 3.2.5 is a "covering result", so we start with the approximation from without. The lemma shows that there exists a sequence $(E_n)_{n\in\mathbb{N}}$ of pairwise disjoint nails such that $E \subset \bigcup_{n=1}^\infty E_n$, $\sum_{n=1}^\infty \mu(E_n) < \mu(E) + \varepsilon/2$. For each $n \in \mathbb{N}$, let E_n' be an open interval such that $E_n \subset E_n'$, $\mu(E_n') < \mu(E_n) + \varepsilon\cdot 2^{-n-1}$. Put $O := \bigcup_{n=1}^\infty E_n'$, then O is open, $O \supset E$, and

$$\mu(O) \le \sum_{n=1}^\infty \mu(E_n') \le \sum_{n=1}^\infty \{\mu(E_n) + \varepsilon\cdot 2^{-n-1}\} < \mu(E) + \varepsilon .$$

Hence $\mu(O\backslash E) = \mu(O) - \mu(E) < \varepsilon$. So E can be approximated from without by open sets.

For the approximation from within, assume first that E is bounded. Choose a bounded closed interval $F \supset E$. The set $F\backslash E$ is measurable and has finite measure. By the result of the preceding paragraph there exists an open set $G \supset F\backslash E$ with $\mu(G\backslash(F\backslash E)) < \varepsilon$. Put $C := F\backslash G$, then C is compact, $C \subset E$, and $\mu(E\backslash C) < \varepsilon$ since $E\backslash C \subset G\backslash(F\backslash E)$. If E is not bounded, consider the sets $(E_n)_{n\in\mathbb{N}}$ where $E_n := E \cap [-n,n]$. Obviously, the E_n increase to E, so $\mu(E_n) \to \mu(E)$. Hence it is possible to choose n so large that $\mu(E_n) > \mu(E) - \varepsilon/2$. Apply the result for bounded sets to such an E_n. This gives a compact set C with $C \subset E_n$, $\mu(E_n \setminus C) < \varepsilon/2$, so C satisfies the requirement.

The results obtained can be rephrased as

$$\mu(E) = \sup\{\mu(C) \mid C \text{ compact, } C \subset E\} ,$$

$$\mu(E) = \inf\{\mu(O) \mid O \text{ open, } E \subset O\} .$$

Thus formulated the results hold even without the condition that $\mu(E)$ be finite (Exercise 3.3-4).

3.3.10. Most of this section concerned the question of whether for a general measure space (S,Γ,μ) the class Σ_μ is much larger than $\sigma(\Gamma)$. The answer was that measurable sets of σ-finite measure are almost equal to members of $\sigma(\Gamma)$. But we do not know, nor is it intuitively clear, how large Σ_μ is. It is even conceivable that Σ_μ is equal to the class of all subsets of S. In fact, simple examples of this phenomenon are easy to find.

What about Lebesgue measure on the real line? Here the class of Borel sets is already very extensive, for, loosely speaking, any easily described subset of $\mathbb{R}$ is a Borel set. (This is related to the fact noted in 2.1.14 that all easily described real-valued functions on $\mathbb{R}$ are Baire functions. We return to this connection in Section 4.4.) Thus it is plausible that unusual methods are needed for the construction of a non Lebesgue measurable subset of $\mathbb{R}$. The tool we shall use is the axiom of choice. This axiom has been discussed fiercely since its explicit formulation at the beginning of this century because of some startling consequences of accepting it. The fight does not seem so hot anymore.

So let μ denote Lebesgue measure on $\mathbb{R}$. Consider $\mathbb{R} \pmod{\mathbb{Q}}$. This set is formed in the following way. Two members of $\mathbb{R}$ are called *equivalent* if their difference is rational. It is easy to see that this really is an equivalence relation (it is reflexive, symmetric and transitive). Now $\mathbb{R} \pmod{\mathbb{Q}}$ consists of the equivalence classes that $\mathbb{R}$ is split into by this relation. That is, members x and y of $\mathbb{R}$ are in the same class if and only if $x - y \in \mathbb{Q}$. Choose in each equivalence class an element (a representative of that class) that belongs to $[0,1]$. Denote the set of all the representatives thus obtained by A. (In the two preceding sentences, innocent as they may look, the axiom of choice was involved.) The set A is not measurable, for we shall show that the assumption that A is measurable leads to a contradiction.

So assume A is measurable. Let $(q_n)_{n\in\mathbb{N}}$ be an enumeration of $\mathbb{Q} \cap [-1,1]$, and for $n \in \mathbb{N}$ put $A_n := \{a + q_n \mid a \in A\}$. For each $n \in \mathbb{N}$ the set A_n is measurable, and $\mu(A_n) = \mu(A)$, since Lebesgue measure is invariant under translations (Exercise 1.2-4). Moreover, the A_n are pairwise disjoint. Hence $B := \bigcup_{n=1}^\infty A_n$ is measurable, and $\mu(B) = \sum_{n=1}^\infty \mu(A_n)$. Now either $\mu(A) > 0$ or $\mu(A) = 0$. If $\mu(A) > 0$, then $\mu(B) = \infty$, which is impossible because B is contained in $[-1,2]$. On the other hand, $\mu(A) = 0$ implies $\mu(B) = 0$. But this is impossible too. In fact, every $x \in [0,1]$ may be written as

$x = a + q$ for some $a \in A$, $q \in \mathbb{Q} \cap [-1,1]$. Therefore $B \supset [0,1]$, so $\mu(B) \geq 1$. Hence A cannot be Lebesgue measurable.

Exercises Section 3.3

1. Give the proof of Theorem 3.3.7.

2. Give an example of a non-σ-finite measure space (S,Γ,μ) with an $A \in \Sigma_\mu$ such that for any $B \in \sigma(\Gamma)$ we have $\mu(A\backslash B) = \infty$ if $B \subset A$ and $\mu(B\backslash A) = \infty$ if $A \subset B$.

3. This exercise lines out the extension procedure of *Carathéodory* in its original form; here a measure μ defined on a ring is extended to at least the σ-algebra generated by the initial ring (in fact we will see that μ is extended to Σ_μ). The details are as follows. Let (S,Γ,μ) be a measure space where Γ is a ring (with a little more effort the case that Γ is only a semiring can be treated as well). Follow the line $\mathcal{R} \to \mathcal{D} \to \mathcal{A} \to \mathcal{N}(B,I,\|\cdot\|) \to \mathcal{L}(L,I)$. Define a set function μ^* by

$$\mu^*(A) := \inf \left\{ \sum_{n=1}^{\infty} \mu(E_n) \mid A \subset \bigcup_{n=1}^{\infty} E_n,\ E_n \in \Gamma\ (n \in \mathbb{N}) \right\}$$

for $A \subset S$ (as usual, $\inf \emptyset = \infty$). This μ^* is called the *exterior measure*.

(i) Show that $\|\chi_A\| = \mu^*(A)$ for $A \subset S$. (Hint. Lemma 3.3.5.)

A set $E \subset S$ is called μ-measurable if

$$\mu^*(A) = \mu^*(A \cap E) + \mu^*(A \cap E^*)$$

for every $A \subset S$. Let Λ be the class of μ-measurable sets.

(ii) Show that Λ is stable with respect to $\cap$, $\backslash$, *.

(iii) Let $F_k \in \Lambda$ $(k = 1,\ldots,n)$ be pairwise disjoint. Show that $F = \bigcup_{k=1}^n F_k \in \Lambda$ and that for $A \subset S$

$$\mu^*(A \cap F) = \sum_{k=1}^{n} \mu^*(A \cap F_k) .$$

(iv) Let $F_k \in \Lambda$ $(k \in \mathbb{N})$ be pairwise disjoint, let $F = \bigcup_{k=1}^\infty F_k$ and let $A \subset S$. Show that $\mu^*(A \cap F) + \mu^*(A \cap F^*) \geq \mu^*(A)$, and prove the reverse inequality by using the Fatou property of $\|\cdot\|$ (cf. 1.2.9).

(v) Show that $\Lambda = \Sigma_\mu$. (Hint. Exercise 2.1-7.)

(vi) Let $E \in \Lambda$ and define $\mu(E) := \mu^*(E)$. Show that this extension of μ agrees with the μ of Subsection 3.2.10.

4. Consider $S = \mathbb{R}$ with Lebesgue measure μ and let $E \in \Sigma_\mu$. Show that

$$\mu(E) = \sup\{\mu(C) \mid C \text{ compact}, C \subset E\} ,$$

$$\mu(E) = \inf\{\mu(O) \mid O \text{ open}, E \subset O\} .$$

5. Let E be as in Exercise 4. Show that there exists a sequence $(O_n)_{n\in\mathbb{N}}$ of open sets such that $E \subset O_n$ $(n \in \mathbb{N})$, $\mu(\bigcap_{n=1}^\infty O_n \backslash E) = 0$. Show that there exists a sequence $(C_n)_{n\in\mathbb{N}}$ of compact sets such that $C_n \subset E$ $(n \in \mathbb{N})$, $\mu(E \backslash \bigcup_{n=1}^\infty C_n) = 0$.

6. Let E be as in Exercise 4 and assume that $\mu(E) > 0$. Let $0 < \alpha < 1$. Show that there exists a bounded open interval I such that $\mu(E \cap I) > \alpha\mu(I)$.

7. Let E be as in Exercise 4. Let V be a countable, dense set in $\mathbb{R}$ and put $X := E + V := \{s + v \mid s \in E, v \in V\}$. Show that either $\mu(X) = 0$ or $\mu(\mathbb{R}\backslash X) = 0$ according as $\mu(E) = 0$ or $\mu(E) > 0$. (Hint. Assume $\mu(E) > 0$ and $\mu(\mathbb{R}\backslash X) > 0$, and take bounded open intervals I and J such that $\mu(E \cap I) > \frac{3}{4}\mu(I)$, $\mu((\mathbb{R}\backslash X) \cap J) > \frac{3}{4}\mu(J)$. Show that $\mu(J \cap X) > \frac{1}{2}\mu(J)$ and derive a contradiction.)

8. Let A be the non-measurable set constructed in 3.3.10. Let E be a measurable subset of $\mathbb{R}$ with $\mu(E) > 0$. Show that for at least one $q \in \mathbb{Q}$ the set $E \cap A_q$ is non-measurable. Here $A_q := \{q + a \mid a \in A\}$. (Hint. Use Exercise 7.) Hence, E contains a non-measurable set.

9. We shall construct a function f defined on $S_0 := [0,1]$ that exhibits several strange features. Let

$$J_{k\ell} := \left(\frac{\ell}{3^k}, \frac{\ell+1}{3^k}\right) \qquad \text{for } \ell = 0,1,\ldots,3^k-1 , \quad k \in \mathbb{N},$$

and let $f_0(x) := x$. For $k \in \mathbb{N}$ we define f_k as the continuous, piecewise linear function that is linear on all intervals $J_{k\ell}$ and whose function values at the end-points of the $J_{k\ell}$'s are given by

$$f_k\left(\frac{3\ell}{3^k}\right) = f_{k-1}\left(\frac{\ell}{3^{k-1}}\right) , \qquad f_k\left(\frac{3\ell+3}{3^k}\right) = f_{k-1}\left(\frac{\ell+1}{3^{k-1}}\right) ,$$

$$f_k\left(\frac{3\ell+1}{3^k}\right) = f_k\left(\frac{3\ell+2}{3^k}\right) = f_{k-1}\left(\frac{\ell+\frac{1}{2}}{3^{k-1}}\right)$$

for $\ell = 0,1,\ldots,3^{k-1}$. If C is the Cantor set constructed in Exercise 3.2-6, show that $f(x) := \lim_{k\to\infty} f_k(x)$ exists for $x \in S_0\backslash C$, that f is non-decreasing on $S_0\backslash C$ and that

$f(S_0\backslash C)$ is dense in $[0,1]$. Show that there exists a unique continuous, non-decreasing extension of f to all of S_0. This extended f is called *Lebesgue's singular function*. Note that f is non-decreasing and that $f'(x) = 0$ for $x \in S_0\backslash C$, i.e. for almost every $x \in [0,1]$.

10. Let f be Lebesgue's singular function, and let $g(x) := x + f(x)$ $(x \in [0,1])$. Show that g is continuous, strictly increasing and that $g(S_0) = [0,2]$. Show that $g(S_0\backslash C)$ is Lebesgue measurable and has measure 1. Hence, $g(C)$ has measure 1. Let A be a non-measurable subset of $g(C)$ (cf. Exercise 8) and let $B := g^{-1}(A)$. Show now that the composition $h \circ \varphi$ of a measurable function h and a continuous function φ (in that order, cf. Theorem 2.1.15) need not be measurable.

11. Let $f : \mathbb{R} \to \mathbb{R}$ be continuously differentiable. Show that $f(E)$ is Lebesgue measurable if $E \subset \mathbb{R}$ is Lebesgue measurable.

12. Show that the class of Borel sets of $\mathbb{R}$ is the smallest σ-algebra generated by the open (closed, compact) sets.

13. Let Σ be a σ-algebra of subsets of S and assume that $g : S \to \mathbb{R}^*$ is Σ-measurable. Show that $\{g^{\leftarrow}(A) \mid A \in \mathcal{B}\}$ is a σ-algebra contained in Σ. ($\mathcal{B}$ is the class of Borel sets of $\mathbb{R}$.)

14. This is not a book on set theory, and we use things like the axiom of choice just to give examples that suggest the boundaries of the theory. In some of the exercises we shall need what is usually called the long line whose construction we describe below. We start from the assertion that $\mathbb{R}$ can be well-ordered, i.e. there exists an order relation $<$ for $\mathbb{R}$ such that any non-empty subset of $\mathbb{R}$ contains a minimal element. (This assertion, which can be proved using the axiom of choice, was one of the reasons for the controversies around this axiom.) Now let $<$ be a well-ordering for $\mathbb{R}$, and consider $X := \{x \in \mathbb{R} \mid x \text{ has uncountably many predecessors}\}$. If X is empty we add a symbol Ω to $\mathbb{R}$ and thus obtain a set Y with the properties

(i) Y is uncountable,
(ii) Y is well-ordered (treat Ω as the last element of Y),
(iii) every $x \in Y\backslash\{\Omega\}$ has countably many predecessors.

If X is non-empty, let Ω be the least member of X, and put $Y := \{\Omega\} \cup (\mathbb{R}\backslash X)$. This Y has also properties (i), (ii), (iii). The set Y thus obtained is called the *long line*. Show the following: if $A \subset Y\backslash\{\Omega\}$ is countable, then $\sup A < \Omega$, i.e. there is $b \in Y\backslash\{\Omega\}$ such that $a \le b$ for all $a \in A$. (Hint. Let $A = \{a_n \mid n \in \mathbb{N}\}$ and consider the set $A_n := \{x \in Y \mid x \le a_n\}$ for $n \in \mathbb{N}$. Note that A_n is countable and consider $\bigcup_{n=1}^{\infty} A_n$.)

15. This exercise is meant to show that the class $\mathcal{B}$ of Borel sets of $\mathbb{R}$ is a proper subset of the class Σ_μ of Lebesgue measurable sets of $\mathbb{R}$. We use the long line Y of the preceding exercise. For every $y \in A := Y\backslash\{\Omega\}$ we define a class $\mathcal{B}_y$ inductively as follows. Let y_0 be the least element of Y and define $\mathcal{B}_{y_0}$ as the set of all open subsets of $\mathbb{R}$. Let $y \in A$ and assume that $\mathcal{B}_z$ has been defined for all $z \in A$ with $z < y$. Let $\mathcal{B}_y$ be the set of all sets of the form $\bigcup_{n=1}^{\infty} A_n$ and $\bigcap_{n=1}^{\infty} A_n^*$ where each $A_n \in \mathcal{B}_z$ for some $z < y$.

(i) Show that $\mathcal{B}_y \subset \mathcal{B}$ for every $y \in A$. (Hint. Use that Y is well-ordered.)

(ii) Show that $\bigcup_{y\in A} \mathcal{B}_y$ is a σ-algebra. (Hint. Exercise 14.)

(iii) Show that $\mathcal{B} = \bigcup_{y\in A} \mathcal{B}_y$.

(iv) Show that each $\mathcal{B}_y$ has the power of $\mathbb{R}$. (Hint. Exercise 14.)

(v) Show that $\mathcal{B}$ has the power of $\mathbb{R}$. (Hint. $\mathbb{R} \times A$ and $\mathcal{B}$ have the same power and the power of A is at most that of $\mathbb{R}$; use that $\mathbb{R}$ and $\mathbb{R}^2$ have the same power.)

(vi) Let C be the Cantor set of Exercise 3.2-6. Show that every subset of C is Lebesgue measurable and that C contains a subset that is not a Borel set. (Hint. C has the power of $\mathbb{R}$.)

3.4. The measure generated by an integral

A situation $L(L,I)$ on a set S leads to a class Σ of measurable subsets of S and a set function μ defined on Σ (Section 2.1). In the terminology of the present chapter the triple (S,Σ,μ) is a measure space (Example 3.1.10). The construction of the preceding section applied to the measure space (S,Σ,μ) gives a new integral. Question: how is the original integral I related to integration with respect to μ? Answer: if M is satisfied, then every $f \in L$ is integrable with respect to μ with $I(f) = \int f\,d\mu$.

3.4.1. Let S be a non-empty set, let L be a Riesz function space of real-valued functions on S, and let I be a positive linear functional defined on L such that $L(L,I)$ is satisfied. Introduce the measurable sets as in Section 2.1, and assume that M holds, that is, assume that the class Σ of measurable sets is a σ-algebra. Let μ be the measure on Σ derived from I as in Section 2.1. There are now two methods to introduce a norm on Φ, the class of $\mathbb{R}^*$-valued functions defined on S. The first method is described by

$$L \to D \to A \to N(B_1,I_1,\|\cdot\|_1) ,$$

and the second one by

$$L \to R(\Sigma,\mu) \to D \to A \to N(B_2,I_2,\|\cdot\|_2) .$$

As described in Section 1.1, the resulting N-systems lead to L-systems, $L(L_1,\bar{I}_1)$ and $L(L_2,\bar{I}_2)$, respectively. It is remarkable that the two norms $\|\cdot\|_1$ and $\|\cdot\|_2$, the spaces L_1 and L_2, and the extensions $\bar{I}_1$ and $\bar{I}_2$ coincide. Before giving the formal statement and proof of these assertions, we derive two lemmas.

3.4.2. Lemma. Under the conditions stated in 3.4.1, $B_2 \subset L_1$ and I_2 is the restriction of $\bar{I}_1$ to B_2.

Proof. By definition, B_2 consists of all functions $\sum_{n=1}^N \alpha_n \chi_{A_n}$, where $\alpha_n \in \mathbb{R}$ and the A_n are pairwise disjoint measurable sets of finite measure, and for such a function

$$I_2\left(\sum_{n=1}^N \alpha_n \chi_{A_n}\right) = \sum_{n=1}^N \alpha_n \mu(A_n) .$$

The functions χ_{A_n} that occur are all in L_1, and satisfy $\mu(A_n) = \bar{I}_1(\chi_{A_n})$. Hence $B_2 \subset L_1$, and I_2 is the restriction of $\bar{I}_1$ to B_2. □

3.4.3. Lemma. Under the conditions stated in 3.4.1, let $b \in B_2^+$ and $\varepsilon > 0$. Then there exists a sequence $(b_n)_{n\in\mathbb{N}}$ in B_2^+ such that $b \le \sum_{n=1}^\infty b_n$ and $\sum_{n=1}^\infty I_2(b_n) < I_1(b) + \varepsilon$.

Proof. Fix $0 < \alpha < 1$. For each integer n, let

$$A_n := \{s \in S \mid \alpha^n < b(s) \le \alpha^{n-1}\}$$

and $c_n := \alpha^{n-1}\chi_{A_n}$. Since M is satisfied for the pair (L,I), and $B_1 = L$ by definition, each A_n is measurable, and therefore each c_n belongs to B_2^+. Moreover,

$$\alpha \sum_{n=-\infty}^{\infty} c_n \le b \le \sum_{n=-\infty}^{\infty} c_n .$$

By the preceding lemma and the monotone convergence theorem 1.1.18,

$$\sum_{n=-\infty}^{\infty} I_2(c_n) = \sum_{n=-\infty}^{\infty} \bar{I}_1(c_n) = \bar{I}_1\left(\sum_{n=-\infty}^{\infty} c_n\right) \le \alpha^{-1}\bar{I}_1(b) = \alpha^{-1}I_1(b) .$$

If one takes α sufficiently near to 1, then the sequence $(b_n)_{n\in\mathbb{N}}$ defined by $b_{2n-1} := c_n$, $b_{2n} := c_{1-n}$ $(n \in \mathbb{N})$ satisfies the requirements. □

3.4.4. Theorem. Start from $L(L,I)$ and assume that M is satisfied. Perform the processes indicated in 3.4.1. Then

(i) $\|\varphi\|_1 = \|\varphi\|_2$ for $\varphi \in \Phi$,

(ii) $L_1 = L_2$, and $\bar{I}_1 f = \bar{I}_2 f$ for $f \in L_1$.

Proof. (i) Let $\varphi \in \Phi$. To show that $\|\varphi\|_1 \leq \|\varphi\|_2$, we may assume $\|\varphi\|_2 < \infty$. By definition

$$\|\varphi\|_2 = \inf \left\{ \sum_{n=1}^{\infty} \alpha_n \mu(A) \mid \alpha_n \geq 0,\ A_n \in \Sigma\ (n \in \mathbb{N}),\ |\varphi| \leq \sum_{n=1}^{\infty} \alpha_n \chi_{A_n} \right\} .$$

For any admissible choice of sequences $(\alpha_n)_{n \in \mathbb{N}}$, $(A_n)_{n \in \mathbb{N}}$ we have, noting that $\|\chi_{A_n}\|_1 = \mu(A_n)$ for all n,

$$\|\varphi\|_1 \leq \| \sum_{n=1}^{\infty} \alpha_n \chi_{A_n} \|_1 \leq \sum_{n=1}^{\infty} \alpha_n \|\chi_{A_n}\|_1 = \sum_{n=1}^{\infty} \alpha_n \mu(A_n) .$$

Hence $\|\varphi\|_1 \leq \|\varphi\|_2$.

For the proof of the reverse inequality, we may assume $\|\varphi\|_1 < \infty$. Then

$$\|\varphi\|_1 = \inf \left\{ \sum_{n=1}^{\infty} I_1(b_n) \mid b_n \in B_1^+\ (n \in \mathbb{N}),\ |\varphi| \leq \sum_{n=1}^{\infty} b_n \right\} .$$

Now let $\varepsilon > 0$, and choose a sequence $(b_n)_{n \in \mathbb{N}}$ in B_1^+ such that $|\varphi| \leq \sum_{n=1}^{\infty} b_n$ and $\sum_{n=1}^{\infty} I_1(b_n) < \|\varphi\|_1 + \varepsilon/2$. For each $n \in \mathbb{N}$, apply Lemma 3.4.3 (with b_n and $\varepsilon \cdot 2^{-n-1}$ instead of b and ε) to conclude the existence of a sequence $(\alpha_{nm})_{m \in \mathbb{N}}$ of non-negative real numbers and a sequence $(A_{nm})_{m \in \mathbb{N}}$ of measurable sets such that

$$b_n \leq \sum_{m=1}^{\infty} \alpha_{nm} \chi_{A_{nm}} , \qquad \sum_{m=1}^{\infty} \alpha_{nm} \mu(A_{nm}) \leq I_1(b_n) + \varepsilon \cdot 2^{-n-1} .$$

Since

$$|\varphi| \leq \sum_{n=1}^{\infty} b_n \leq \sum_{n=1}^{\infty} \sum_{m=1}^{\infty} \alpha_{nm} \chi_{A_{nm}} ,$$

it follows that

$$\|\varphi\|_2 \leq \sum_{n=1}^{\infty} \sum_{m=1}^{\infty} \alpha_{nm} \mu(A_{nm}) \leq \sum_{n=1}^{\infty} (I_1(b_n) + \varepsilon \cdot 2^{-n-1}) < \|\varphi\|_1 + \varepsilon .$$

Therefore $\|\varphi\|_2 \leq \|\varphi\|_1$.

(ii) Let $f \in L_2$ and $\varepsilon > 0$. By definition there exists $b \in B_2$ such that $\|f - b\|_2 < \varepsilon/2$. Then $\|f - b\|_1 < \varepsilon/2$ by (i). Since $B_2 \subset L_1$, there exists $b' \in B_1$ such that $\|b' - b\|_1 < \varepsilon/2$, and therefore $\|f - b'\|_1 < \varepsilon$. This shows that $f \in L_1$. Also, if $\varepsilon > 0$, and b is as before, then

$$|\bar{I}_1 f - \bar{I}_2 f| = |\bar{I}_1 f - I_2 b + I_2 b - \bar{I}_2 f| \leq \|f - b\|_1 + \|b - f\|_2 < \varepsilon .$$

Hence $\bar{I}_1 f = \bar{I}_2 f$.

On the other hand, let $f \in L_1^+$. Fix $\varepsilon > 0$. Just as in the last paragraph of (i) there exists a sequence $(b_n)_{n\in\mathbb{N}}$ in B_2^+ such that $f \leq \sum_{n=1}^{\infty} b_n$ and $\sum_{n=1}^{\infty} I_2(b_n) \leq \bar{I}_1(f) + \varepsilon$. By the Beppo-Levi theorem 1.1.17 there exists $b \in L_2$ such that $b = \sum_{n=1}^{\infty} b_n$ (a.e.) and $\bar{I}_2(b) = \sum_{n=1}^{\infty} I_2(b_n)$. By the preceding paragraph, $b \in L_1$ and $\bar{I}_1(b) = \bar{I}_2(b)$. Hence

$$\|b - f\|_2 = \|b - f\|_1 = \bar{I}_1(|b - f|) = \bar{I}_1(b - f) =$$

$$= \bar{I}_1(b) - \bar{I}_1(f) = \bar{I}_2(b) - \bar{I}_1(f) < \varepsilon .$$

Therefore $b \in L_2$.

Finally, if $f \in L_1$, then f can be written as the difference of two members of L_1^+, and therefore $f \in L_2$. Hence $L_1 \subset L_2$. □

3.4.5. The result just derived can also be interpreted as a representation theorem: any abstract integral in the sense of Section 1.1 that satisfies M is (a restriction of) an integral with respect to a measure in the sense of Section 3.2.

Theorem. Start from $L(L,I)$ and assume M. Then there exists a σ-algebra Σ of subsets of S and a measure μ defined on Σ such that every $f \in L$ is integrable with respect to μ, while $I(f) = \int f\,d\mu$.

3.4.6. How final are the space of integrable functions and the integral obtained from a measure space? A more carefully formulated version of this question can be answered by using Theorem 3.4.5.

First we formulate the problem more precisely. Start from a measure space (S,Γ,μ), that is, start from $R(\Gamma,\mu)$. Apply the processes

$$R \to D \to A \to N \to L(L,I) \to R(\Sigma,\mu)$$

of Sections 3.2 and 2.1. The measure space (S,Σ,μ) is an extension of (S,Γ,μ). Now start from $R(\Sigma,\mu)$:

$$R(\Sigma,\mu) \to D \to A \to N \to L(L_2,I_2) .$$

The question is: How are the pairs (L,I) and (L_2,I_2) related?

The answer is as satisfactory as one might desire: (L,I) is equal to (L_2,I_2). To see this, consider the auxiliary process in the circle line

$$L(L,I) \to D \to A \to N \to L(L_1,I_1) .$$

Theorem 3.4.4 shows that (L_2,I_2) and (L_1,I_1) are equal, while (L,I) and (L_1,I_1) are equal by 1.4.7 and 1.4.8.

CHAPTER FOUR

INTEGRATION ON LOCALLY COMPACT HAUSDORFF SPACES

Lebesgue had based his definition of an integral on the familiar notions of length in the one-dimensional case and of volume in the multi-dimensional case, and thus obtained an extension of the Riemann integral. In 1913, Radon used more general set functions on Euclidean space as a starting point, which led to extensions of the Riemann-Stieltjes integral. Eventually, Radon's approach culminated in a rich theory of integration on locally compact Hausdorff spaces. The present chapter contains our rendering of a minor part of this theory.

As usual, the point of departure is a class B of basic functions together with a positive linear functional I on B. The class B now consists of the continuous real-valued functions of compact support (that is, each member of B vanishes outside some compact set) on a locally compact Hausdorff space S. It turns out that the pair (B,I) satisfies the condition $\mathcal{D}'$ (and hence $\mathcal{D}$) of Section 1.3. The transitions $\mathcal{D}' \to A \to N$ and $\mathcal{D}' \to \mathcal{D} \to A \to N$ lead to two norms, denoted by $\|\cdot\|_1$ and $\|\cdot\|_2$, respectively. Thus we obtain two integrals, I_1 and I_2, on two spaces of integrable functions, L_1 and L_2. Both I_1 and I_2 are extensions of the original functional I. Moreover, the norms are comparable in the sense that

$$\|\varphi\|_1 \leq \|\varphi\|_2 \qquad (\varphi \in \Phi) ,$$

so $L_2 \subset L_1$, and I_1 is an extension of I_2. The direct route $\mathcal{D}' \to A$, which leads to L_1, gives the larger class of integrable functions. Therefore it is generally preferred over $\mathcal{D}' \to \mathcal{D} \to A$. In Section 4.1 we consider in some detail the question whether L_1 is much larger than L_2.

In the basic space B we have $\inf(b,\chi_S) \in B$ whenever $b \in B^+$. Consequently, both systems of integrable functions satisfy condition M. Hence, by Theorem 3.4.5, both I_1 and I_2 are integrals with respect to measures in the sense of the preceding chapter. There seems therefore to be no particular reason for this topological approach to integration. However, on a topological space the continuous real-valued functions play a dominant role, and since these functions become measurable in the $\mathcal{D}'$-case, it makes sense to incorporate them into the theory right from the start. (In the $\mathcal{D}$-case the situation is not so nice, at least if the topology does not have a countable base, which is another reason to prefer the direct $\mathcal{D}' \to A$ route.)

Since I_1 is generated by a measure, μ say, there is an interplay between the topological and the measure theoretic aspects of subsets of S, and of real-valued functions on S. Some results in this direction are presented in Section 4.2. We show, for instance, that an integrable function is in some sense "almost" a continuous function (Lusin's theorem).

The measure μ of the preceding paragraph is defined at least on all open subsets of S, and hence on the σ-algebra generated by these sets. The members of this σ-algebra are called Borel sets. A measure defined on the Borel sets that is finite on compact sets is called a Borel measure. The Riesz representation theorem is the assertion that there is a one-to-one correspondence between positive linear functionals on B and certain Borel measures. This important result is treated in Section 4.3.

The final section of the chapter deals with Baire sets and Baire functions. The Baire functions appeared in Section 2.1 already, in the discussion of the measurability of composite functions. Here the point of view is somewhat different, because Baire functions arise in a natural way in the process $\mathcal{D}' \to \mathcal{D} \to A$.

4.1. The assumptions $\mathcal{T}$ and the two routes towards an integral

The Riemann integral for continuous functions on a bounded interval in $\mathbb{R}$ may be used as a starting point for an extension process; this was shown in Example 1.2.4. The present section deals with a more general situation: we start with a positive linear functional defined on the continuous real-valued functions of compact support on a locally compact Hausdorff space. The system (B,I) thus obtained satisfies the conditions $\mathcal{D}$ and $\mathcal{D}'$ introduced in Section 1.3. Hence, there are two ways to obtain a class of auxiliary functions with an appropriate J-functional, and each of these leads to a norm. These norms may be different, which may result in different classes of integrable functions, L_1 (derived from $\mathcal{D}' \to A \to N \to L$) and L_2 (derived from $\mathcal{D}' \to \mathcal{D} \to A \to N \to L$).

How are L_1 and L_2 related? Often there is no difference at all. A simple condition on the topology guaranteeing that L_1 is equal to L_2 is contained in Theorem 4.1.6; this condition is satisfied by the Euclidean spaces $\mathbb{R}^n$. In general, however, L_1 is bigger, for it contains L_2. It is not so much bigger, though: each $f \in L_1$ can be written as $f = g + h$, where $g \in L_2$ and $h \in L_1$ is a null function (Theorem 4.1.8). Adding more conditions, we refine this comparison in Theorem 4.1.9. Part of this last result is in terms of local null functions; it may be ignored by those who skipped Section 2.4 on the local norm.

4.1.1. Let S be a locally compact Hausdorff space with topology T. Let B denote the class of real-valued continuous functions f on S that have compact support, that is, for which $\{s \in S \mid f(s) \neq 0\}$ has compact closure. It is easy to see that B is a Riesz function space. Let I be a positive linear functional on B. We denote this situation by $\mathcal{T}(T,I)$, or simply by $\mathcal{T}$.

If there exists a sequence $(b_n)_{n\in\mathbb{N}}$ in B^+ such that $\sum_{n=1}^{\infty} b_n > 0$, then, as usual, we call the system σ-finite, and write $\mathcal{T}_\sigma(T,I)$ or $\mathcal{T}_\sigma$. The σ-finiteness property is purely topological, of course, and it has nothing to do with I. It can also be formulated in terms of compactness: the system $\mathcal{T}(T,I)$ is σ-finite if and only if S is

σ-*compact*, that is, if S is the union of countably many compact sets (Exercise 4.1-1). Well-known examples of σ-compact spaces are the spaces $\mathbb{R}^n$ with their usual topology.

We note that in the present case B has the property that $\inf(b,\chi_S) \in B$ for every $b \in B^+$. According to Exercise 2.1-1 this guarantees that condition $\mathcal{M}$ of Section 2.1 is satisfied by any class of integrable functions for which B acts as class of basic functions.

4.1.2. Examples. (i) Take $S := \mathbb{R}$ with the usual topology, and put $I(b) := \int_{-\infty}^{\infty} b(x)dx$ for $b \in B$ (the integral is the ordinary Riemann integral).

(ii) More generally, take $S := \mathbb{R}^n$ (where $n \in \mathbb{N}$) with the usual topology. Let h be any non-negative continuous real-valued function on $\mathbb{R}^n$, and put $I(b) := \int_{\mathbb{R}^n} b(x)h(x)dx$ for $b \in B$.

(iii) Let $(S,\mathcal{T})$ be any locally compact Hausdorff space. Let $x_1,x_2,\ldots,x_n$ be points of S. Put $I(b) := \sum_{k=1}^n b(x_k)$ for $b \in B$.

In each of these cases it is obvious that I is positive and linear. Moreover, in (i) and (ii) the system is σ-finite.

4.1.3. We are now going to show that for a system $\mathcal{T}(T,I)$ the pair (B,I) satisfies the conditions $\mathcal{D}'$ and $\mathcal{D}$ of Section 1.3. This will enable us to enter the circle line, and thus to obtain an integral. The following lemma proves slightly more than needed. (Note that the result is purely topological: the functional I does not really play a role here.)

Lemma. Assume $\mathcal{T}(T,I)$. Let $V \subset B^+$ with $\inf_{b\in V} b = 0$. Then there exists $h \in B^+$ with the following property: For every $\varepsilon > 0$ there exists a finite subset $W \subset V$ such that $\inf_{b\in W} b \leq \varepsilon h$.

Proof. Let c be any member of V, and let C denote the support of c. Urysohn's lemma (Theorem 0.2.16) shows that there exists $h \in B^+$ such that $0 \leq h \leq 1$, $h(s) = 1$ if $s \in C$. Fix $\varepsilon > 0$. For each $s \in C$ there exists $b_s \in V$ satisfying $b_s(s) < \varepsilon$, and because of the continuity of b_s, there even exists a neighborhood U_s of s such that $b_s(t) < \varepsilon$ for all $t \in U_s$. Since C is compact, C may be covered by finitely many of these neighborhoods, $U_{s_1},U_{s_2},\ldots,U_{s_k}$, say. Now $W := \{c,b_{s_1},b_{s_2},\ldots,b_{s_k}\}$ does the trick. □

4.1.4. Theorem. Assume $\mathcal{T}(T,I)$. Then the pair (B,I) satisfies $\mathcal{D}'$ and $\mathcal{D}$.

Proof. It follows easily from the preceding lemma that $\mathcal{D}'$ is satisfied. Since $\mathcal{D}'$ implies $\mathcal{D}$, as noted in 1.3.7, the proof is complete. □

4.1.5. The routes $T \to \mathcal{D}' \to A \to N$ and $T \to \mathcal{D}' \to \mathcal{D} \to A \to N$ give two norms on Φ. One can construct examples where these norms are in fact different (see Exercise 4.1-5). There is a simple condition on the topology, however, that ensures that the norms are equal. This condition is that (S,T) satisfies the *second axiom of countability*, that is, there exists a countable subset T_1 of T such that every member of T is a union of members of T_1, so that T_1 is a *countable base* for the topology T. Obviously, $\mathbb{R}^n$ satisfies this condition (for instance, the class of open cubes all of whose vertices have rational coordinates is a countable base for the Euclidean topology).

The following lemma is an important step in the proof of the equality of norms in the special case mentioned.

Lemma. Assume $T(T,I)$, and let (S,T) satisfy the second axiom of countability. Then for every subset V of B^+ there exists a countable set $W \subset V$ such that $\sup_{b\in V} b = \sup_{c\in W} c$.

Proof. Let $(O_n)_{n\in\mathbb{N}}$ be a countable base for the topology. For each pair $(n,r) \in \mathbb{N} \times \mathbb{Q}$, define $V_{nr} := \{b \in V \mid b(s) > r \text{ if } s \in O_n\}$, choose an arbitrary element $v_{nr} \in V_{nr}$ if $V_{nr} \neq \emptyset$, and put $v_{nr} := 0$ (the zero element of B) otherwise. The family $W := \{v_{nr} \mid (n,r) \in \mathbb{N} \times \mathbb{Q}\}$ is obviously countable; we shall show that it also has the right supremum.

Let $s \in S$, and put $\alpha := \sup_{b\in V} b(s)$, $\beta := \sup_{b\in W} b(s)$. Since $W \subset V$, the inequality $\beta \le \alpha$ is trivial, so we need only prove $\beta \ge \alpha$. Choose $r \in \mathbb{Q}$ such that $r < \alpha$. Then there exists $b \in V$ with $b(s) > r$, and by continuity of b there exists a neighborhood U of s such that $b(t) > r$ for all $t \in U$. Now U contains an O_n, so for this pair (n,r) we have $V_{nr} \neq \emptyset$, and therefore $v_{nr}(s) > r$. This shows that $\beta \ge r$. Letting r increase to α, we obtain $\beta \ge \alpha$. □

4.1.6. Theorem. Assume $T(T,I)$, and let (S,T) satisfy the second axiom of countability. Then the routes $T \to \mathcal{D}' \to A(A_1,J_1) \to N(B,I,\|\cdot\|_1)$ and $T \to \mathcal{D}' \to \mathcal{D} \to A(A_2,J_2) \to N(B,I,\|\cdot\|_2)$ lead to the same norm.

Proof. Since the $\mathcal{D}$-system is derived from the $\mathcal{D}'$-system, we have $A_2 \subset A_1$ and $J_2(h) = J_1(h)$ if $h \in A_2$, and therefore $\|\varphi\|_1 \le \|\varphi\|_2$ if $\varphi \in \Phi$. So the rest of the proof consists in deriving the reverse inequality $\|\varphi\|_2 \le \|\varphi\|_1$.

Fix $\varphi \in \Phi$. We may assume $\|\varphi\|_1 < \infty$. Let $\varepsilon > 0$, and choose a member $h \in A_1$ such that $|\varphi| \le h$ and $J_1(h) \le \|\varphi\|_1 + \varepsilon$ (see 1.3.12). Let $V := \{b \in B^+ \mid b \le h\}$. By Lemma 4.1.5 there exists a countable subset W of V such that $h = \sup_{b\in W} b$. Let $(b_n)_{n\in\mathbb{N}}$ be an enumeration of W. For $n \in \mathbb{N}$, let $c_n := \sup(b_1,\dots,b_n)$. The sequence $(c_n)_{n\in\mathbb{N}}$ thus obtained is increasing, and $h = c_1 + \sum_{n=1}^{\infty}(c_{n+1} - c_n)$. Hence

$$\|h\|_2 \le \|c_1\|_2 + \sum_{n=1}^{\infty} \|c_{n+1} - c_n\|_2 = I(c_1) + \sum_{n=1}^{\infty} \{I(c_{n+1}) - I(c_n)\} =$$

$$= \lim_{n\to\infty} I(c_n) \le J_1(h) \; .$$

It follows that $\|\varphi\|_2 \le \|h\|_2 \le \|\varphi\|_1 + \varepsilon$. Since $\varepsilon > 0$ is arbitrary, the proof is complete. □

We still want to mention that under the assumption of the theorem the space S is σ-compact (Exercise 4.1-2).

4.1.7. The next few subsections deal with the question of how the final results of the two routes are related. We compare the processes

$$\mathcal{T} \to \mathcal{D}'(B,I) \to \mathcal{A}(B,I,A_1,J_1) \to \mathcal{N}(B,I,\|\cdot\|_1) \to \mathcal{L}(L_1,I_1) \; ,$$

$$\mathcal{T} \to \mathcal{D}'(B,I) \to \mathcal{D}(B,I) \to \mathcal{A}(B,I,A_2,J_2) \to \mathcal{N}(B,I,\|\cdot\|_2) \to \mathcal{L}(L_2,I_2) \; .$$

We note in passing that in both cases the norm can be described by integrable functions, so the condition $\mathcal{A}5$ of Subsection 1.2.6 is satisfied. This follows immediately from the fact that any class of auxiliary functions derived from a $\mathcal{D}$- or a $\mathcal{D}'$-system has this property (see 1.3.9 and 1.3.10).

The main result in comparing the two processes is the following one.

4.1.8. Theorem. Assume $\mathcal{T}(T,I)$ and follow the routes indicated in 4.1.7. Then:

(i) $L_2 \subset L_1$, and $I_2(f) = I_1(f)$ if $f \in L_2$.

(ii) For each $f \in L_1$ there exists $g \in L_2$ such that $\|f - g\|_1 = 0$.

Proof. Since $\|\varphi\|_1 \le \|\varphi\|_2$ for all $\varphi \in \Phi$ (see the first line of the proof of Theorem 4.1.6), everything follows from Theorem 1.1.20. □

4.1.9. The next theorem gives more precise information under an additional condition.

Theorem. Assume $\mathcal{T}(T,I)$ and follow the routes indicated in 4.1.7. For $i = 1,2$, let M_i denote the class of measurable functions derived from L_i by the procedure of Section 2.1. Assume $A_1 \subset M_2$. Then we have:

(i) If $\varphi \in \Phi$ and $\|\varphi\|_2 < \infty$, then $\|\varphi\|_1 = \|\varphi\|_2$.

(ii) If $\varphi \in \Phi$ and $\|\varphi\|_1 = 0$, then $\|\varphi\|_{2,loc} = 0$ (see Section 2.4 for the definition of the local norm).

(iii) If $f \in L_1$, then there exists $g \in L_2$ such that

$$\|f - g\|_1 = \|f - g\|_{2,loc} = 0 \quad \text{and} \quad I_2(g) = I_1(f) \; .$$

(iv) If $f \in L_1$ and $\|f\|_2 < \infty$, then $f \in L_2$.

(v) $M_1 = M_2$.

Proof. (i) Let $\varphi \in \Phi$ satisfy $\|\varphi\|_2 < \infty$. We know already that $\|\varphi\|_1 \le \|\varphi\|_2$, so we need only prove the reverse inequality $\|\varphi\|_2 \le \|\varphi\|_1$.

Since $\|\varphi\|_2 < \infty$, and the system $A(B,I,A_2,J_2)$ satisfies $A5$, Corollary 1.2.8 guarantees the existence of an $f \in L_2^+$ such that $|\varphi| \le f$ almost everywhere in the sense of $\|\cdot\|_2$. Choose $h \in A_1$ such that $|\varphi| \le h$, and put $g := (h)_f$. Since $h \in M_2$ by assumption, we have $g \in L_2$. Hence, $g \in L_1$ and $I_1(g) = I_2(g)$ by the preceding theorem, and therefore

$$J_1(h) = \|h\|_1 \ge \|g\|_1 = I_1(g) = I_2(g) = \|g\|_2 \ge \|\varphi\|_2 .$$

Now Theorem 1.3.12(ii) shows that $\|\varphi\|_1 \ge \|\varphi\|_2$.

(ii) Let $\varphi \in \Phi$ satisfy $\|\varphi\|_1 = 0$. Let $f \in L_2^+$. Then $|(\varphi)_f| \le |f|$, so $\|(\varphi)_f\|_2 \le \|f\|_2 < \infty$. Hence, by (i), $\|(\varphi)_f\|_2 = \|(\varphi)_f\|_1 \le \|\varphi\|_1 = 0$. Therefore $\|\varphi\|_{2,loc} = 0$.

(iii) This assertion follows immediately from Theorem 4.1.8 and (ii) above.

(iv) Let $f \in L_1$ satisfy $\|f\|_2 < \infty$. Take $g \in L_2$ such that $\|f - g\|_1 = 0$. Since $\|f - g\|_2 \le \|f\|_2 + \|g\|_2 < \infty$, (i) implies $\|f - g\|_2 = 0$. Hence $f \in L_2$.

(v) To show $M_1 \subset M_2$, let $h \in M_1$ and $f \in L_2^+$. We must prove that $(h)_f \in L_2$. By 4.1.8(i), $f \in L_1^+$, so by the definition of measurability, $(h)_f \in L_1$. According to 4.1.8(ii) there exists $g \in L_2$ such that $\|g - (h)_f\|_1 = 0$. Since $\|g - (h)_f\|_2 \le \|g\|_2 + \|f\|_2 < \infty$, (i) shows that $\|g - (h)_f\|_2 = 0$. Hence $(h)_f \in L_2$.

For the proof of the reverse inclusion, let $g \in M_2$ and $f \in L_1^+$. Choose $h \in L_2^+$ such that $\|h - f\|_1 = 0$. Then $(g)_f = (g)_h$ almost everywhere with respect to $\|\cdot\|_1$. Now $(g)_h \in L_2 \subset L_1 \subset M_1$, and therefore $(g)_f \in M_1$. Hence $M_2 \subset M_1$. □

4.1.10. We now give two conditions, each of which implies that the cardinal assumption $A_1 \subset M_2$ of the preceding theorem is satisfied. The first one (in Theorem 4.1.11) still has a mixed character, for it relates topological and measure theoretic properties of the space; the second one (in Theorem 4.1.12) is entirely in topological terms.

The following result, which is also interesting in its own right, will be needed.

Proposition. An $h \in \Phi^+$ belongs to A_1 if and only if for every $a \in \mathbb{R}$ the set $O(a) := \{s \in S \mid h(s) > a\}$ is open.

Proof. First let O be an open subset of S. We show that $\chi_O \in A_1$. For each $s \in O$ there exists by Urysohn's lemma 0.2.16 a function $b_s \in B^+$ such that $0 \le b_s \le 1$, $b_s(t) = 0$ $(t \in S\setminus O)$, $b_s(s) = 1$. It follows that $\chi_O = \sup_{s\in O} b_s$. Therefore $\chi_O \in A_1$.

Next let $h \in \Phi^+$ be such that $O(a)$ is open for every $a \in \mathbb{R}$. The result of the preceding paragraph shows that for each $a \in \mathbb{R}$ there exists a subset $W(a)$ of B^+ satisfying $\chi_{O(a)} = \sup_{b \in W(a)} b$. Now it is easy to see that $h = \sup_{a>0} \sup_{b \in W(a)} (ab)$, that is, $h \in A_1$.

On the other hand, if $h \in A_1$, then $h = \sup_{b \in W} b$ for some subset W of B^+. If $a \in \mathbb{R}$, then $O(a) = \bigcup_{b \in W} \{s \in S \mid b(s) > a\}$, so $O(a)$ is a union of open sets, and therefore is itself open. □

A function $\varphi : S \to \mathbb{R}$ such that $\{s \in S \mid \varphi(s) > a\}$ is open for every $a \in \mathbb{R}$ is often called *lower semi-continuous*. Thus the members of A_1 are precisely the non-negative lower semi-continuous functions on S.

4.1.11. Theorem. Assume $\mathcal{T}(T,I)$ and follow the routes indicated in 4.1.7. For $i = 1,2$ let M_i denote the class of measurable functions derived from L_i by the procedure of Section 2.1. Then the following two assumptions are equivalent:

(i) $\chi_O \in M_2$ for every open $O \subset S$,

(ii) $\chi_C \in M_2$ for every compact $C \subset S$;

and each of (i) and (ii) implies that $A_1 \subset M_2$.

Proof. Assume (i). Let C be a compact subset of S. Since S is a Hausdorff space, C is closed. Therefore $S \backslash C$ is open, so $\chi_{S \backslash C} \in M_2$ by assumption. As already stated at the end of 4.1.1, $\mathcal{M}$ is satisfied, that is, $\chi_S \in M_2$. Hence $\chi_C \in M_2$.

Conversely, assume (ii) holds. Let $O \subset S$ be open. It will suffice to show that $(\chi_O)_b \in L_2^+$ for every $b \in B^+$ (see Exercise 2.1-1). Fix $b \in B^+$. Let C denote the support of b. Let $F := S \backslash O$, then F is closed. Now note that $(\chi_F)_b(s)$ is equal to $\inf\{b(s),1\}$ if $s \in C \backslash O$, and equal to zero otherwise. This proves $(\chi_F)_b = (b)_{\chi_S} \cdot \chi_{C \backslash O}$. The function $(b)_{\chi_S}$ is obviously in B, and therefore in M_2. Also, $C \backslash O = C \cap F$ is a closed subset of a compact set, and therefore compact (see 0.2.8). Hence $\chi_{C \backslash O} \in M_2$, and therefore $(\chi_F)_b \in M_2$ (see Example 2.1.16(ii)). Now note that $0 \leq (\chi_F)_b \leq b$ and use Theorem 2.1.17 to get $(\chi_F)_b \in L_2$. It follows that $\chi_F \in M_2$, and so $\chi_O \in M_2$ as before. Thus (i) and (ii) are equivalent.

To show that (i) implies $A_1 \subset M_2$, use Proposition 4.1.10, Theorem 2.1.12, and the fact that $\mathcal{M}$ holds. □

4.1.12. Theorem. Besides the conditions of Theorem 4.1.11, assume that for every compact $C \subset S$ there exists a sequence $(O_n)_{n \in \mathbb{N}}$ of open sets such that $C = \bigcap_{n \in \mathbb{N}} O_n$. Then $A_1 \subset M_2$. In particular, if the topology T can be derived from a metric, then $A_1 \subset M_2$.

Proof. We show that assumption (ii) of 4.1.11 is satisfied. Let C be a compact subset of S and assume $C = \bigcap_{n\in\mathbb{N}} O_n$ where the O_n are open for all n. By Urysohn's lemma, for each $n \in \mathbb{N}$ there exists $b_n \in B$ such that $0 \leq b_n \leq 1$, $b_n(s) = 1$ $(s \in C)$, $b_n(s) = 0$ $(s \in S\backslash O_n)$. Since $\chi_C = \lim_{n\to\infty} b_n$ and each b_n is measurable, Proposition 2.1.3(iv) shows that $\chi_C \in M_2$.

If T can be derived from a metric d, then each compact subset of S is automatically an intersection of countably many open sets. For if C is compact, then the sequence $(O_n)_{n\in\mathbb{N}}$ with $O_n := \{t \in S \mid d(s,t) < 1/n \text{ for some } s \in C\}$ does what is required. □

A subset of S that is the intersection of countably many open sets is often called a G_δ-*set*. So with this terminology the main assumption of the theorem is that each compact set is a G_δ-set.

4.1.13. For continuous real-valued functions there are never measurability problems, for we have the following result.

Proposition. Assume the conditions of Theorem 4.1.11. Let f be a real-valued continuous function defined on S. Then $f \in M_1 \cap M_2$.

Proof. If $b \in B^+$, then $(f)_b \in B^+ \subset L_1 \cap L_2$. Now use the result of Exercise 2.1-1. □

4.1.14. As noted several times already, in general the route $\mathcal{T} \to \mathcal{D}' \to A \to N \to L$ produces more integrable functions than the alternative route using the short-circuit $\mathcal{D}' \to \mathcal{D} \to A$. This is a good reason to prefer the first method, and we shall do so in the remainder of this book, unless stated otherwise. We shall often drop the subscript 1 in $\|\cdot\|_1$, I_1, L_1, J_1, A_1 and M_1, and use the words "integrable" and "measurable" without qualification.

4.1.15. The final result of this section gives another justification for our preference.

Proposition. Under the conditions of Theorem 4.1.11 we have

(i) lower semi-continuous functions are measurable, that is, they belong to M_1;

(ii) open sets and closed sets are measurable, that is, their characteristic functions belong to M_1.

Proof. In view of the definition of lower semi-continuity, and because the measurable sets form a σ-algebra, it is sufficient to show that open sets are measurable (see Theorem 2.1.12). Let O be an open subset of S, and let $b \in B^+$. Proposition 4.1.10

shows that $(\chi_0)_b$ is an auxiliary function. Since $J((\chi_0)_b) \le I(b) < \infty$, Proposition 1.3.9 furnishes the existence of a sequence $(b_n)_{n\in\mathbb{N}}$ in B^+ and a non-negative null function φ such that $(\chi_0)_b = \varphi + \sum_{n=1}^{\infty} b_n$, while $J((\chi_0)_b) = \sum_{n=1}^{\infty} I(b_n)$. Beppo Levi's theorem 1.1.17 implies that $\sum_{n=1}^{\infty} b_n$ converges almost everywhere to an integrable function h. Since $h = (\chi_0)_b$ almost everywhere, $(\chi_0)_b \in L$ (this is L_1 in the old notation). Hence χ_0 is measurable. □

Exercises Section 4.1

1. Let S, T and B be as in 4.1.1. Show that there is a sequence $(b_n)_{n\in\mathbb{N}}$ in B^+ with $\sum_{n=1}^{\infty} b_n(s) > 0$ $(s \in S)$ if and only if there is a sequence $(C_n)_{n\in\mathbb{N}}$ of compact sets with $S = \bigcup_{n=1}^{\infty} C_n$.

2. Assume that T has a countable base. Show that there is a sequence $(C_n)_{n\in\mathbb{N}}$ of compact sets with $S = \bigcup_{n=1}^{\infty} C_n$.

3. With the notations and assumptions of Theorem 4.1.9 show that $\|\varphi\|_{2,loc} \le \|\varphi\|_1$ $(\varphi \in \Phi)$. Give an example where one has $\|\varphi\|_{2,loc} = 0$, $\|\varphi\|_1 = \infty$ for some $\varphi \in \Phi$. (Hint. "Topologize" Exercise 2.4-11.)

4. Take $S = \mathbb{R}$ with the usual topology and let $F : \mathbb{R} \to \mathbb{R}$ be non-decreasing. Put $I(b) = \int_{-\infty}^{\infty} b(x)dF(x)$ for $b \in B$ (ordinary Riemann-Stieltjes integral). Show that the processes $T \to \mathcal{D}' \to A \to N \to L$ and $T \to \mathcal{D}' \to \mathcal{D} \to A \to N \to L$ give the same results.

5. This exercise gives an example where the processes $T \to \mathcal{D}' \to A$ and $T \to \mathcal{D}' \to \mathcal{D} \to A$ yield different results. Let S be an uncountable set and add a point $p \notin S$ to S to obtain the set $S_0 = S \cup \{p\}$. We take on S the discrete topology, and the neighborhoods of p are all sets containing p whose complements are finite. Thus we obtain a compact Hausdorff space.

(i) Let $b \in B$. Show that for any $\varepsilon > 0$ the set $\{s \in S_0 \mid |b(s) - b(p)| \ge \varepsilon\}$ is finite. Show that the set $\{s \in S_0 \mid b(s) \ne b(p)\}$ is countable. Put $I(b) = b(p)$ for $b \in B$ and assume the notation of 4.1.7.

(ii) Show that $\chi_S \in A_1$, $J_1(\chi_S) = 0$. Conclude that $\|\varphi\|_1 = |\varphi(p)|$ for $\varphi \in \Phi$ and determine L_1 and I_1.

(iii) Show that $\|\chi_S\|_2 = \|\chi_{\{p\}}\|_2 = 1$ and that neither S nor $\{p\}$ is a measurable set in the second process.

This example is due to Ralph Howard.

6. This exercise is a refined version of the previous one. (It is due to Jean Dieudonné.) We give it since we shall need it also in Exercise 4.3-2. Let Y be the long line as constructed in Exercise 3.3-14. On Y we take the topology T for which a base is given by $\{\{0\}\} \cup \{\{s \mid a < s \leq b\} \mid a \in Y, b \in Y\}$ (we treat 0 as the first element of Y).

(i) Show that Y is a compact Hausdorff space. (Hint. Given an open covering of Y, let W be the set of $y \in Y$ for which $[y,\Omega]$ can be covered by a finite subcover. Show that $W \neq \emptyset$, and that $\min W = 0$.)

(ii) Let $b \in B$. Show there is an $a \in Y \setminus \{\Omega\}$ with $b(y) = b(\Omega)$ for $y \geq a$. (Hint. Exercise 3.3-14; compare Exercise 4.1-5(i).)

Let $I(b) := b(\Omega)$ for $b \in B$ and assume the notation of 4.1.7.

(iii) Let $\varphi \in \Phi$. For every $\varepsilon > 0$ there is an $a \in Y\setminus\{\Omega\}$ such that $|\varphi(y)| \leq \|\varphi\|_2 + \varepsilon$ for $y \geq a$. Conversely, if m is such that $|\varphi(y)| \leq m$ for $y \geq a$ (where $a \in Y\setminus\{\Omega\}$), then $\|\varphi\|_2 \leq m$. (Hint. Construct $b_1 \in B$, $b_2 \in B$ with $b_1(\Omega) = 0$, $b_1(y) = 1$ $(y \leq a)$, $b_2(y) = 1$ $(y \geq a)$.)

(iv) Let $f \in \Phi_F$. Show that $f \in L_2$ if and only if there is $c \in \mathbb{R}$, $a \in Y\setminus\{\Omega\}$ such that $f(y) = c$ for $y \geq a$.

(v) Let $f \in \Phi$. Show that $f \in M_2$ if and only if there is $c \in \mathbb{R}^*$, $a \in Y\setminus\{\Omega\}$ such that $f(y) = c$ for $y \geq a$.

4.2. Approximation properties of measurable sets and measurable functions

In a measure space (S,Γ,μ), measurable sets of σ-finite measure can be approximated by sets from the σ-algebra generated by the semiring Γ (see Theorem 3.3.7), and there is a corresponding property for measurable functions (see Theorem 3.3.8). The present section contains similar results for the case of an integral on a locally compact Hausdorff space. Now the approximation is carried out by topologically nice objects, such as open sets and (semi-)continuous functions.

4.2.1. Let S be a locally compact Hausdorff space. Let S, T, B, I, J, A and so on have the meaning of the preceding section (recall that we follow the route $\mathcal{D}' \to A$ rather than $\mathcal{D}' \to \mathcal{D} \to A$, as agreed in 4.1.14). Let Σ denote the class of measurable subsets of S. The last proposition of the preceding section asserts that open sets and closed sets belong to Σ. Define the set function μ on Σ as follows: for $E \in \Sigma$, let $\mu(E) := I(\chi_E^-)$ if $\chi_E \in L$, $\mu(E) := \infty$ otherwise. According to 3.2.10, the set function μ is a measure on the σ-algebra Σ. We note that the compact subsets of S, which are closed and hence measurable, have finite measure (Exercise 4.2-1).

Here is the approximation result for sets.

Theorem. Assume $\mathcal{T}(T,I)$.

(i) Let $E \in \Sigma$. Then

$$\mu(E) = \inf\{\mu(O) \mid O \text{ open}, O \supset E\} .$$

(ii) Let $O \subset S$ be open. Then

$$\mu(O) = \sup\{\mu(C) \mid C \text{ compact}, C \subset O\} .$$

(iii) Let $E \in \Sigma$ have σ-finite measure, that is, assume that E is contained in the union of countably many measurable sets of finite measure. Then

$$\mu(E) = \sup\{\mu(C) \mid C \text{ compact}, C \subset E\} .$$

Proof. (i) If $\mu(E) = \infty$, there is nothing to prove, so assume $\mu(E)$ finite. Let $\varepsilon > 0$. By the definition of norm and integral (see also 1.3.12), there exists $h \in A$ such that $\chi_E \le h$, $J(h) \le \mu(E) + \varepsilon/2$. For each $n \in \mathbb{N}$, let

$$O_n := \{s \in S \mid h(s) > 1 - \tfrac{1}{n}\} .$$

The sets O_n are open by Proposition 4.1.10, while $O_n \supset E$, and

$$(1 - \tfrac{1}{n})\mu(O_n) = I((1 - \tfrac{1}{n})\chi_{O_n}) \le I(h) \le J(h) < \mu(E) + \tfrac{\varepsilon}{2} .$$

Hence $\mu(O_n) < \mu(E) + \varepsilon$ if n is large enough.

(ii) Let $0 < c < \mu(O)$. Since

$$\mu(O) = I(\chi_O) = \|\chi_O\| = J(\chi_O) = \sup\{I(b) \mid 0 \le b \le \chi_O, b \in B\} ,$$

there exists $b \in B^+$ such that $I(b) > c$ and $b \le \chi_O$. If C is the support of b, then $\mu(C) = I(\chi_C) \ge I(b) > c$. Hence $\mu(O) \le \sup\{\mu(C) \mid C \text{ compact}, C \subset O\}$. The reverse inequality is trivial.

(iii) First assume $\mu(E) < \infty$. Fix $\varepsilon > 0$. Using (i), choose an open set O such that $E \subset O$, $\mu(O) < \mu(E) + \varepsilon/3$. Then $\mu(O\backslash E) < \varepsilon/3$. Use (ii) to choose a compact set $C \subset O$ such that $\mu(C) > \mu(O) - \varepsilon/3$, and again use (i) to obtain an open set $U \supset O\backslash E$ such that $\mu(U) < 2\varepsilon/3$. (Draw a picture to keep track of all these sets!) Let $C_1 := C\backslash U$, then C_1 is compact as a closed subset of a compact set, and

$$\mu(C_1) \ge \mu(C) - \mu(U) > \mu(O) - \frac{\varepsilon}{3} - \frac{2\varepsilon}{3} \ge \mu(E) - \varepsilon .$$

Also $C_1 \subset O$, while $C_1 \cap (O\backslash E) \subset C_1 \cap U = \emptyset$. and therefore $C_1 \subset E$. This proves the result for sets of finite measure.

Now assume $\mu(E)$ is infinite, but that E has σ-finite measure. Let $(F_n)_{n\in\mathbb{N}}$ be a sequence of measurable sets of finite measure such that $E \subset \bigcup_{n=1}^{\infty} F_n$. Define $E_1 := F_1 \cap E$, $E_{n+1} := (E \cap F_{n+1})\setminus E_n$ for $n \in \mathbb{N}$. The sets E_n are pairwise disjoint, and $E = \bigcup_{n=1}^{\infty} E_n$. Hence $\sum_{n=1}^{\infty} \mu(E_n) = \mu(E) = \infty$. Fix $c > 0$, and choose $N \in \mathbb{N}$ so large that

$$\mu\left(\bigcup_{n=1}^{N} E_n\right) = \sum_{n=1}^{N} \mu(E_n) > c + 1 .$$

The result of the preceding paragraph shows that there exists a compact set $C \subset \bigcup_{n=1}^{N} E_n$ such that $\mu(C) > \mu(\bigcup_{n=1}^{N} E_n) - 1 > c$. Hence the result for this case. □

4.2.2. The preceding theorem says roughly that measurable sets are not very different from certain topologically well-behaved sets. There are similar results for integrable and measurable functions. For instance, by Theorem 1.1.19(i), any integrable function is the almost everywhere limit of a sequence of basic functions, which are continuous in the case at hand. Our next result, commonly known as Lusin's theorem, is in the same direction: it is concerned with continuity properties of integrable functions.

Theorem (Lusin's theorem). Assume $\mathcal{T}(T,I)$. Let f be integrable, and assume that $S_0 := \{s \in S \mid f(s) \neq 0\}$ has finite measure. Then for every $\varepsilon > 0$ there exists a basic function b such that

$$\mu(\{s \in S \mid f(s) \neq b(s)\}) < \varepsilon ,$$

$$\|b\|_{\infty} \leq \|f\|_{\infty} \quad \text{(notation explained in 2.3.1).}$$

Proof. First assume $0 \leq f \leq 1$. Once this special case has been treated, the rest is easy. By the preceding theorem, choose a compact set $C \subset S_0$ such that $\mu(S_0\setminus C) < \varepsilon/4$, and an open set $U \supset C$ such that $\mu(U\setminus C) < \varepsilon/4$. Also arrange that the closure $\bar{U}$ of U is compact. Obviously $\mu(S_0\setminus U) < \varepsilon/2$.

Now define a sequence $(E_n)_{n\in\mathbb{N}}$ recursively. Put $E_1 := \{s \in U \mid f(s) > \frac{1}{2}\}$ and when $E_1, E_2, \ldots, E_n$ have been defined, put

$$E_{n+1} := \left\{s \in U \mid f(s) - \sum_{k=1}^{n} c_k(s)2^{-k} > 2^{-n-1}\right\} ,$$

where $c_k := \chi_{E_k}$ $(k \in \mathbb{N})$. It is easy to see that each E_n is measurable (use Theorem 2.1.12), and that $f(s) = \sum_{k=1}^{\infty} c_k(s)2^{-k}$ if $s \in U$. Moreover, $U \cap S_0 = \bigcup_{n=1}^{\infty} E_n$. For each $n \in \mathbb{N}$, choose a compact set C_n and an open set U_n such that $C_n \subset E_n \subset U_n \subset U$ and $\mu(U_n\setminus C_n) < \varepsilon\cdot 2^{-n-1}$ (again by Theorem 4.2.1). Application of Urysohn's lemma gives a sequence $(b_n)_{n\in\mathbb{N}}$ in B such that $0 \leq b_n \leq 1$, $b_n(s) = 1$ $(s \in C_n)$, $b_n(s) = 0$ $(s \in S\setminus U_n)$ for $n \in \mathbb{N}$.

Put $b_0 := \sum_{n=1}^{\infty} 2^{-n} b_n$. The series converges uniformly on S, so b_0 is continuous (Exercise 0.2-14), and $b_0(s) = 0$ if $s \in S\backslash U$, so the support of b_0 is contained in the compact set $\bar{U}$. Hence b_0 is a basic function. Also $b_0(s) = f(s)$ if $s \in \bigcup_{n=1}^{\infty} C_n$, and therefore

$$\mu(\{s \in S \mid f(s) \neq b_0(s)\}) \le \mu\left(S_0 \setminus \bigcup_{n=1}^{\infty} C_n\right) .$$

Obviously, $S_0\backslash\bigcup_{n=1}^{\infty} C_n$ is contained in the union of $S_0\backslash U$ and $\bigcup_{n=1}^{\infty}(E_n\backslash C_n)$. Hence

$$\mu(\{s \in S \mid f(s) \neq b_0(s)\} \le \mu(S_0\backslash U) + \sum_{n=1}^{\infty} \mu(E_n\backslash C_n) < \varepsilon .$$

Now put $b := \inf(b_0, \|f\|_\infty)$ to obtain a function that is also small enough. It is useful to note that $b \ge 0$ in this case.

Secondly, assume $f \ge 0$. Let $\varepsilon > 0$ and choose $n \in \mathbb{N}$ such that $\mu(\{s \in S \mid f(s) > n\}) < \varepsilon/2$. By the special case already treated, there exists a member $b' \in B^+$ such that

$$\mu(\{s \in S \mid \inf(1, n^{-1} f(s)) \neq b'(s)\}) < \frac{\varepsilon}{2} ,$$

$$\|b'\|_\infty \le n^{-1} \|f\|_\infty .$$

Hence $b := nb'$ will do in this case.

Finally, in the general case, write $f = \sup(f,0) - \sup(-f,0)$, and apply the result of the preceding paragraph to $\sup(f,0)$ and $\sup(-f,0)$ separately. (It is in checking the inequality $\|b\|_\infty \le \|f\|_\infty$ that we employ the possibility of choosing a non-negative b' in the second step.) □

4.2.3. The final theorem of this section gives an analogue of Theorem 4.2.1(i) for integrable functions. It is usually named after Vitali and Carathéodory. An interesting application of this result occurs in Theorem 5.4.16.

Theorem (Vitali-Carathéodory theorem). Assume $\mathcal{T}(T,I)$. Let f be integrable. Then for every $\varepsilon > 0$ there exists an integrable lower semi-continuous function φ with $\varphi(s) \ge -A$ for some $A \ge 0$, while $f \le \varphi$ and $I(\varphi) < I(f) + \varepsilon$.

Proof. Let $\varepsilon > 0$ be fixed. Choose $b \in B$ such that $\|f - b\| < \varepsilon$. By the definition of the norm and Proposition 1.3.12(ii), there exists an auxiliary function h such that $|f - b| \le h$, $\|h\| \le J(h) < \varepsilon$. Since h is measurable (4.1.15(i)) and has finite norm, it is integrable with $I(h) = \|h\|$. Put $\varphi := b + h$, then φ satisfies the requirements. □

Exercises Section 4.2

1. Assume the notation of 4.2.1. If C is compact, then $\mu(C) < \infty$.

2. Let E be measurable with $\mu(E) < \infty$. Show there is a sequence $(O_n)_{n\in\mathbb{N}}$ of open sets and a sequence $(C_n)_{n\in\mathbb{N}}$ of compact sets such that

$$F := \bigcup_{n=1}^{\infty} C_n \subset E \subset \bigcap_{n=1}^{\infty} O_n =: G \quad \text{and} \quad \mu(F) = \mu(E) = \mu(G) .$$

3. Give an example of a situation $\mathcal{T}(T,I)$ where there is a measurable set E with $\mu(E) > \sup\{\mu(C) \mid C \subset E,\ C \text{ compact}\}$. (Hint. Consider the situation of Exercise 2.3-11 where (with the proper topology) one has that C is compact if $C \cap O_\alpha \neq \emptyset$ for only finitely many $\alpha \in [0,1]$.)

4.3. The Riesz representation theorem

Every abstract integral satisfying $\mathcal{M}$ is an integral with respect to a measure (Theorem 3.4.5). In the situation of a locally compact Hausdorff space S this leads to the Riesz representation theorem, which connects the positive linear functionals on the class B with the Borel measures on S. The correspondence is even one-to-one if certain regularity conditions for the measure are added.

4.3.1. As before, let (S,T) be a locally compact Hausdorff space, and let B be the class of continuous real-valued functions on S that have compact support. If I is a positive linear functional on B, then after following the route $\mathcal{T} \to \mathcal{D}' \to \mathcal{A} \to \mathcal{N} \to \mathcal{L}$ the open sets in S become measurable. This motivates the following definition.

Definition. The class of *Borel sets* of S is the smallest σ-algebra of subsets of S that contains the open sets; it is denoted by $\mathcal{B}$. A *Borel measure* on S is a measure defined on $\mathcal{B}$ that is finite on compact sets. A function $f : S \to \mathbb{R}^*$ is called *Borel measurable* if it is $\mathcal{B}$-measurable, that is, if $\{s \in S \mid f(s) < \alpha\} \in \mathcal{B}$ for all $\alpha \in \mathbb{R}$. A Borel measure ν on S is called *regular* if

(i) $\nu(O) = \sup\{\nu(C) \mid C \subset O,\ C \text{ compact}\}$ if $O \subset S$ is open, and

(ii) $\nu(E) = \inf\{\nu(O) \mid E \subset O,\ O \text{ open}\}$ if $E \subset S$ is a Borel set.

A Borel measurable function is often called a *Borel function* for short. Obviously, the continuous real-valued functions on S are Borel functions, and so are the lower semi-continuous functions.

4.3.2. Borel measures generate positive linear functionals on B in a simple way.

<u>Theorem</u>. Let (S,T) and B be as before. Let ν be a Borel measure on S. Then every member of B is ν-integrable. If I is defined by

$$I(b) := \int b\,d\nu \qquad (b \in B) ,$$

then I is a positive linear functional on B.

<u>Proof</u>. Let $b \in B$. Since b is continuous, it is a Borel function, and therefore it is ν-measurable (Theorem 3.3.4(ii)). Also, if C is the support of b, then b is dominated by the ν-integrable function $\|b\|_\infty \cdot \chi_C$. Hence b is ν-integrable (Theorem 2.1.17). The rest of the proof is trivial. □

4.3.3. We now come to the converse of this result. It is usually called the Riesz representation theorem, after F. Riesz, who established a first version of it in 1909 (see Exercise 4.3-5), thereby virtually marking the beginning of functional analysis.

<u>Theorem</u> (Riesz representation theorem). Let (S,T) and B be as before, and let I be a positive linear functional on B. Then there exists a unique regular Borel measure ν on S such that $I(b) = \int b\,d\nu$ for all $b \in B$.

<u>Proof</u>. Most of the work has been done already. Follow the route $\mathcal{T} \to \mathcal{D}' \to \mathcal{A} \to \mathcal{N} \to \to \mathcal{L}(L,I)$. Let Σ denote the class of measurable subsets of S, and let μ be the measure defined on Σ in the usual way. According to Theorem 3.4.5, each $b \in B$ is μ-integrable and satisfies $I(b) = \int b\,d\mu$. Since Σ is a σ-algebra and all open sets belong to Σ (Proposition 5.1.15(ii)), Σ contains $\mathcal{B}$. Let ν be the restriction of μ to $\mathcal{B}$. Then ν is a Borel measure, for μ is finite on compact sets. Theorem 4.2.1 shows that ν is regular, and by Exercise 4.3-1 this regular Borel measure represents I in the sense that $I(b) = \int b\,d\mu$ for $b \in B$.

It remains to show that there is only one regular Borel measure of the kind required. Start from two regular Borel measures ν and ν', both representing I. First we show that ν and ν' coincide on compact sets. Let $C \subset S$ be compact, and let $(U_n)_{n\in \mathbb{N}}$ be a sequence of open sets such that $C \subset \bar{U}_{n+1} \subset U_n$ and $\bar{U}_n$ is compact for all $n \in \mathbb{N}$, while $\nu(U_n) \to \nu(C)$ (Exercise 4.3-2). For each $n \in \mathbb{N}$, let φ_n be a basic function with $\varphi_n(s) = 1$ $(s \in \bar{U}_{n+1})$, $\varphi_n(s) = 0$ $(s \in S\setminus U_n)$, $0 \le \varphi_n \le 1$. It is obvious that $\nu'(C) \le \nu'(U_{n+1}) \le I(\varphi_n) \le \nu(U_n)$ for all $n \in \mathbb{N}$, so $\nu'(C) \le \nu(C)$. Interchanging the roles of ν and ν', we find $\nu(C) \le \nu'(C)$. Hence $\nu(C) = \nu'(C)$. It follows immediately that ν and ν' coincide on the open sets by condition (i) of regularity, and then also that ν and ν' coincide on all Borel sets, by condition (ii). □

If one drops the requirement that ν be regular, ν need not be uniquely determined (see Exercises 4.3-3 and 4.3-4). There are simple conditions on the topology, though, that make ν unique even without this condition of regularity (see Exercise 4.3-6). Otherwise stated: on such spaces all Borel measures are automatically regular.

Exercises Section 4.3

1. Let S, T, B, I be as in Theorem 4.3.3, let μ be the measure obtained in the process $\mathcal{T} \to \mathcal{D}' \to \mathcal{A} \to \mathcal{N} \to \mathcal{L} \to \mathcal{R}$ and let ν be the restriction of μ to the Borel sets $\mathcal{B}$. Show that $I(b) = \int b\,d\mu$ for $b \in B$.

2. Let S, T, B, I and ν be as in the proof of Theorem 4.3.3 and let C be compact. For every $\varepsilon > 0$ there is a $U \in T$ with $\bar{U}$ compact and $C \subset U$, $\nu(C) \geq \nu(U) - \varepsilon$.

3. If condition (i) in Definition 4.3.1 is replaced by

$$\nu(O) = \sup\{\nu(C) \mid C \subset O,\ C \text{ compact}\} \text{ if } O \subset S \text{ is open, } \nu(O) < \infty ,$$

then the Borel measure ν in Theorem 4.3.3 need not be unique.

4. This exercise is meant to show that the Borel measure ν in Theorem 4.3.3 need not be unique if we drop condition (ii) in the definition of regularity (cf. 4.3.1). Take S = Y where Y is the long line of Exercise 3.3-14 and take on Y the topology of Exercise 4.1-6. Let $I(b) := b(\Omega)$ for $b \in B$.

(i) Let $\mu_1(A) := 1$ or 0 according as A does or does not contain Ω $(A \in \mathcal{B})$. Show that μ_1 is a regular Borel measure with $I(b) = \int b\,d\mu_1$ for $b \in B$.

Call a subset E of Y unbounded if for every $y \in Y\backslash\{\Omega\}$ there exists an $a \in E$ with $y < a < \Omega$.

(ii) Let $(E_n)_{n\in\mathbb{N}}$ be a sequence of closed, unbounded sets in Y. Show that $\bigcap_{n=1}^{\infty} E_n$ is closed and unbounded. (Hint. Let $y \in Y\backslash\{\Omega\}$. There exist $\alpha_{nk} \in E_k$ with $\alpha_{nk} < \Omega$ $(k = 1,\dots,n;\ n \in \mathbb{N})$ such that $y < \alpha_{11} < \alpha_{21} < \alpha_{22} < \alpha_{31} < \alpha_{32} < \alpha_{33} < \alpha_{41} < \dots$. Consider $\sup \alpha_{nk}$.)

(iii) Let $\mu_2(A) := 1$ or 0 according as A does or does not contain a closed, unbounded subset of Y $(A \in \mathcal{B})$. Show that μ_2 is a Borel measure with $I(b) = \int b\,d\mu_2$ for $b \in B$. (Hint. Use (ii) and Exercise 3.1-7.)

(iv) Show that μ_2 satisfies condition (i) of Definition 4.3.1.

(v) Show that $\mu_2 \neq \mu_1$, whence μ_2 does not satisfy condition (ii) of Definition 4.3.1. (Hint. Show that $\mu_2(\{\Omega\}) = 0$.)

5. Let $S = [0,1]$ with the usual topology. Let I be a positive linear functional on B. For $x \in S$, define

$$g(x) := \sup\{I(b) \mid b \in B^+, b(s) = 0 \ (s \geq x), b(s) \leq 1 \ (s < x)\} .$$

Show that g is non-decreasing and real-valued on S and that

$$I(b) = \int_0^1 b(x)dg(x)$$

for $b \in B$ (ordinary Riemann-Stieltjes integral).

6. Let (S,T) be a compact Hausdorff space and assume that every open set in S is a countable union of compact sets. Show that every Borel measure ν is regular. (Hint. Show that the class of all Borel sets E for which

$$\nu(E) = \sup\{\nu(C) \mid C \subset E, C \text{ compact}\} = \inf\{\nu(O) \mid E \subset O, O \text{ open}\}$$

is a σ-algebra containing the open sets.) Generalize to σ-compact spaces.

7. Show that every Borel set is measurable in the process $T \to \mathcal{D}' \to A \to N \to L$. Show that if f is integrable, there is a Borel function f_0 such that $f = f_0$ (a.e.). If the system is σ-compact and g is measurable, show that there is a Borel function g_0 such that $g = g_0$(a.e.).

8. Let $f : \mathbb{R} \to \mathbb{R}$ and $g : \mathbb{R} \to \mathbb{R}$ be Borel measurable. Show that $f \circ g$ is Borel measurable.

9. Let I be a linear functional defined on B and assume that for every compact set C there is an $M > 0$ such that $|I(b)| \leq M$ if $b \in B$, $0 \leq b \leq 1$, $b(x) = 0$ $(x \notin C)$. Show there exist Borel measures ν_1 and ν_2 such that $I(b) = \int b\,d\nu_1 - \int b\,d\nu_2$ for $b \in B$. (Hint. Exercise 0.1-1.)

In the following exercises we deal with representation of positive linear functionals defined on certain spaces of continuous functions on a locally compact Hausdorff space (S,T). If I is a positive linear functional on B, then μ_I will denote the unique Borel measure ν satisfying (i) and (ii) of Definition 4.3.1 (cf. Theorem 4.3.3).

10. Assume that there is an $M > 0$ such that $I(b) \leq M$ for $b \in B^+$, $0 \leq b \leq 1$. Show that μ_I is bounded.

11. Let B_0 be the set of continuous real-valued functions b on S for which $\{s \mid |b(s)| \geq c\}$ is compact if $c > 0$. Let I_0 be a positive linear functional on B_0.

Show that there is a bounded Borel measure μ such that $I_0(b) = \int b\,d\mu$ $(b \in B_0)$ as follows.

(i) Let $I(b) := I_0(b)$ for $b \in B$ and show that μ_I is bounded. (Hint. If μ_I is unbounded, there are compact sets C_n with $\mu_I(C_n) \geq 4^n$ $(n \in \mathbb{N})$. Construct $b_n \in B^+$ with $0 \leq b_n \leq 2^{-n}$, $b_n(s) = 2^{-n}$ $(s \in C_n)$ and consider $b := \sum_{n=1}^{\infty} b_n$.)

(ii) Let $b \in B_0^+$. Show there is a sequence $(b_n)_{n\in\mathbb{N}}$ in B^+ with $b_n \uparrow b$ uniformly, $I(b_n) \to I_0(b)$. (Hint. There is a sequence $(b_n)_{n\in\mathbb{N}}$ in B^+ with $b_n \uparrow b$ uniformly. Take $n_k \in \mathbb{N}$ $(k \in \mathbb{N})$ such that $\|b - b_{n_k}\|_\infty < 2^{-k}$ and let $\varphi := \sum_{k=1}^{\infty}(b - b_{n_k})$. Show that $\varphi \in B_0^+$, and use $\sum_{k=1}^{\infty} I_0(b - b_{n_k}) \leq I_0(\varphi) < \infty$ $(k \in \mathbb{N})$ to show that $I(b_{n_k}) \to I_0(b)$ $(k \to \infty)$.)

(iii) Let $b \in B_0$. Show that $I_0(b) = \int b\,d\mu$.

12. Let B_1 be the class of all bounded continuous functions and let I_1 be a positive linear functional defined on B_1. Let $I(b) := I_1(b)$ for $b \in B$ and show that μ_I is bounded. Give an example where $I = 0$ and $I_1 \neq 0$. (Hint. Let $S := Y\backslash\{\Omega\}$, where Y is the long line, and take on S the relative topology (cf. Exercise 4.1-6). If $A \in \mathcal{B}$, put $\mu_1(A) := 1$ or 0 according as A does or does not contain a closed set E with $\sup E > s$ for any $s \in S$. Show that μ_1 is a bounded Borel measure. Let $I_1(b) = \int b\,d\mu_1$ for $b \in B_1$. Show that $I_1 \neq 0$, $I = 0$.)

13. Let $S = \mathbb{R}$ with the usual topology and let I_2 be a positive linear functional on the set B_2 of all continuous functions. Let $I(b) := I_2(b)$ for $b \in B$. Show there is a compact set C with $\mu_I(S\backslash C) = 0$. (Hint. If such a C does not exist there is a sequence $(C_n)_{n\in\mathbb{N}}$ of compact sets with $\mu_I(C_n) > 0$, $C_n \cap [-\alpha_n, \alpha_n] = \emptyset$ where $\alpha_1 = 0$, $\alpha_n = 1 + \max\{|s| \mid s \in \bigcup_{k=1}^{n-1} C_k\}$ for $n \in \mathbb{N}$. Construct a $b \in B_2^+$ with $b(s) \geq (\mu_I(C_n))^{-1}$ if $s \in C_n$.)

4.4. Baire sets and Baire functions

The σ-algebra of the Borel sets appears in a natural way in this chapter's development of integration on locally compact Hausdorff spaces. In the literature, one finds several other σ-algebras in this context. Unfortunately, the terminology is very confusing. Quite often, different authors use different names for the same object, which is bad enough, or the same name for different objects, which is even worse. We shall consider only one other σ-algebra, namely $\mathcal{B}_0$, the smallest σ-algebra making all basic functions measurable. The class $\mathcal{B}_0$ also appears quite naturally: we cannot do with a smaller σ-algebra in discussing B. Exercise 4.4-10 shows that $\mathcal{B}_0$ is still large enough to be meaningful.

4.4.1. Again, let (S,T) be a locally compact Hausdorff space, and let B be as before.

Definition. The *class of Baire sets* of S is the smallest σ-algebra with respect to which all members of B are measurable; it is denoted by $\mathcal{B}_0$. A *Baire measure* is a measure defined on $\mathcal{B}_0$ that is finite on compact Baire sets. A *Baire function* on S is an $\mathbb{R}^*$-valued function that is $\mathcal{B}_0$-measurable.

A few comments are in order. First, the existence of such a class $\mathcal{B}_0$ must be proved (Exercise 4.4-1). Second, we defined Baire functions on $\mathbb{R}$ in 2.1.14 already; is there no danger of confusion? No, see Proposition 4.4.3. Finally, although all basic functions are Baire functions, on certain spaces S there may exist continuous real-valued functions that are not Baire functions (Exercise 4.4-2).

4.4.2. The reader may have noted a difference between the definition of the Borel sets and that of the Baire sets: the former involves open sets only, whereas the latter is formulated in terms of continuous functions. The next result is connected with this observation; it gives a characterization of $\mathcal{B}_0$ in terms of certain compact sets. Recall first that a G_δ-set in a topological space is the intersection of countably many open sets.

Proposition. Let (S,T) be a locally compact Hausdorff space. Let Σ be the σ-algebra generated by the compact G_δ's. Then $\mathcal{B}_0 = \Sigma$.

Proof. Let C be a compact G_δ, and assume $C = \bigcap_{n=1}^{\infty} O_n$, where each O_n is open. Since S is locally compact, it can be arranged that each $\bar{O}_n$ is compact (Exercise 4.3-2). For each $n \in \mathbb{N}$, choose a member $\varphi_n \in B$ such that $\varphi_n(s) = 1$ if $s \in C$, $\varphi_n(s) = 0$ if $s \in S\backslash O_n$, $0 \le \varphi_n \le 1$. Obviously, each φ_n is a Baire function, and so is χ_C because $\varphi_n \to \chi_C$ (see 3.3.3(iii)). Hence C belongs to the σ-algebra $\mathcal{B}_0$. Since Σ is contained in any σ-algebra that contains all compact G_δ's, it follows that $\Sigma \subset \mathcal{B}_0$.

On the other hand, if $f \in B$, then f is Σ-measurable. To see this, assume first $f \ge 0$. Let $c \in \mathbb{R}$, and $E_c := \{s \in S \mid f(s) \ge c\}$. We have to show that $E_c \in \Sigma$. This is obvious if $c \le 0$. And if $c > 0$, then E_c is a compact G_δ. For,

$$E_c = \bigcap_{n=1}^{\infty} \{s \in S \mid f(s) > c - \tfrac{1}{n}\} ,$$

so E_c is a G_δ, and also E_c is a closed subset of the (compact) support of f, and therefore compact. Hence f is Σ-measurable in this special case. For a general $f \in B$, write $f = \sup(f,0) - \sup(-f,0)$, and use 3.3.3(ii). So all members of B are Σ-measurable. From the definition of $\mathcal{B}_0$ it follows that $\mathcal{B}_0 \subset \Sigma$. □

Exercise 4.4-4 gives a similar characterization of $\mathcal{B}_0$.

4.4.3. If S is σ-compact, then the Baire functions on S can be described in a way similar to the definition in Section 2.1 for the case $S = \mathbb{R}$. (As to the existence of the class X mentioned below, we refer to the corresponding discussion in 2.1.14.)

Proposition. Let S be a locally compact Hausdorff space which is σ-compact. Let Y denote the class of Baire functions on S. Let X denote the smallest class of $\mathbb{R}^*$-valued functions that contains all continuous real-valued functions on S and that is closed under pointwise sequential limits. Then X = Y.

Proof. Y is closed under sequential limits by 3.3.3(iii). Moreover, since S is σ-compact, there exists a sequence $(b_n)_{n\in\mathbb{N}}$ in B^+ such that $b_n \to \chi_S$. So if $f : S \to \mathbb{R}$ is continuous, then $f\cdot b_n \in B \subset Y$ for each $n \in \mathbb{N}$, and therefore $f = \lim_{n\to\infty} f\cdot b_n \in Y$ again by 3.3.3(iii). Hence $X \subset Y$.

The proof of the reverse inclusion is more intricate. First we show the following: If f and g are finite-valued members of X, then $f + g \in X$. Fix $f_0 : S \to \mathbb{R}$ and assume f_0 is continuous. Let $Z_0 := \{g \in X \mid f_0 + g \in X\}$. It is easy to see that Z_0 contains the continuous real-valued functions and that it is closed under pointwise sequential limits. Hence $Z_0 = X$. Now repeat the same reasoning with an $f_0 \in X$ which is finite-valued to find that $f_0 + g_0 \in X$ for any two finite-valued members of X. Argue likewise for scalar multiples.

Next we show that $\chi_E \in X$ for any Baire set E. Let $\Gamma := \{E \in \mathcal{B}_0 \mid \chi_E \in X\}$. This class Γ is a σ-algebra. For, if E and F belong to Γ, then $\chi_{E\cap F} = \lim_{n\to\infty}\{\frac{1}{2}(\chi_E + \chi_F)\}^n$ belongs to X by the result of the preceding paragraph. Also, Γ is closed under complementation, so it is an algebra. Since X is closed under pointwise sequential limits, Γ is even an σ-algebra. If C is a compact G_δ, then there exists a sequence $(b_n)_{n\in\mathbb{N}}$ in B^+ such that $b_n \to \chi_C$ (see the proof of Proposition 4.4.2). So $C \in \Gamma$. Hence Γ is a σ-algebra contained in $\mathcal{B}_0$ that contains the compact G_δ-sets. Since $\mathcal{B}_0$ is the smallest σ-algebra of this kind, $\Gamma = \mathcal{B}_0$. Hence $\chi_E \in X$ if $E \in \mathcal{B}_0$.

Finally, by the preceding two paragraphs, X contains all simple Baire functions (the finite linear combinations of characteristic functions of Baire sets). By 2.1.13, X contains all Baire functions. □

Exercises Section 4.4

1. Show that the definition of $\mathcal{B}_0$ in 4.4.1 makes sense.

2. Give an example of a locally compact Hausdorff space with a continuous function on it that is not Baire measurable. (Hint. In view of Exercises 5 and 9 look for an example where the topology does not have a countable base and the space itself is not σ-compact.)

3. If $S = \mathbb{R}$ with the usual topology then $\mathcal{B} = \mathcal{B}_0$.

4. Show that $\mathcal{B}_0$ is the σ-algebra generated by all open F_σ's (an F_σ is a countable union of closed sets).

5. If (S,T) is a locally compact Hausdorff space with a countable base, then $\mathcal{B}_0 = \mathcal{B}$.

6. Assume that S is a σ-compact metric space. Show that $\mathcal{B}_0 = \mathcal{B}$.

7. Give an example of a compact Hausdorff space where $\mathcal{B}_0 \neq \mathcal{B}$.

8. Give an example of a locally compact metric space where $\mathcal{B}_0 \neq \mathcal{B}$.

9. If S is σ-compact every continuous real-valued function on it is Baire measurable.

10. Let I be a positive linear functional defined on B. Show that every Baire set is measurable in the process $\mathcal{T} \to \mathcal{D}' \to \mathcal{D} \to \mathcal{A} \to \mathcal{N} \to \mathcal{L}$ and that every compact Baire set has finite measure. Show that if f is integrable there is a Baire function f_0 such that $f = f_0$ (a.e.). Show that if the system is σ-compact and g is measurable there is a Baire function g_0 such that $g = g_0$ (a.e.). Compare with Exercise 2.

11. Let I be a positive linear functional defined on B and follow the line $\mathcal{T} \to \mathcal{D}' \to \mathcal{D} \to \mathcal{A} \to \mathcal{N} \to \mathcal{L}$. Let Σ be the σ-algebra of measurable sets with measure μ. If $E \in \Sigma$, show that

$$(*) \qquad \mu(E) = \inf\{\mu(O) \mid O \text{ open } F_\sigma,\ O \supset E\} .$$

12. Let I be a positive linear functional defined on B. Show there is a Baire measure μ such that $I(b) = \int b\,d\mu$ for every $b \in B$. Show that μ is unique if it satisfies $(*)$ of the previous exercise for $E \in \mathcal{B}_0$ and that μ need not be unique if this condition is dropped.

13. Let $\mathcal{B}_1$ be the σ-algebra generated by all compact sets (this is called by Hahn the class of Borel sets). Show that $\mathcal{B}_0 \subset \mathcal{B}_1 \subset \mathcal{B}$ and give examples where the inclusions are proper.

CHAPTER FIVE

THE RADON-NIKODYM THEOREM AND ITS CLASSICAL ORIGIN

The main subjects of calculus are differentiation and integration of functions on $\mathbb{R}$ or $\mathbb{R}^n$. Now we have treated integration in a more general set-up, it is time that we turn to differentiation in a similar context. We present differentiation of measures first, and discuss its relation to ordinary differentiation on the real line afterwards.

In the first section of this chapter we define signed measures. A signed measure behaves like the measures of Chapter 3 in being a σ-additive set function, but it is defined from the outset on a σ-algebra, and it takes values either in $(-\infty,\infty]$ or in $[-\infty,\infty)$. Signed measures and measures are closely connected. For instance, if μ_1 and μ_2 are measures on the same σ-algebra Σ and if at least one of μ_1 and μ_2 is finite, then $\mu_1 - \mu_2$ is a signed measure on Σ. It is a remarkable fact that every signed measure on Σ may be obtained in this way. This is the content of the Jordan decomposition theorem, which is often useful, because it enables us to reduce questions about signed measures to questions about measures.

Signed measures can also be obtained by integration. Let (S,Σ,μ) be a measure space, where Σ is a σ-algebra. Let f be a μ-integrable function on S, and put

$$(*) \qquad \nu(E) := \int_E f\,d\mu \qquad (E \in \Sigma)\ .$$

Then ν is a signed measure on Σ. Now, how general is this construction? That is, which signed measures on Σ are obtained in this way? The results of Section 5.2 give an answer to this question. They say that if μ is a σ-finite measure and ρ a finite signed measure on a σ-algebra Σ, then there is a μ-integrable function f on S such that $\rho = \nu + \tau$, where ν is the signed measure given by $(*)$ and τ is as far removed from being representable by $(*)$ as one can imagine. The signed measure ν has the property that $\nu(E) = 0$ whenever $E \in \Sigma$ and $\mu(E) = 0$; we say that ν is absolutely continuous with respect to μ. If ρ itself is absolutely continuous with respect to μ, then the signed measure τ is absent, so ρ coincides with ν and is given by an integral expression as in $(*)$. This is the Radon-Nikodym theorem.

In Section 5.3 we apply the Radon-Nikodym theorem to discuss the bounded linear functionals on L^p-spaces with $1 < p < \infty$. (Another application of the same theorem is in Section 6.5.)

The rest of the chapter treats classical questions of which the Radon-Nikodym theorem is an outgrowth. To introduce the problem, we go back to the fundamental theorem of calculus, which says that under appropriate conditions integration is inverse to differentiation. For example, if f is a differentiable real-valued function on a bounded interval $[a,b]$ and if f' is Riemann integrable, then

$$(**) \qquad \int_a^b f'(x)dx = f(b) - f(a) .$$

The practical use of this formula is that it enables us to evaluate integrals by primitivization: if we want to know what $\int_a^b g(x)dx$ is, we seek a function f (a primitive of g) such that f' = g, and apply (**).

There is another interpretation of (**) that is more interesting for us at this point. After rewriting (**) as

$$f(t) = f(a) + \int_a^t f'(x)dx \qquad (t \in [a,b]) ,$$

we see that the formula enables us to recover f from f'. One of the important points of Lebesgue's integral is that this last formula also holds if f is everywhere differentiable with a finite derivative f' which is integrable in the Lebesgue sense. The restriction that f has a finite derivative everywhere does not fit nicely into a theory where things happening on null sets usually are irrelevant. One would hope that integrability of f' would suffice. Unfortunately, things are not so easy. For instance, there exists a simple continuous function on [0,1] that is strictly increasing with f(0) = 0, f(1) = 1, and which has almost everywhere a derivative equal to 0. So one must compensate for less restrictions on f' by restrictions on f itself. The appropriate condition for f is called absolute continuity. As its name suggests, this notion is related to the property for measures mentioned a little while ago in connection with the Radon-Nikodym theorem. In the description of the relationship, the Lebesgue-Stieltjes measure generated by f turns up. Since Lebesgue-Stieltjes measures themselves are connected to functions of bounded variation, this gives us an opportunity to discuss these, too. One of the classical results we prove is that such functions are almost everywhere differentiable with a (Lebesgue) integrable derivative. (Here we follow the treatment of D. Austin in Proc. Amer. Math. Soc. 16 (1965), 220-221.)

The final section of this chapter is also concerned with practical matter. It treats the formula for change of variable in integrals,

$$\int_{g(a)}^{g(b)} f(x)dx = \int_a^b f(g(x))\, g'(x)dx .$$

There is, of course, a corresponding formula in more dimensions, where Jacobians play a role. We do not treat this more general result, which is best discussed in the context of manifolds where it has its main applications.

5.1. Signed measures; Hahn and Jordan decompositions

One often meets σ-additive set functions that take both positive and negative values. Although functions of this kind are not measures in the sense of Chapter 3, they are closely related to these. They are called signed measures. In this section we define the new concept, illustrate it by some examples, and consider its connection with measures.

5.1.1. Let S be a non-empty set, let Σ be a σ-algebra of subsets of S, and let ν be an $\mathbb{R}^*$-valued function defined on Σ.

Definition. ν is called a *signed measure* if it satisfies the following conditions:

(i) $\nu(\emptyset) = 0$,

(ii) ν takes at most one of the values ∞ and $-\infty$,

(iii) if $(E_n)_{n\in\mathbb{N}}$ is a sequence of pairwise disjoint members of Σ, then

$$\nu\left(\bigcup_{n=1}^{\infty} E_n\right) = \lim_{N\to\infty} \sum_{n=1}^{N} \nu(E_n) .$$

A signed measure is called *finite* if it takes finite values only, and σ-*finite* if S can be covered by countably many sets of finite measure.

Some points about this definition must be noted. First, signed measures live on σ-algebras, not on semirings like measures; the reason is explained in Exercise 5.1-1. Second, condition (ii) is to prevent us from meeting two disjoint sets one of which has measure ∞ and the other $-\infty$. Finally, in the partial sums $\sum_{n=1}^{N} \nu(E_n)$ of condition (iii) there may occur infinite terms. If they do occur, (ii) ensures that they are all of the same kind of infinity, so that there is no difficulty in interpreting these partial sums or the limit. Hence, there is no danger in replacing $\lim_{N\to\infty} \sum_{N=1}^{N} \nu(E_n)$ in (iii) by $\sum_{n=1}^{\infty} \nu(E_n)$, and we shall do so in the future.

Now assume ν is a signed measure as defined above. According to (iii), ν is σ-additive, and (i) and (iii) taken together show that ν is finitely additive: if $E_1,\ldots,E_n$ are disjoint sets in Σ with union E, then $\nu(E) = \sum_{k=1}^{n} \nu(E_k)$. Together with (ii) this leads to the following proposition: Any member of Σ contained in a set of finite measure has itself finite measure. To prove this, let $E,F \in \Sigma$, $\nu(F)$ finite, $E \subset F$. Assume $\nu(E) = \infty$. Since $\nu(F) = \nu(E) + \nu(F\backslash E)$ and $\nu(F\backslash E)$ is either finite or ∞, it follows that $\nu(F) = \infty$: contradiction. Similarly, $\nu(E) = -\infty$ is impossible.

The result of the preceding paragraph can be strengthened as follows. If $(E_n)_{n\in\mathbb{N}}$ is a sequence of pairwise disjoint elements of Σ, and $E := \bigcup_{n=1}^{\infty} E_n$ has finite measure, then $\sum_{n=1}^{\infty} |\nu(E_n)|$ is convergent (to a real number). For, let F_+ be the union of those E_n that satisfy $\nu(E_n) \geq 0$, and F_- the union of the remaining E_n's. As we have seen, both $\nu(F_+)$ and $\nu(F_-)$ are finite, and since

$$\nu(F_+) = \sum_{n=1}^{\infty} \max\{\nu(E_n),0\} , \quad \nu(F_-) = \sum_{n=1}^{\infty} \min\{\nu(E_n),0\} ,$$

the result follows. On the other hand, if $\nu(E)$ is not finite, then $\sum_{n=1}^{\infty} \nu(E_n)$ cannot be absolutely convergent, since the series does not even converge to a real number. However, if $\nu(E) = \infty$, say, then $\sum_{n=1}^{\infty} \max\{\nu(E_n,0\} = \infty$, while $\sum_{n=1}^{\infty} \min\{\nu(E_n),0\}$ is finite.

5.1.2. Example. Let ν_1 and ν_2 be measures on the σ-algebra Σ, and assume that at least one of them is finite. Then $\nu := \nu_1 - \nu_2$, defined by

$$\nu(E) := \nu_1(E) - \nu_2(E) \qquad (E \in \Sigma) ,$$

is a signed measure. This example is typical: every signed measure on Σ can be obtained in this way. We prove this in theorem 5.1.6.

5.1.3. Example. Let (S,Γ,μ) be a measure space (so Γ need only be a semiring). Let Σ be the σ-algebra of measurable subsets of S obtained by the procedure of Chapter 3. Let f be a μ-integrable function, and define

$$\nu(E) := \int_E f\,d\mu \qquad (E \in \Sigma) .$$

Then ν is a signed measure on Σ. For the proof we check the conditions stated in 5.1.1. The first two of these are trivially satisfied, and the last one follows from Lebesgue's dominated convergence theorem. For, if E_n and E are as in 5.1.1(iii), then $f\cdot\sum_{k=1}^{n} \chi_{E_k} \to f\cdot\chi_E$ as $n \to \infty$, and the terms of the sequence are dominated by the integrable function $|f|$. Hence

$$\nu(E) = \int f\cdot\chi_E\,d\mu = \lim_{n\to\infty} \int f\cdot \sum_{k=1}^{n} \chi_{E_k}\,d\mu = \lim_{n\to\infty} \sum_{k=1}^{n} \nu(E_k) .$$

In the present example μ and ν are related in a peculiar way: if $E \in \Sigma$ and $\mu(E) = 0$, then $\nu(E) = 0$. We say that ν is *absolutely continuous* with respect to μ. In the next section we shall see that, conversely, this property is sufficient to reconstruct f from μ.

5.1.4. Again, let ν be a signed measure on the σ-algebra Σ of subsets of S. A set $E \in \Sigma$ is called *positive* (*with respect to* ν) if $\nu(F) \geq 0$ whenever $F \subset E$, $F \in \Sigma$, and *strongly positive* if it is positive and $\nu(E) > 0$. *Negative* and *strongly negative* sets are defined similarly.

Our first result shows that the space S can be split into a positive and a negative set.

Theorem (Hahn decomposition theorem). Let ν be a signed measure on the σ-algebra Σ. Then S can be decomposed as $S = S_1 \cup S_2$, where S_1 and S_2 are disjoint members of Σ, while S_1 is positive and S_2 is negative with respect to ν.

Proof. Without loss of generality we assume that ν takes both positive and negative values, and that $\nu(E) > -\infty$ for all $E \in \Sigma$. First we show that any $E \in \Sigma$ with $\nu(E) < 0$ contains a strongly negative set F.

So let $E \in \Sigma$, $\nu(E) < 0$. Let $E_0 := \emptyset$. Define a sequence $(E_n)_{n\in\mathbb{N}}$ recursively: once $E_0, E_1, \ldots, E_n$ have been chosen, let

$$\alpha_{n+1} := \sup\left\{\nu(G) \mid G \in \Sigma,\ G \subset E \setminus \bigcup_{k=0}^{n} E_k\right\}$$

and choose $E_{n+1} \in \Sigma$, $E_{n+1} \subset E \setminus \bigcup_{k=0}^{m} E_k$, such that $\nu(E_{n+1}) \geq \min\{1, \frac{1}{2}\alpha_{n+1}\}$. This process generates a pairwise disjoint sequence of members of Σ, all contained in E, and all having non-negative measure. So if $F := \bigcup_{n=1}^{\infty} E_n$, then $\nu(F) \geq 0$, and since $F \subset E$ and $\nu(E)$ is finite, $\nu(F)$ is finite, too. Hence $\sum_{n=1}^{\infty} \nu(E_n)$ is convergent (to a real number), and therefore $\alpha_n \to 0$.

We assert that $E\setminus F$ is strongly negative. In the first place, $\nu(E\setminus F) = \nu(E) - \nu(F) \leq \nu(E) < 0$. Also, if $G \in \Sigma$ and $G \subset E\setminus F$, then G is disjoint from all sets $\bigcup_{k=0}^{n} E_k$, so $\nu(G) \leq \alpha_{n+1}$ for all $n \in \mathbb{N}$, and therefore $\nu(G) \leq 0$. Hence $E\setminus F$ is strongly negative. In particular, strongly negative sets do exist.

Put

$$\beta := \inf\{\nu(E) \mid E \in \Sigma,\ E \text{ strongly negative}\}\ ,$$

and choose a sequence $(E_n)_{n\in\mathbb{N}}$ of strongly negative subsets of S such that $\nu(E_n) \to \beta$. Let $S_2 := \bigcup_{n=1}^{\infty} E_n$, then S_2 is strongly negative, and $\nu(S_2) = \beta$ (Exercise 5.1-2), so β is finite. Finally, $S_1 := S\setminus S_2$ is a positive set for ν. In fact, if $F \subset S_1$ and $\nu(F) < 0$, then F contains a strongly negative subset G by the first part of the proof. But then $S_2 \cup G$ is strongly negative and has measure $\nu(S_2) + \nu(G)$, which is less than β since β is finite. This contradicts the definition of β. □

A decomposition of S as given in the theorem is called a *Hahn decomposition of* S *with respect to* ν.

5.1.5. Hahn decompositions are unique up to null sets. To be precise, let S and ν be as in the theorem, and let S have decomposition by S_1 and S_2 on the one hand and by S_1' and S_2' on the other hand, where S_1 and S_1' are the positive sets. Then any $E \in \Sigma$ contained in the symmetric difference of S_1 and S_1' (or of S_2 and S_2') has $\nu(E) = 0$. The easy proof is left to the reader.

5.1.6. The final result of this section is the promised counterpart to Example 5.1.2. To formulate it we need a new concept. Assume μ_1 and μ_2 are measures on the same σ-algebra Σ, and assume that there exists $E \in \Sigma$ such that $\mu_1(E) = \mu_2(S\backslash E) = 0$. Then we say that μ_1 and μ_2 are *mutually singular*.

Theorem. Let ν be a signed measure on the σ-algebra Σ. Then there exists a unique pair of mutually singular measures ν_1 and ν_2 on Σ such that $\nu = \nu_1 - \nu_2$. If ν does not take the value ∞, then ν_1 is a finite measure, and if ν does not take the value $-\infty$, then ν_2 is finite.

Proof. Let S have a Hahn decomposition with respect to ν, consisting of the positive set S_1 and the negative set S_2. Define

$$\nu_1(E) := \nu(S_1 \cap E) \qquad (E \in \Sigma) ,$$

$$\nu_2(E) := \nu(S_2 \cap E) \qquad (E \in \Sigma) .$$

It is easy to check that ν_1 and ν_2 are mutually singular and that $\nu = \nu_1 - \nu_2$. The remaining assertions are also easily proved. □

A decomposition as described in the theorem is usually called the *Jordan decomposition* of ν. It is related to the decomposition of a function of bounded variation on $\mathbb{R}$ into two monotone functions (see 0.5.11), which is not unique, however. Exercise 5.1-3 gives another way to single out the Jordan decomposition of ν among all possible decompositions $\nu = \nu_1 - \nu_2$.

Exercises Section 5.1

1. Give an example of a set S with a semiring Γ of subsets of S and a mapping $\nu : \Gamma \to \mathbb{R}$ satisfying $\nu(\bigcup_{n=1}^{\infty} E_n) = \sum_{n=1}^{\infty} \nu(E_n)$ for every sequence $(E_n)_{n\in\mathbb{N}}$ of pairwise disjoint elements of Γ with $\bigcup_{n=1}^{\infty} E_n \in \Gamma$, such that ν cannot be extended to a signed measure defined on the σ-algebra generated by Γ. (Hint. $\mathbb{Z}$ with $\pm$ counting measure.)

2. Let ν be a signed measure on the σ-algebra Σ. Put

$$\beta := \inf\{\nu(E) \mid E \in \Sigma,\ E \text{ strongly negative}\} .$$

If $(E_n)_{n\in\mathbb{N}}$ is a sequence of strongly negative elements of Σ with $\nu(E_n) \downarrow \beta$, then $\bigcup_{n=1}^{\infty} E_n$ is strongly negative and $\nu(\bigcup_{n=1}^{\infty} E_n) = \beta$.

3. This exercise is meant to establish the result of Proposition 5.1.6 without using the Hahn decomposition theorem. Let ν be a signed measure on Σ.

(i) Let $E \in \Sigma$, $|\nu(E)| < \infty$. Show that $\Upsilon_{F \in \Sigma, F \subset E}\, \nu(F)$ is bounded as follows. Suppose that $\Upsilon_{F \in E, F \subset E}\, \nu(F)$ is unbounded.

(a) Show that there is an $F \in \Sigma$, $F \subset E$ such that $\nu(F) > 1$, $\Upsilon_{G \in \Sigma, G \subset E \setminus F}\, \nu(G)$ unbounded. (Hint. If such an F does not exist, one can find $F_n \in \Sigma$ with $\nu(F_1) > 1$, $\nu(F_{n+1}) > \nu(F_n) + 1$, $E \supset F_n \supset F_{n+1}$. Now if $F = \bigcap_{n=1}^{\infty} F_n$, then

$$\nu(F_1) = \nu(F) + \sum_{n=1}^{\infty} \nu(F_n \setminus F_{n+1}) = -\infty \, .)$$

(b) Show that one can find $E_n \in \Sigma$, $E_n \subset E$ such that $\nu(E_n) > 1$ and E_n pairwise disjoint. Derive a contradiction.

(ii) Define for $E \in \Sigma$

$$\nu_+(E) := \sup\{\nu(F) \mid F \in \Sigma,\ F \subset E\} \, ,$$

$$\nu_-(E) := \sup\{-\nu(F) \mid F \in \Sigma,\ F \subset E\} \, .$$

Show that ν_+ and ν_- are measures on Σ.

(iii) Let $E \in \Sigma$. Show there is an $F \subset E$ such that $\nu_+(E) = \nu(F)$. (Hint. Let $(F_n)_{n \in \mathbb{N}}$ be a sequence in Σ with $F_n \subset E$ such that $\nu(F_n) \geq \nu_+(E) - 2^{-n}$ if $\nu_+(E) < \infty$ and $\nu(F_n) \geq 1 + \nu(\bigcup_{k=1}^{n-1} F_k)$ if $\nu_+(E) = \infty$. Consider $\bigcap_{m=1}^{\infty} \bigcup_{n=m}^{\infty} F_n$.)

(iv) Show that for $E \in \Sigma$ at least one of the numbers $\nu_+(E)$ and $\nu_-(E)$ is finite, and show that $\nu = \nu_+ - \nu_-$.

(v) Assume that $\nu(E) < \infty$ $(E \in \Sigma)$. Show that ν_+ is bounded.

(vi) Show that ν_+ and ν_- are mutually singular. (Hint. Assume $\nu(E) < \infty$ $(E \in \Sigma)$, and take an $F \in \Sigma$ with maximal value of $\nu_+(F)$.)

4. Let $-\infty < a < b < \infty$, and let $g : (a,b] \to \mathbb{R}$ have bounded variation. Show that there is exactly one signed measure ν defined on the σ-algebra Σ of Borel sets contained in $(a,b]$ such that

$$(*) \qquad \nu((\alpha,\beta]) = g(\beta + 0) - g(\alpha + 0)$$

for $a \leq \alpha \leq \beta \leq b$ (we have $g(b + 0) := g(b)$). Conversely, if ν is a finite signed measure on Σ, then there is a $g : (a,b] \to \mathbb{R}$ of bounded variation such that $(*)$ holds for $a \leq \alpha \leq \beta \leq b$.

5. Let S be a set, Γ a semiring of subsets of S, and let ν_1, ν_2 be two σ-finite signed measures defined on the σ-algebra Σ generated by Γ. If $\nu_1(A) = \nu_2(A)$ $(A \in \Gamma)$, then $\nu_1(A) = \nu_2(A)$ $(A \in \Sigma)$.

5.2. The Radon-Nikodym theorem

Given a measure μ on a σ-algebra Σ and a μ-integrable function f, we constructed in Example 5.1.3 a signed measure ν on Σ by putting

$$(*) \qquad \nu(E) := \int_E f\,d\mu \qquad (E \in \Sigma) .$$

In this section we seek a converse: given a measure μ on a σ-algebra Σ and a signed measure ν on Σ, can ν be expressed by (*) for some f? We show first that under a pretty mild assumption ν can be decomposed in a unique way as $\nu = \nu_a + \nu_s$, where ν_a is absolutely continuous with respect to μ, while ν_s and μ are mutually singular. As suggested by the terminology, there is nothing that can be done to ν_s in terms of μ: ν_s and μ are completely unrelated. However, the main result of the section says that ν_a may be expressed as in (*); this is the Radon-Nikodym theorem.

In most of this section we can restrict our attention to measures, because of the Jordan decomposition theorem 5.1.6.

5.2.1. We have already met the notion of absolute continuity in a special context. Now we give the definition in a more general situation.

Let Σ be a σ-algebra of subsets of S, let μ be a measure on Σ, and let ν be a signed measure on Σ. Then ν is called *absolutely continuous with respect to* μ (written $\nu \ll \mu$) if $\nu(E) = 0$ whenever $E \in \Sigma$ and $\mu(E) = 0$.

5.2.2. Our first result is another decomposition theorem.

Proposition (Lebesgue decomposition theorem). Let Σ be a σ-algebra of subsets of S, let μ and ν be measures on Σ, and let ν be σ-finite. Then there exists a unique pair ν_a, ν_s of measures on Σ such that $\nu = \nu_a + \nu_s$, $\nu_a \ll \mu$, ν_s and μ mutually singular.

Proof. Let $(E_n)_{n\in\mathbb{N}}$ be a pairwise disjoint sequence in Σ such that $S = \bigcup_{n=1}^\infty E_n$, $\nu(E_n) < \infty$ $(n \in \mathbb{N})$. For each $n \in \mathbb{N}$, let $\nu_n(E) := \nu(E \cap E_n)$ $(E \in \Sigma)$. Obviously, the ν_n's are finite measures on Σ.

Now let $n \in \mathbb{N}$ be fixed. Put

$$\Sigma_n := \{E \in \Sigma \mid E \subset E_n,\ \mu(E) = 0\} \quad \text{and} \quad \alpha_n := \sup\{\nu_n(F) \mid F \in \Sigma_n\} .$$

Choose a sequence $(F_{nm})_{m\in\mathbb{N}}$ in Σ_n such that $\nu_n(F_{nm}) \to \alpha_n$ as $m \to \infty$. Put $F_n := \bigcup_{m=1}^\infty F_{nm}$. Since $\nu_n(F_{nm}) \le \nu_n(F_n)$ for each m, we have $\nu_n(F_n) \ge \alpha_n$, and since $F_n \in \Sigma_n$, we even have $\nu_n(F_n) = \alpha_n$. The point is, of course, that now E_n has been split into sets F_n and $E_n \backslash F_n$ where $\mu(F_n) = \nu_n(E_n \backslash F_n) = 0$.

Let $F := \bigcup_{n=1}^\infty F_n$, $G := S\backslash F$, and define the measures ν_a and ν_s by

$$\nu_a(E) := \nu(E \cap G) , \quad \nu_s(E) := \nu(E \cap F) \qquad (E \in \Sigma) .$$

To show that ν_a and ν_s satisfy the conditions, note first that $\mu(F) = 0$ and $\nu_s(G) = 0$, so that μ and ν_s are mutually singular. To show that $\nu_a \ll \mu$, let $E \in \Sigma$, $\mu(E) = 0$. For each $n \in \mathbb{N}$, $F_n \cup (E \cap E_n \cap G) \in \Sigma_n$, so

$$\alpha_n \geq \nu_n(F_n \cup (E \cap E_n \cap G)) = \nu_n(F_n) + \nu_n(E \cap E_n \cap G) =$$

$$= \alpha_n + \nu_n(E \cap E_n \cap G) ,$$

that is, $\nu_n(E \cap E_n \cap G) = 0$. Hence

$$\nu_a(G) = \nu(E \cap G) = \sum_{n=1}^{\infty} \nu(E \cap E_n \cap G) = \sum_{n=1}^{\infty} \nu_n(E \cap E_n \cap G) = 0 .$$

Since the relation $\nu = \nu_a + \nu_s$ is trivial, we have shown the existence of a decomposition of the kind required.

The uniqueness proof is left to the reader as an exercise. □

5.2.3. The absolutely continuous part in the Lebesgue decomposition which we just derived can be decomposed further (under an additional σ-finiteness assumption) by means of the next result, known as the Radon-Nikodym theorem. The theorem is not only theoretically important, but has many applications as well, one of which will be treated in the next section and another one in Section 6.5.

Theorem (Radon-Nikodym). Let Σ be a σ-algebra of subsets of S, let μ and ν be σ-finite measures on Σ with $\nu \ll \mu$. Then there exists a non-negative real-valued Σ-measurable function h such that

$$(*) \qquad \nu(E) = \int_E h\,d\mu \qquad (E \in \Sigma) .$$

If h_1 is another function with the same properties, then $h = h_1$ a.e. with respect to μ.

Proof. First assume that both μ and ν are finite measures. Let H be the class of Σ-measurable non-negative $\mathbb{R}^*$-valued functions that satisfy

$$\int_E h\,d\mu \leq \nu(E) \qquad (E \in \Sigma) ,$$

and let $\alpha := \sup\{\int h\,d\mu \mid h \in H\}$. Call an $h \in H$ with $\int h\,d\mu = \alpha$ an extremal function. Since $\alpha \leq \nu(S)$, each extremal function is integrable. We shall show that extremal functions actually exist, and that any extremal function satisfies (*).

In the first place, let $h_1, h_2 \in H$, and $h := \sup(h_1, h_2)$. Put

$$A := \{s \in S \mid h_1(s) \geq h_2(s)\} \quad \text{and} \quad B := S \backslash A .$$

Both A and B are members of Σ, and for each $E \in \Sigma$ we have

$$\int_E h\,d\mu = \int_{A\cap E} h_1\,d\mu + \int_{B\cap E} h_2\,d\mu \le \nu(A \cap E) + \nu(B \cap E) = \nu(E) .$$

Hence $h \in H$. Also, if $(h_n)_{n\in\mathbb{N}}$ is a pointwise increasing sequence in H tending to a function h, then the monotone convergence theorem shows that

$$\int_E h\,d\mu = \lim_{n\to\infty} \int_E h_n\,d\mu \le \nu(E)$$

for any $E \in \Sigma$, and therefore $h \in H$.

Now choose a sequence $(g_n)_{n\in\mathbb{N}}$ in H such that $\int g_n\,d\mu \to \alpha$, and put $h_n := \sup(g_1,\ldots,g_n)$ $(n \in \mathbb{N})$. Repeated application of the first part of the preceding paragraph shows that the h_n belong to H. Hence they increase to an $h \in H$, and the inequalities

$$\int g_n\,d\mu \le \int h_n\,d\mu \le \int h\,d\mu \le \alpha \qquad (n \in \mathbb{N})$$

show that $\int h\,d\mu = \alpha$. We conclude that extremal functions do exist.

Let h be any extremal function. Define $\sigma(E) := \nu(E) - \int_E h\,d\mu$ for $E \in \Sigma$. Clearly σ is a measure. We assert that σ is the zero measure, that is, h does what we want it to do. Assume σ is not the zero measure. Choose $F \in \Sigma$ with $\sigma(F) > 0$. Then there exists $\beta > 0$ with $\sigma(F) > \beta\mu(F)$. Let σ_0 be the signed measure $\sigma - \beta\mu$, and choose a Hahn decomposition of S with respect to σ_0: $S = A \cup B$, say, where A is the positive set. Put $F_0 := F \cap A$. Then $\mu(F_0) > 0$. For, $\mu(F_0) = 0$ would imply $\nu(F_0) = 0$, so $\sigma(F_0) = 0$, and therefore $\sigma(F) = \sigma(F\setminus F_0) + \sigma(F_0) \le 0$. The function $h + \beta\chi_{F_0}$ is non-negative and Σ-measurable. Also, if $E \in \Sigma$, then

$$\int_E (h + \beta\chi_{F_0})\,d\mu = \int_{E\cap F_0} (h + \beta)\,d\mu + \int_{E\cap(S\setminus F_0)} h\,d\mu$$

$$\le \nu(E \cap F_0) - \sigma(E \cap F_0) + \beta\mu(E \cap F_0) + \nu(E \cap (S\setminus F_0))$$

$$\le \nu(E \cap F_0) + \nu(E \cap (S\setminus F_0)) = \nu(E) .$$

Hence $h + \beta\chi_{F_0} \in H$. But now $\int (h + \beta\chi_{F_0})\,d\mu = \alpha + \beta\mu(F_0) > \alpha$, which contradicts the definition of α. So σ is the zero measure, as asserted, and ν is represented by h as in (*). Clearly we may choose a real-valued extremal function.

Now that the case of finite measures is finished, the rest is easy. Let ν and μ be σ-finite measures. Write $S = \bigcup_{n=1}^{\infty} E_n$ where the E_n are pairwise disjoint elements of Σ with $\nu(E_n)$ and $\mu(E_n)$ finite for all n. For each $n \in \mathbb{N}$, let

$$\mu_n(E) := \mu(E \cap E_n) , \quad \nu_n(E) := \nu(E \cap E_n) \qquad (E \in \Sigma) .$$

Obviously, μ_n and ν_n are finite measures on Σ with $\nu_n \ll \mu_n$ for all $n \in \mathbb{N}$. Hence, there exist Σ-measurable real-valued and non-negative functions h_n such that

$$\nu_n(E) = \int_E h_n \, d\mu_n \qquad (n \in \mathbb{N},\ E \in \Sigma) ,$$

and the monotone convergence theorem shows that $h := \sum_{n=1}^{\infty} h_n \cdot \chi_{E_n}$ is a function with property (*).

The proof that h is unique up to null functions is easy. □

5.2.4. The Radon-Nikodym theorem for the case that ν is a signed measure follows from the Jordan decomposition theorem. Although the formulation of the result is rather lengthy, the proof is short.

Corollary. Let Σ be a σ-algebra of subsets of S, let μ be a σ-finite measure and let ν be a σ-finite signed measure on Σ, $\nu \ll \mu$. Let $\nu = \nu_1 - \nu_2$ be the Jordan decomposition of ν. Then $\nu_1 \ll \mu$ and $\nu_2 \ll \mu$, and there exist non-negative real-valued and Σ-measurable functions h_1 and h_2, unique up to null functions, such that

$$\nu_i(E) = \int_E h_i \, d\mu \qquad (E \in \Sigma,\ i = 1,2) .$$

Also, h_1 or h_2 is μ-integrable according as ν does not take the value ∞ or $-\infty$. If ν is finite, then both h_1 and h_2 are μ-integrable.

Proof. We need only prove that ν_1 and ν_2 are σ-finite measures, absolutely continuous with respect to μ, and this follows immediately from the construction of ν_1 and ν_2 in the proof of 5.1.6. □

5.2.5. The equality (*) in the Radon-Nikodym theorem 5.2.3 may also be written as $\int f \, d\nu = \int f \cdot h \, d\mu$ for those f that are characteristic functions of sets in Σ. The following result is concerned with extensions of this relationship to more general functions f.

Theorem. Let Σ, μ, ν and h be as in 5.2.3. As usual, let L_μ be the space of real-valued functions integrable with respect to μ, and let M_μ be the corresponding space of measurable $\mathbb{R}^*$-valued functions. Let L_ν and M_ν be defined similarly. Then the following assertions hold.

(i) $M_\mu \subset M_\nu$.

(ii) If $f \in M_\nu$ then $f \cdot h \in M_\mu$.

(iii) If $f \in M_\nu^+$, then

(**) $$\int f\,d\nu = \int f\cdot h\,d\mu \ .$$

(iv) If f is a real-valued function on S, then $f \in L_\nu$ if and only if $f\cdot h \in L_\mu$. Also, if $f \in L_\nu$, then (**) holds.

Proof. Let Σ_μ be the σ-algebra of μ-measurable subsets of S (see 3.2.10), and let Σ_ν be defined similarly. Let $P := \{s \in S \mid h(s) > 0\}$. As a preliminary result we show that if $E \in \Sigma_\nu$ and $\nu(E) = 0$, then $E \cap P \in \Sigma_\mu$ and $\mu(E \cap P) = 0$. Take such a set E. By Theorem 3.3.6 there exists $F \in \Sigma$ such that $E \subset F$ and $\nu(F) = 0$. If $\mu(F \cap P) > 0$, then $\nu(F \cap P) = \int_{F\cap P} h\,d\mu > 0$, which is impossible. Hence $0 = \mu(F \cap P) \geq \mu(E \cap P) \geq 0$, and therefore $E \cap P \in \Sigma_\mu$ and $\mu(E \cap P) = 0$.

(i) Let $f \in M_\mu$. By Theorem 3.3.8 there exists a Σ-measurable g such that f = g except possibly on a set of μ-measure zero. Since $\nu << \mu$, this means that f = g almost everywhere with respect to ν. Hence, since $g \in M_\nu$ by Theorem 3.3.4(ii), $f \in M_\nu$.

(ii) Now let $f \in M_\nu$. Again by Theorem 3.3.8 there exists a Σ-measurable g such that f = g almost everywhere with respect to ν. Let $E := \{s \in S \mid f(s) \neq g(s)\}$. Then $\nu(E) = 0$. Therefore $f\cdot h = g\cdot h$ except on $E \cap P$ (recall that $0\cdot\infty = 0$), where $\mu(E \cap P) = 0$. Since $g\cdot h$ is Σ-measurable, we have $f\cdot h \in \Sigma_\mu$.

(iii) First, let g be a simple function of the form $\sum_{k=1}^n \alpha_k \chi_{E_k}$ where $\alpha_k \geq 0$, $E_k \in \Sigma$ $(1 \leq k \leq n)$. We know already that $\int \chi_{E_k} d\nu = \int \chi_{E_k}\cdot h\,d\mu$ for each k, and therefore $\int g\,d\nu = \int g\cdot h\,d\mu$ for such special functions g.

If $f \in M_\nu^+$, then there exists a sequence $(g_n)_{n\in\mathbb{N}}$ of simple functions such that $g_n \uparrow f$ a.e. with respect to μ; this follows from the result of 2.1.13. As in (ii) we see that $g_n\cdot h \uparrow f\cdot h$ (a.e. with respect to μ), and two applications of the monotone convergence theorem yield

$$\int f\,d\nu = \lim_{n\to\infty} \int g_n\,d\nu = \lim_{n\to\infty} \int g_n\cdot h\,d\mu = \int f\cdot h\,d\mu \ .$$

(iv) The last assertion from the theorem follows from (iii), first for functions f that are non-negative, and then for arbitrary functions by the decomposition $f = \sup(f,0) - \sup(-f,0)$. □

Exercises Section 5.2

1. Show that ν_a and ν_s in the decomposition of ν (Proposition 5.2.2) are uniquely determined by ν.

2. Take $S = \mathbb{R}$, Σ the σ-algebra of Borel sets, μ ordinary Lebesgue measure, $\nu(A)$ = the number of elements of A or ∞ according as $A \in \Sigma$ is finite or not. Show that this ν has no decomposition as in Proposition 5.2.2.

3. Let ν be a finite signed measure on Σ. Show that $\nu \ll \mu$ if and only if for every $\varepsilon > 0$ there is a $\delta > 0$ such that $|\nu(E)| < \varepsilon$ for every $E \in \Sigma$ with $\mu(E) < \delta$. (Hint. Use Theorem 5.2.4.)

4. Take $S = \mathbb{R}$, Σ the σ-algebra of Borel sets, μ ordinary Lebesgue measure, and let ν be the Lebesgue-Stieltjes measure generated by $\Upsilon_x\, x^3$. Show that $\nu \ll \mu$. Is it true that for every $\varepsilon > 0$ there is a $\delta > 0$ such that $|\nu(E)| < \varepsilon$ for every $E \in \Sigma$ with $\mu(E) < \delta$?

5. Take $S = \mathbb{R}$, Σ the σ-algebra of Borel sets, $\mu(A)$ = the number of elements of A or ∞ according as A is finite or not, ν ordinary Lebesgue measure. Show that $\nu \ll \mu$. Does there exist a Σ-measurable function h such that $\nu(E) = {}_E\!\int h\,d\mu$ for all $E \in \Sigma$?

6. Let ν be a σ-finite measure on Σ, and let ν_a, ν_s and h be as in Proposition 5.2.2 and Theorem 5.2.3. Show that

$$\int f\,d\nu = \int f\cdot h\,d\mu + \int f\,d\nu_s$$

for $f \in L_\nu$.

7. Assume $\mu(S) < \infty$ and let ν be a (possibly non σ-finite) measure on (S,Σ) with $\nu \ll \mu$.

(i) Show there is an $E \in \Sigma$ such that

(a) ν_0 is a σ-finite measure on Σ_0. Here Σ_0 is the induced σ-algebra on E, and ν_0 is the induced measure (cf. Exercise 3.1-10),

(b) if $F \in \Sigma$, $F \cap E = \emptyset$ then $\nu(F) = 0$ or $\nu(F) = \infty$,

(c) if $F \in \Sigma$, $F \cap E = \emptyset$ then $\nu(F) = 0 \Rightarrow \mu(F) = 0$. (Hint. Let $\Sigma_1 := \{E \in \Sigma \mid E$ has σ-finite measure with respect to $\nu\}$. Show that Σ_1 is a σ-ring. Let $\alpha := \sup\{\mu(E) \mid E \in \Sigma_1\}$, and show that there is an $E \in \Sigma_1$ such that $\mu(E) = \alpha$. This E satisfies the conditions.)

(ii) Show that there is a Σ-measurable function $f \geq 0$ such that $\nu(A) = {}_A\!\int f\,d\mu$ $(A \in \Sigma)$.

8. Take $S = \mathbb{R}$, Σ the σ-algebra of Borel sets of $\mathbb{R}$, μ ordinary Lebesgue measure. Let ν be a σ-finite measure on Σ, and write $\nu = \nu_a + \nu_s$ with $\nu_a \ll \mu$, ν_s and μ mutually singular. Show there are a non-decreasing continuous function $g : \mathbb{R} \to \mathbb{R}$ and sequences $(\alpha_n)_{n\in\mathbb{N}}$, $(a_n)_{n\in\mathbb{N}}$ with $\alpha_n \geq 0$, $a_n \in \mathbb{R}$, such that

$$\nu_s(A) = \mu_g(A) + \sum_{a_n\in A} \alpha_n \qquad (A \in \Sigma) .$$

Here μ_g denotes the Lebesgue-Stieltjes measure generated by g.

9. We indicate an alternative proof of Theorems 5.2.2 and 5.2.3 for the case that both μ and ν are finite. (With some effort the σ-finite case can be handled as well; we leave the details to the reader.)

(i) Show that for every $r \in \mathbb{Q}$ there is a $W_r \in \Sigma$ such that

$$\nu(E) \leq r\mu(E) \qquad (E \in \Sigma,\ E \subset W_r) ,$$

$$\nu(E) \geq r\mu(E) \qquad (E \in \Sigma,\ E \subset S\backslash W_r) .$$

(Hint. Apply the Hahn decomposition theorem to $\gamma_{E\in\Sigma}(\nu(E) - r\mu(E))$.)

(ii) For $u \in \mathbb{R}$, put

$$F(u) := \bigcup_{r\in\mathbb{Q},r<u} W_r .$$

Show that for $u \in \mathbb{R}$, $v \in \mathbb{R}$ one has

(a) if $E \in \Sigma$, $E \subset F(u)$ then $\nu(E) \leq u\mu(E)$,

(b) if $E \in \Sigma$, $E \subset S\backslash F(u)$ then $\nu(E) \geq u\mu(E)$,

(c) if $u < v$ then $F(u) \subset F(v)$.

(iii) Put

$$N_1 := \bigcap_{u\in\mathbb{R}} F(u) , \quad N_2 := S \setminus \bigcup_{u\in\mathbb{R}} F(u) .$$

Show that $N_1 \in \Sigma$, $N_2 \in \Sigma$, $\mu(N_1) = \mu(N_2) = 0$, $\nu(N_1) \leq 0 \leq \nu(N_2)$.

(iv) For $k \in \mathbb{N}$, put

$$B_{nk} := F\left(\frac{n+1}{k}\right) \setminus F\left(\frac{n}{k}\right) \qquad (n \in \mathbb{Z}),$$

and let

$$t_k := \sum_{n=-\infty}^{\infty} \frac{n}{k} \chi_{B_{nk}} .$$

Show that t_k is Σ-measurable and integrable with respect to μ for $k \in \mathbb{N}$.

(v) Let $N := N_1 \cup N_2$, and for $s \in S$ put

$$f(s) = \begin{cases} \inf\{u \in \mathbb{R} \mid s \in F(u)\} & (s \in S\setminus N) , \\ 0 & (s \in N) . \end{cases}$$

Show that $t_k \leq f \leq t_k + 1/k$ $(k \in \mathbb{N})$, that f is Σ-measurable and that f is integrable with respect to μ.

(vi) Show that

$$\nu(E) = \int_E f\,d\mu + \nu(E \cap N) \qquad (E \in \Sigma) .$$

(Hint. For $k \in \mathbb{N}$ write $E \in \Sigma$ as a union of $E \cap N$ and $\bigcup_{n=-\infty}^{\infty}(E \cap B_{nk})$, use Lebesgue's theorem on dominated convergence and let $k \to \infty$.)

5.3. Continuous linear functionals on L^p-spaces

As an important and interesting application of the Radon-Nikodym theorem we derive a representation theorem for continuous linear functionals on L^p-spaces.

5.3.1. In all of this section (S,Σ,μ) is a σ-finite measure space, with Σ a σ-algebra of subsets of S (not merely a semiring), and p is any number in $[1,\infty)$. The space L^p consists of equivalence classes of functions, which we treat as functions (see the remarks in 2.3.12). With the p-norm as norm, L^p is a Banach space (2.3.11). As we know, a linear functional L on L^p is a real-valued function on L^p such that

$$L(\alpha f + \beta g) = \alpha L(f) + \beta L(g) \qquad (\alpha,\beta \in \mathbb{R} \text{ and } f,g \in L^p) .$$

Such a linear functional L is called *continuous* if $L(f_n) \to 0$ for every sequence $(f_n)_{n\in\mathbb{N}}$ in L^p that satisfies $\|f_n\|_p \to 0$.

We want to describe the continuous linear functionals on L^p in a concrete way, where the meaning of "concrete" is exemplified by the Riesz representation theorem 4.3.3, which describes certain positive linear functionals in terms of regular Borel measures. First we treat an example.

5.3.2. Example. Let q be the conjugate exponent of p as defined in 2.3.2. Let $g \in L^q$, and define

$$L(f) := \int f\cdot g\,d\mu \qquad (f \in L^p) .$$

Obviously L is a linear functional on L^p, and Hölder's inequality 2.3.8 shows that

$$(\dagger) \qquad |L(f)| \le \|g\|_q \cdot \|f\|_p \qquad (f \in L^p) ,$$

which implies that L is continuous.

The interesting point in the construction is that by varying the function g one obtains all continuous linear functionals on L^p. In other words: here is a concrete description of such functionals.

5.3.3. Now we show that the continuous linear functions on L^p are characterized by an inequality as (†) in 5.3.2. (For this reason these linear functionals are often called *bounded* instead of continuous.)

Proposition. Let L be a linear functional on L^p. Then L is continuous if and only if there exists $M > 0$ such that $|L(f)| \le M\|f\|_p$ for all $f \in L^p$.

Proof. The sufficiency of the condition is clear. To prove necessity, argue by contradiction. Assume that L is continuous and that no such M exists. Then for each $n \in \mathbb{N}$ there exists $f_n \in L^p$ with $|L(f_n)| > n\|f_n\|_p$. Put $g_n := (n\|f_n\|_p)^{-1} f_n$, then obviously $|L(g_n)| > 1$ for all n. On the other hand, $\|g_n\|_p = 1/n \to 0$ as $n \to \infty$, so $L(g_n) \to 0$ as $n \to \infty$: contradiction. □

5.3.4. Theorem. Let $1 \le p < \infty$, and let L be a continuous linear functional on L^p. Then there exists a unique $g \in L^q$ such that

$$(*) \qquad Lf = \int f \cdot g \, d\mu \qquad (f \in L^p) .$$

Proof. Write $S = \bigcup_{m=1}^{\infty} S_m$, where $S_m \in \Sigma$, $S_m \subset S_{m+1}$, $\mu(S_m) < \infty$ for all m. For each $m \in \mathbb{N}$ we are going to define an auxiliary signed measure ν_m on Σ to which the Radon-Nikodym theorem will be applied.

Fix $m \in \mathbb{N}$. If $A \in \Sigma$, then $A \cap S_m$ has finite μ-measure, so its characteristic function is in L^p. Hence

$$\nu_m(A) := L(\chi_{A \cap S_m}) \qquad (A \in \Sigma)$$

gives a real-valued set function ν_m on Σ. This ν_m is a signed measure which is absolutely continuous with respect to μ. For, let $(A_n)_{n \in \mathbb{N}}$ be a sequence of disjoint elements of Σ, let $B_N := \bigcup_{n=1}^{N} A_n$ $(N \in \mathbb{N})$, and $A := \bigcup_{n=1}^{\infty} A_n$. Since $\|\chi_{B_N \cap S_m} - \chi_{A \cap S_m}\|_p \to 0$ by dominated convergence, and L is linear,

$$\nu_m(A) = L(\chi_{A \cap S_m}) = \lim_{N \to \infty} L(\chi_{B_N \cap S_m}) = \lim_{N \to \infty} \sum_{n=1}^{N} L(\chi_{A_n \cap S_m}) = \sum_{n=1}^{\infty} \nu_m(A_n) .$$

Also, if $A \in \Sigma$ and $\mu(A) = 0$, then $\mu(A \cap S_m) = 0$ and therefore $\|\chi_{A \cap S_m}\|_p = 0$, which implies $\nu_m(A) = 0$. In particular, $\nu_m(\emptyset) = 0$. By Corollary 5.2.4 to the Radon-Nikodym theorem there is a real-valued, Σ-measurable and μ-integrable g_m such that

$$(**)\qquad \nu_m(A) = \int_A g_m \, d\mu \qquad (A \in \Sigma) .$$

The function g will be built from the g_m's. Note first that if $n > m$, then g_n and g_m coincide μ-almost everywhere on S_m in the sense that $g_m = g_n \cdot \chi_{S_m}$ (μ-a.e.). In fact

$$\nu_m(A) = L(\chi_{A\cap S_m}) = L(\chi_{A\cap S_m \cap S_n}) = \int_{A\cap S_m} g_n \, d\mu = \int_A g_n \cdot \chi_{S_m} \, d\mu \qquad (A \in \Sigma) ,$$

so $g_n \cdot \chi_{S_m}$ does the same as g_m in representing ν_m. According to 5.2.4 this means that g_m and $g_n \cdot \chi_{S_m}$ differ by a null function. Now let

$$D_m := \{s \in S_m \mid g_m(s) \neq g_n(s) \text{ for some } n > m\} .$$

Then $\mu(D_m) = 0$. Let $D := \bigcup_{m=1}^{\infty} D_m$. Then $\mu(D) = 0$, and for each $s \in S_k \setminus D$ the terms of the sequence $(g_m(s))_{m\in\mathbb{N}}$ are eventually constant. Hence, if g is defined by

$$g(s) := \begin{cases} \lim_{m\to\infty} g_m(s) & (s \in S\setminus D) , \\ 0 & (s \in D) , \end{cases}$$

then g is Σ-measurable, and $g \cdot \chi_{S_m} = g_m$ (a.e.) for every $m \in \mathbb{N}$.

The proof that g satisfies (*) is in several steps. First, let $A \in \Sigma$, $\mu(A) < \infty$, and assume $g(s) \geq 0$ on A. Then $\chi_A \in L^p$ and

$$L(\chi_{A\cap S_m}) = \nu_m(A \cap S_m) = \int_{A\cap S_m} g_m \, d\mu = \int g \cdot \chi_{S_m} \cdot \chi_A \, d\mu \qquad (m \in \mathbb{N}) .$$

By monotone convergence,

$$\int g \cdot \chi_{S_m} \cdot \chi_A \, d\mu \to \int g \cdot \chi_A \, d\mu ,$$

$$\|\chi_{A\cap S_m} - \chi_A\|_p \to 0 ,$$

as $m \to \infty$. Hence, by the continuity of L, (*) follows for χ_A. The same proof works for an $A \in \Sigma$ with $\mu(A) < \infty$ and $g(s) < 0$ on A. Combination of the two results shows that (*) holds for all χ_A with $A \in \Sigma$, $\mu(A) < \infty$. Next, let f be a basic function, that is, $f = \sum_{k=1}^{n} \alpha_k \chi_{A_k}$ where $\alpha_k \in \mathbb{R}$, $A_k \in \Sigma$, $\mu(A_k) < \infty$ $(1 \leq k \leq n)$. Then $f \in L^p$, and (*) follows for this f from the preceding step by the linearity of L and of the integral. Finally, suppose that we already know that $g \in L^q$. If f is any member of L^p, then by 2.3.13 there exists a sequence $(f_n)_{n\in\mathbb{N}}$ of basic functions such that $\|f_n - f\|_p \to 0$. Since L is continuous and linear, this means that $Lf_n \to Lf$. On the other hand,

Hölder's inequality implies $\int f_n \cdot g\,d\mu \to \int f \cdot g\,d\mu$. Therefore (*) follows for this general f.

So we are left with the task of proving $g \in L^q$. This is the trickiest part of the proof. Let M be such that $|L(f)| \le M\|f\|_p$ for all $f \in L^p$. We shall show that $\|g\|_q \le M$. The arguments for the cases $p > 1$ and $p = 1$ are somewhat different.

If $p > 1$, let $(f_n)_{n\in\mathbb{N}}$ be a sequence of basic functions such that $0 \le f_n \le f_{n+1}$ $(n \in \mathbb{N})$, $f_n \to |g|^q$ (see 2.1.13). For each $n \in \mathbb{N}$, let $h_n := f_n^{1/p}\,\mathrm{sgn}(g)$, where sgn(g) is the function that takes at $s \in S$ the value 1 if $g(s) > 0$, -1 if $g(s) < 0$, and 0 otherwise. Each h_n is a basic function, with $\|h_n\|_p = \|f_n\|_1^{1/p}$, and

$$\int h_n \cdot g\,d\mu = L(h_n) \le M\|h_n\|_p = M\|f_n\|_1^{1/p}$$

by (*) applied to the basic function h_n. Now

$$f_n = f_n^{1/p} \cdot f_n^{1/q} \le f_n^{1/p} \cdot |g| = h_n \cdot g$$

implies

$$\|f_n\|_1 = \int h_n \cdot g\,d\mu \le M\|f_n\|_1^{1/p} ,$$

and therefore $\|f_n\|_1 \le M^q$ for all n. Since $f_n \to |g|^q$, Fatou's lemma 2.2.12 shows that $\||g|^q\|_1 \le M^q$, that is, $\|g\|_q \le M$.

If $p = 1$, we have $q = \infty$, so we must show that g is essentially bounded. Let $\alpha > M$ and put $A_\alpha := \{s \in S \mid g(s) > \alpha\}$. Suppose $\mu(A_\alpha) > 0$ and take an $E \in \Sigma$ with $E \subset A_\alpha$, $0 < \mu(E) < \infty$. If $f := \chi_E$, then $f \in L$ and

$$L(f) = \int_E g\,d\mu \ge \alpha \int_E d\mu = \alpha \int f\,d\mu > M\|f\|_1 ,$$

which is a contradiction. Therefore $\mu(A_\alpha) = 0$. Since α is arbitrary, it follows that $\{s \in S \mid g(s) > M\}$ is a null set. Similarly, $\{s \in S \mid g(s) < -M\}$ is a null set. So $\|g\|_\infty \le M$.

The proof that there is only one $g \in L^q$ with the assigned properties is left to the reader. □

5.3.5. <u>Remark</u>. The condition $p < \infty$ in the theorem is essential. In fact, using the axiom of choice one can prove the existence of continuous linear functionals on L^∞ (with [0,1] in the role of S and μ ordinary Lebesgue measure) that cannot be written as $\forall_f(\int f \cdot g\,d\mu)$ for any $g \in L^1$. The interested reader is referred to Exercise 5.3-3.

Exercises Section 5.3

1. Show that the g of Theorem 5.3.4 is uniquely determined by L.

2. Let $1 < p \leq \infty$ and let $g \in L^q$. Show that there is an $f \in L^p$, $f \neq 0$ such that $\int f \cdot g \, d\mu = \|f\|_p \cdot \|g\|_q$. Is this always true if $p = 1$?

3. Let $S = [0,1]$, Σ the Borel sets contained in $[0,1]$ and μ ordinary Lebesgue measure. Define $L(f) := f(0)$ for every continuous function f defined on $[0,1]$. This L can be extended by the Hahn-Banach theorem (see for instance, A.L. Brown, A. Page, Elements of Functional Analysis (Van Nostrand Reinhold, London, 1970), Sect. 8.3) to a continuous linear functional L_0 of L^∞. Show that there is no $g \in L^1$ such that $L_0(f) = \int f \cdot g \, d\mu$ $(f \in L^\infty)$.

4. Let Σ_0 be a σ-algebra contained in Σ, and let f be Σ-measurable and integrable with respect to μ. Show that there exists a Σ_0-measurable function f_0, integrable with respect to μ, such that

$$\int_A f_0 \, d\mu = \int_A f \, d\mu \qquad (A \in \Sigma_0) .$$

This f_0 is unique, apart from Σ_0-measurable null functions. (If $\mu(S) = 1$ (probability space) this f_0 is called the *conditional expectation* of f with respect to Σ_0.)

(i) Show that $\|f_0\|_p \leq \|f\|_p$ for $f \in L^p$ $(1 \leq p \leq \infty)$.

(ii) Let g be Σ_0-measurable, $f \in L^p$, $g \in L^q$ (with $1 \leq p \leq \infty$). Show that $(f \cdot g)_0 = f_0 \cdot g$.

(iii) Determine f_0 if $S = (0,1]$, $n \in \mathbb{N}$, Σ_0 the σ-algebra generated by the semiring

$$\Gamma := \left\{ \left(\frac{k}{n}, \frac{k+1}{n} \right] \;\middle|\; k = 0, \ldots, n-1 \right\} ,$$

Σ the σ-algebra of Borel sets and f Lebesgue integrable over $(0,1]$.

(iv) Let $S_0 \in \Sigma$, and let Σ_0 be the σ-algebra generated by $\{E \in \Sigma \mid E \subset S_0\}$. Determine f_0 if f is integrable over S.

5.4. Lebesgue integration on $\mathbb{R}$ as the inverse of differentiation

We now derive some classical results on the relation between integration and differentiation on the real line. The first one is about functions of bounded variation on a bounded interval $[a,b]$. We show that such functions are differentiable a.e. on $[a,b]$ with a Lebesgue integrable derivative. However, if f has bounded variation on $[a,b]$, then

$$(\dagger) \qquad \int_a^b f'(x)dx = f(b) - f(a)$$

need not hold. A necessary and sufficient condition on f for $(\dagger)$ to hold is that f be absolutely continuous. Therefore we are also interested in recognizing functions with this property. The final result of the section is in this direction, for it states a condition on the derivative f' of the differentiable function f which ensures that f is absolutely continuous.

In all of this section μ denotes Lebesgue measure on a fixed bounded interval $[a,b]$, and a.e. means almost everywhere with respect to μ.

5.4.1. In the sequel it is sometimes convenient to use geometrical language, based on the connection between derivatives of functions and tangents to curves known from calculus. Note that our concepts are always defined analytically; the reference to geometry is only to help our intuition.

The first notion we introduce is related to what one usually calls the length of a curve. Let f be a real-valued function on the bounded interval $[a,b]$. For each partition $\pi = [x_0, x_1, \ldots, x_n]$ of the interval $[a,b]$ (see 0.5.1), let

$$\ell_\pi(f) := \sum_{k=1}^{n} \{(x_k - x_{k-1})^2 + (f(x_k) - f(x_{k-1}))^2\}^{\frac{1}{2}} .$$

Now put

$$\ell(f) := \sup\{\ell_\pi(f) \mid \pi \text{ a partition of } [a,b]\} .$$

It is not at all obvious why $\ell(f)$ should be finite, and in fact there are even continuous functions f on $[a,b]$ with $\ell(f) = \infty$ (see Exercise 5.4-1). Nevertheless, we have met the functions with finite $\ell(f)$ before, under the guise of functions of bounded variation (see 0.5.10 for the definition).

5.4.2. Proposition. Let $f : [a,b] \to \mathbb{R}$. Then $\ell(f)$ is finite if and only if f has bounded variation on $[a,b]$.

Proof. Let $\pi := [x_0,x_1,\ldots,x_n]$ be any partition of $[a,b]$. Then

$$\sum_{k=1}^{n} |f(x_k) - f(x_{k-1})| \le \sum_{k=1}^{n} \{(x_k - x_{k-1})^2 + (f(x_k) - f(x_{k-1}))^2\}^{\frac{1}{2}}$$

$$\le \sum_{k=1}^{n} \{|x_k - x_{k-1}| + |f(x_k) - f(x_{k-1})|\}$$

$$= b - a + \sum_{k=1}^{n} |f(x_k) - f(x_{k-1})| ,$$

and the assertion follows easily. □

5.4.3. Again, let $\pi = [x_0,x_1,\ldots,x_n]$ be a partition of $[a,b]$, and let $y = (y_0,y_1,\ldots,y_n) \in \mathbb{R}^{n+1}$. Define the piecewise linear continuous function $p_{\pi,y}$ by

$$p_{\pi,y}(x_k) := y_k \qquad (0 \le k \le n) ,$$

$$p_{\pi,y} \text{ is linear on } [x_{k-1},x_k] \qquad (1 \le k \le n) .$$

It is easy to see that $\ell(p_{\pi,y})$ is just the ordinary length

$$\sum_{k=1}^{n} \{(x_k - x_{k-1})^2 + (y_k - y_{k-1})^2\}^{\frac{1}{2}}$$

of the graph of $p_{\pi,y}$. (Thus the number $\ell(f)$ of 5.4.1 is also the supremum of the lengths of the polygonal arcs inscribed in the graph of f. The last supremum is usually called the *length of the graph of* f.)

Now $\ell(p_{\pi,y})$ is completely determined by two ordered sets, namely $[\ell_1,\ldots,\ell_n]$ and $[m_1,\ldots,m_n]$, where $\ell_k := x_k - x_{k-1}$, $m_k := y_k - y_{k-1}$ $(1 \le k \le n)$, for

$$\ell(p_{\pi,y}) = \sum_{k=1}^{n} \{\ell_k^2 + m_k^2\}^{\frac{1}{2}} .$$

If β is any permutation of $\{1,2,\ldots,n\}$, then the ordered sets $[\ell_{\beta(1)},\ldots,\ell_{\beta(n)}]$ and $[m_{\beta(1)},\ldots,m_{\beta(n)}]$ give rise to a new partition π' of $[a,b]$, and to a new $y' \in \mathbb{R}^{n+1}$ where $x_0' := a$, $y_0' := y_0$, $x_k' - x_{k-1}' = \ell_{\beta(k)}$, $y_k' - y_{k-1}' = m_{\beta(k)}$ for $1 \le k \le n$. If $p^\beta := p_{\pi',y'}$, then obviously $\ell(p^\beta) = \ell(p)$. This simple observation comes in handy in the proof of the following lemma.

5.4.4. Lemma. Let q be linear and non-decreasing on $[a,b]$. Let p be a piecewise linear continuous function on $[a,b]$ with $p(a) = q(a)$, $p(b) = q(b)$. Let $\alpha > 0$, and assume that there is a finite collection J of pairwise disjoint open intervals (c_i,d_i) in $[a,b]$ with $p(d_i) - p(c_i) \le \alpha(c_i - d_i)$. If d is the sum of the lengths of the members of J, then

$$\ell(p) \geq \ell(q) + d((1 + \alpha^2)^{\frac{1}{2}} - 1) .$$

Proof. We may assume that $p = p_{\pi,y}$ for a partition π of $[a,b]$ which contains all points c_i and d_i, $\pi = [x_0,x_1,\ldots,x_n]$, say. By a suitable permutation β of $\{1,2,\ldots,n\}$ we can achieve that p^β (as described at the end of 5.4.3) satisfies $p^\beta(a + d) + - p^\beta(a) \leq -\alpha d$. Since $p^\beta(a) = p(a)$, $p^\beta(b) = p(b)$ and $\ell(p^\beta) = \ell(p)$, it suffices to consider p^β.

Without loss of generality we assume $p^\beta(a) = 0$, and then we have the situation depicted in the figure below.

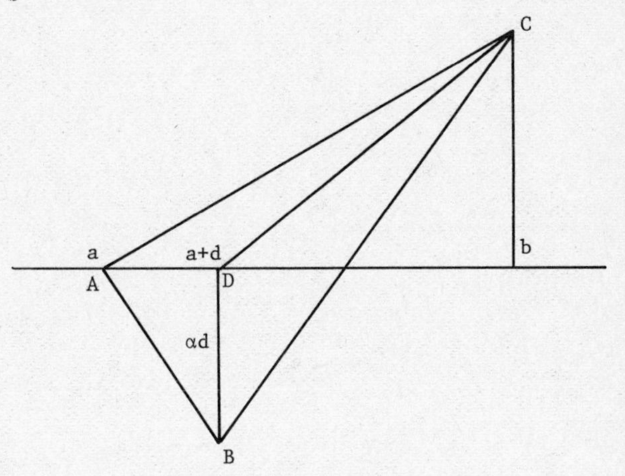

It is easy to see that

$$\ell(p^\beta) \geq |AB| + |BC| \geq |AB| + |DC| \geq |AB| - |AD| + |AC| \geq$$
$$\geq d((1 + \alpha^2)^{\frac{1}{2}} - 1) + \ell(q) . \qquad \square$$

There is, of course, a similar result if q is non-increasing and $\alpha < 0$.

5.4.5. In almost all approaches to differentiation theory one needs a covering lemma, where one replaces certain collections of intervals by collections of pairwise disjoint intervals. The one we use is the simplest of this kind of results.

Lemma. Let J be a finite collection of bounded open intervals in $\mathbb{R}$. Then there exists a subcollection J' of pairwise disjoint intervals from J such that

$$\mu\left(\bigcup_{I \in J'} I\right) \geq \frac{1}{3}\mu\left(\bigcup_{I \in J} I\right) .$$

Proof. The proof is by induction on the number of intervals in J. If J consists of just one interval, $J' := J$ does the trick. Now let $n > 1$ be the number of intervals in J, and assume that the statement of the lemma has been proved for all collections with less than n members. Let I_1 be a member of J with maximal length, and put

$J_1 := \{I \in J \mid I \cap I_1 = \emptyset\}$. By the induction hypothesis, there exists a disjoint collection $J_1' \subset J_1$ such that

$$\mu\left(\bigcup_{I \in J_1'} I\right) \geq \frac{1}{3}\mu\left(\bigcup_{I \in J_1} I\right).$$

Moreover, all members of $J \backslash J_1$ are contained in an open interval concentric with I_1 and having length $3\mu(I_1)$. Hence

$$3\mu(I_1) \geq \mu\left(\bigcup_{I \in J \backslash J_1} I\right),$$

and therefore $J_1' \cup \{I_1\}$ satisfies the condition. □

5.4.6. Let f be a real-valued function on [a,b], and let $x \in [a,b]$. We "split the derivative of f at x into four" as follows:

$$D^+f(x) := \limsup_{h \downarrow 0} h^{-1}\{f(x+h) - f(x)\},$$

$$D_+f(x) := \liminf_{h \downarrow 0} h^{-1}\{f(x+h) - f(x)\},$$

$$D^-f(x) := \limsup_{h \uparrow 0} h^{-1}\{f(x+h) - f(x)\},$$

$$D_-f(x) := \liminf_{h \uparrow 0} h^{-1}\{f(x+h) - f(x)\}.$$

Whether f is differentiable at x or not, these numbers are always defined, though they may be infinite. They are called the *Dini derivates* of f at x, and one has the obvious inequalities $D^+f(x) \geq D_+f(x)$ and $D^-f(x) \geq D_-f(x)$.

The function f is *differentiable at* x if its four Dini derivates at this point are equal and finite. That is, f is differentiable at x if $\lim_{h \to 0} h^{-1}\{f(x+h) - f(x)\}$ exists as a finite number.

5.4.7. The following proposition is essential in the proof of the differentiation theorem.

Proposition. Let $f : [a,b] \to \mathbb{R}$ be non-decreasing. Then $D^+f(x) \leq D_-f(x)$ for almost all $x \in [a,b]$.

Proof. First we show that D^+f and D_-f are measurable on [a,b]. As to D^+f, this will follow from the representation

$$(*) \qquad D^+f(x) = \lim_{n \to \infty} \sup_{0 < r_i < 1/n} r_i^{-1}\{f(x + r_i) - f(x)\},$$

where $x \in (a,b)$ and $(r_i)_{i\in\mathbb{N}}$ is any enumeration of $\mathbb{Q}$ (see 2.1.3(iv) and (v)).

Let $x \in (a,b)$. Choose $n \in \mathbb{N}$, $\varepsilon > 0$. There exists $k \in (0,1/n)$ such that

$$\frac{f(x+k)-f(x)}{k} \geq D^+f(x) - \frac{\varepsilon}{2} .$$

Since f is non-decreasing, there exists $i \in \mathbb{N}$ with $k < r_i < 1/n$ such that

$$\frac{f(x+r_i)-f(x)}{r_i} \geq D^+f(x) - \varepsilon .$$

Hence

$$\sup_{0<r_i<1/n} \frac{f(x+r_i)-f(x)}{r_i} \geq D^+f(x) ,$$

and therefore

$$D^+f(x) = \limsup_{h\downarrow 0} \frac{f(x+h)-f(x)}{h} \geq \lim_{n\to\infty} \sup_{0<r_i<1/n} \frac{f(x+r_i)-f(x)}{r_i} \geq D^+f(x) ,$$

which proves (*).

The measurability of D_-f could be established similarly. There is an easier way, however: apply the result already proved to the function $\biguplus_{x\in[-b,-a]}[-f(-x)]$.

The rest of the proof is an argument by contradiction; it is rather long, but not really difficult. Assume that the assertion of the proposition is false. Then there exist $\alpha > 0$ and $\beta \in \mathbb{R}$ such that

$$E_{\alpha,\beta} := \{x \in (a,b) \mid D^+f(x) > \beta + \alpha,\ D_-f(x) < \beta - \alpha\}$$

has positive measure. Since f has only countably many discontinuities (see Exercise 0.5-7), $E := \{x \in E_{\alpha,\beta} \mid f \text{ continuous at } x\}$ has positive measure, too.

Put $g := \biguplus_{x\in[a,b]}[f(x) - \beta x]$. Clearly g has bounded variation on $[a,b]$. Choose a polygonal arc inscribed in the graph of g, represented by a piecewise linear continuous function q based on the partition $\pi := [x_0,x_1,\ldots,x_n]$ such that

$$(**) \qquad \ell(q) > \ell(g) - \frac{1}{6}\mu(E)((1+\alpha^2)^{\frac{1}{2}} - 1) .$$

Choose a compact set $C \subset E$ such that $\mu(C) \geq \frac{1}{2}\mu(E)$ (see 3.3.9 or Theorem 4.2.1) and such that C does not contain points of π. Now note that in each point $x \in C$ we have $D^+g(x) > \alpha$, $D_-g(x) < -\alpha$, while g is continuous at x. It follows easily that for each $x \in C$ there exists an interval $(a_x,b_x) \subset (a,b)$ that contains x while q is linear on (a_x,b_x) and

$$\frac{g(b_x) - g(a_x)}{b_x - a_x} < -\alpha \qquad \text{if } q(b_x) \geq q(a_x) ,$$

$$\frac{g(b_x) - g(a_x)}{b_x - a_x} > \alpha \qquad \text{if } q(b_x) < q(a_x) .$$

Since C is compact, some finite collection J of these intervals suffices to cover C. Use 5.4.5 to choose a subcollection J' of pairwise disjoint members of J of total measure at least 1/3 $\mu(C)$, that is, at least 1/6 $\mu(E)$. Let π' be the partition of [a,b] obtained from π by adding the endpoints of the members of J to it, and let p be the corresponding piecewise linear approximation to g. Apply Lemma 5.4.4 to each interval $[x_{k-1}, x_k]$ that contains intervals from J', and add the inequalities thus obtained to get

$$\ell(g) \geq \ell(p) \geq \ell(q) + \frac{1}{6}\mu(E)((1 + \alpha^2)^{\frac{1}{2}} - 1) .$$

However, this is absurd in view of (**). □

5.4.8. Now we are in a position to prove our first main result on differentiation.

Theorem. Let $f : [a,b] \to \mathbb{R}$ have bounded variation on [a,b]. Then f is differentiable a.e. on [a,b], and the derivative f' is integrable. Moreover, if f is non-decreasing, then f' is non-negative a.e. and

$$\int_a^b f'(x)dx \leq f(b) - f(a) .$$

Proof. Since f can be written as the difference of two non-decreasing functions (Theorem 0.5.11), we may assume f non-decreasing.

Apply Proposition 5.4.7 to the function $\forall_{x\in[-b,-a]}[-f(-x)]$ to get $D_+f(x) \geq D^-f(x)$ almost everywhere in (a,b) (be careful with the signs!). Together with 5.4.7 applied to f itself this gives

$$D^+f(x) \leq D_-f(x) \leq D^-f(x) \leq D_+f(x) \leq D^+f(x)$$

for almost all $x \in (a,b)$. For an x where the Dini derivates are equal, denote the common value by f'(x) (so f'(x) may be ∞ at some points). Then f' is defined for almost all $x \in (a,b)$, it is non-negative, and it is measurable by the first part of the proof of 5.4.7.

Put $f(x) := f(b)$ for $x \geq b$. If $n \in \mathbb{N}$, then

$$\int_a^b n\left\{f(x + \tfrac{1}{n}) - f(x)\right\}dx = n\int_b^{b+1/n} f(x)dx - n\int_a^{a+1/n} f(x)dx .$$

The right-hand side of this equality tends to $f(b) - f(a + 0)$ as $n \to \infty$ (Exercise 5.4-4), and the non-negative integrands on the left-hand side tend to f' a.e. So by Fatou's lemma 2.2.12

$$\int_a^b f'(x)dx \le f(b) - f(a + 0) \le f(b) - f(a) ,$$

which also shows that f is finite a.e. on $[a,b]$. □

5.4.9. The remainder of this section is concerned with the question whether the inequality in Theorem 5.4.8 can be replaced by an equality. Now it is already obvious from the proof of 5.4.8 that for a non-decreasing f the value of $\int_a^b f'(x)dx$ is strictly less than $f(b) - f(a)$ if f is not continuous at a, but even if f is continuous on $[a,b]$ the inequality may be strict (see Exercise 5.4-2).

We shall need the following simple but useful result.

Lemma. If $f : [a,b] \to \mathbb{R}^*$ is integrable, and if $\int_\alpha^\beta f(x)dx = 0$ for all α, β with $a \le \alpha \le \beta < b$, then $f = 0$ (a.e.).

Proof. This is Exercise 3.2-9. □

5.4.10. Theorem. Let $f : [a,b] \to \mathbb{R}^*$ be integrable over $[a,b]$, and define

$$F := \forall_{x\in[a,b]}\left[\int_a^x f(t)dt\right] .$$

Then F is differentiable almost everywhere on $[a,b]$, and $F'(x) = f(x)$ for almost all $x \in (a,b)$.

Proof. F has bounded variation over $[a,b]$. For, if $[x_0,x_1,\dots,x_n]$ is any partition of $[a,b]$, then

$$\sum_{k=1}^n |F(x_k) - F(x_{k-1})| = \sum_{k=1}^n \left|\int_{x_{k-1}}^{x_k} f(t)dt\right| \le$$

$$\le \sum_{k=1}^n \int_{x_{k-1}}^{x_k} |f(t)|dt = \int_a^b |f(t)|dt < \infty .$$

So, by Theorem 5.4.8, F is differentiable a.e. on (a,b), and its derivative F' is integrable.

To show that $F' = f$ (a.e.), assume first that f is bounded, $|f(x)| \le K$, say, for all $x \in [a,b]$. Define $F(x) := F(b)$ $(x \ge b)$, and put

$$g_n := \curlyvee_{x\in[a,b]}[n\{F(x+\tfrac{1}{n}) - F(x)\}] \qquad (n \in \mathbb{N}).$$

Then $|g_n| \leq K$ for all n, and $g_n \to F'$ (a.e.). Choose an interval $[\alpha,\beta) \subset [a,b]$. By Lebesgue's dominated convergence theorem,

$$\int_\alpha^\beta g_n(x)dx \to \int_\alpha^\beta F'(x)dx$$

as $n \to \infty$. Since F is continuous from the right at α and β (see Exercise 1.2-6),

$$\int_\alpha^\beta g_n(x) = n\int_\beta^{\beta+1/n} F(x)dx - n\int_\alpha^{\alpha+1/n} F(x)dx \to F(\beta) - F(\alpha)$$

(see Exercise 5.4-4), and therefore $\int_\alpha^\beta \{F'(x) - f(x)\}dx = 0$. By Lemma 5.4.9, $F' = f$ (a.e.) in this special case.

Now drop the assumption that f is bounded. Without loss of generality, assume $f \geq 0$. For $n \in \mathbb{N}$, put $f_n := \curlyvee_{x\in[a,b]}\inf\{f(x),n\}$, and define

$$F_n := \curlyvee_{x\in[a,b]} \int_a^x f_n(t)dt .$$

Apply the result of the preceding paragraph to f_n to get $F_n' = f_n$ (a.e.). Since $F_n' \leq F'$ (a.e.) (to see this, apply 5.4.8 to $f - f_n$) and $F_n' = f_n \to f$ (a.e.), we have $F' \geq f$ (a.e.). Let

$$E := \{x \in [a,b] \mid F'(x) > f(x)\} .$$

Clearly

$$\int_a^b F'(x)dx > \int_a^b f(x)dx ,$$

unless $\mu(E) = 0$. But Theorem 5.4.8 implies that

$$\int_a^b F'(x)dx \leq F(b) - F(a) = \int_a^b f(x)dx ,$$

so E must have zero measure. Hence $F' = f$ (a.e.). □

5.4.11. In the language of calculus, the function F of 5.4.10 is an indefinite integral of f. If we call (extending the language of calculus) g a *primitive of* f whenever $g' = f$ (a.e.), then 5.4.10 says that the indefinite integral of an integrable function is a primitive of that function. We have already hinted at the fact that the converse is not true. There is a neat restricted converse, however. To formulate it, we need the concept of absolute continuity of functions. (Exercise 5.4.9 elucidates its relation to the concept of the same name for measures.)

Definition. A real-valued function f defined on [a,b] is called *absolutely continuous on* [a,b] if for every $\varepsilon > 0$ there exists $\delta > 0$ such that $a \le a_1 \le b_1 \le a_2 \le \dots \le \le a_n \le b_n \le b$ and $\sum_{k=1}^{n} (b_k - a_k) < \delta$ implies $\sum_{k=1}^{n} |f(b_k) - f(a_k)| < \varepsilon$.

5.4.12. The next proposition shows that there are many absolutely continuous functions. The point of the fundamental theorem 5.4.15 will be that (apart from additive constants) they are the only ones.

Proposition. Let $h : [a,b] \to \mathbb{R}^*$ be integrable, and let

$$f := \Upsilon_{x\in[a,b]}\left[\int_a^x h(t)dt\right].$$

Then f is absolutely continuous on [a,b].

Proof. Let $\varepsilon > 0$, and let g be any bounded integrable function on [a,b] such that

$$\int_a^b |h(x) - g(x)|dx < \frac{\varepsilon}{2}.$$

If $|g(x)| \le K$ on [a,b], then $\delta := \varepsilon/2K$ does what it should do according to Definition 5.4.11. □

5.4.13. We now list some simple properties of absolutely continuous functions. The proofs, which consist of juggling with partitions, are left to the reader as Exercise 5.4-7.

Proposition. (i) The absolutely continuous functions on [a,b] form a Riesz function space.

(ii) An absolutely continuous function on [a,b] has bounded variation on [a,b].

(iii) If f is absolutely continuous on [a,b], and $g := \Upsilon_{x\in[a,b]}[\mathrm{var}(f;[a,x])]$ (see 0.5.10 for the definition), then g is absolutely continuous on [a,b].

5.4.14. Our next result plays a crucial role in the proof of the fundamental theorem 5.4.15. Note that f' below exists a.e. by 5.4.13(ii) and 5.4.8.

Lemma. Let f be an absolutely continuous non-decreasing function on [a,b] with $f' = 0$ (a.e.). Then f is constant.

Proof. Extend the function f to the interval $[a-1,b+1]$ by putting $f(x) := f(a)$ for $x \in [a-1,a)$, $f(x) := f(b)$ for $x \in (b,b+1]$. Fix $\varepsilon > 0$, and choose $\delta > 0$ corresponding to ε as in Definition 5.4.11. Let $E := \{x \in [a,b] \mid f'(x) = 0\}$ and $F := [a,b]\setminus E$. Since $\mu(F) = 0$, there exists a sequence $(I_n)_{n\in\mathbb{N}}$ of open intervals such that $F \subset \bigcup_{n=1}^{\infty} I_n$ and $\sum_{n=1}^{\infty} \mu(I_n) < \delta$. We may assume that each I_n is contained in $[a-1,b+1]$. For each $x \in E$ choose $\delta_x > 0$ such that $|f(y) - f(x)| \le \varepsilon|y - x|$ whenever $y \in [a,b]$ and $|y - x| \le \delta_x$, and put $J_x := (x-\delta_x, x+\delta_x)$.

The I_n's together with the J_x's constitute an open covering of the compact set $[a,b]$. Hence, there exists a covering of $[a,b]$ by finitely many I_n's and J_x's. Take such a finite subcovering, and assume that it is minimal in the set theoretic sense, that is, none of the intervals may be left out without the covering property being destroyed. (Another way of saying this is that none of the intervals is contained in the union of the others.)

At this point it is convenient to change the notation. Denote the I-intervals in the subcovering by $I_1,\ldots,I_k$ and the J-intervals by $J_1,\ldots,J_\ell$, where $I_p = (a_p,b_p)$ $(1 \le p \le k)$, $J_q = (c_q,d_q)$ $(1 \le q \le \ell)$, $a_1 \le a_2 \le \ldots \le a_k$, $c_1 \le c_2 \le \ldots \le c_\ell$. The minimality assumption implies that $a_1 < a_2 < \ldots < a_k$, $b_1 < b_2 < \ldots < b_k$, and similarly for the c_q's and the d_q's. Now consider the classes $\{I_p \mid p \text{ even}\}$, $\{I_p \mid p \text{ odd}\}$, $\{J_q \mid q \text{ even}\}$, $\{J_q \mid q \text{ odd}\}$. The point is that in each class the intervals are disjoint. For instance, $I_1 \cap I_3 \neq \emptyset$ would obviously imply $I_2 \subset I_1 \cup I_3$, which was ruled out beforehand, and all other cases are similar to this one. Now it is easy to see that

$$(*) \qquad f(b) - f(a) \le \sum_{p=1}^{k} \{f(b_p) - f(a_p)\} + \sum_{q=1}^{\ell} \{f(d_q) - f(c_q)\} .$$

The first of the sums on the right of $(*)$ is dominated by

$$\sum_{p=1}^{k} \varepsilon(b_p - a_p) = \left(\sum_{p \text{ odd}} + \sum_{p \text{ even}} \right) \varepsilon(b_p - a_p) \le 2\varepsilon(b - a + 2) .$$

The second sum on the right of $(*)$ is also split into two parts: the terms with q odd are taken together in one sum, and those with q even in another. In each of these sums the intervals are non-overlapping and have total length less than δ. Hence, each contributes less than ε. It follows that $f(b) - f(a) < \varepsilon(2b - 2a + 4)$. Since $\varepsilon > 0$ is arbitrary, this shows that $f(b) \le f(a)$. Since f is non-decreasing, f is constant. □

5.4.15. Here is the main result of the present section.

Theorem. Let $f : [a,b] \to \mathbb{R}$ be absolutely continuous. Then f is differentiable a.e. on $[a,b]$, f' is integrable on $[a,b]$, and

$$f(\beta) - f(\alpha) = \int_\alpha^\beta f'(x)dx$$

for all $[\alpha,\beta] \subset [a,b]$.

Proof. Using the decomposition from the proof of Theorem 0.5.11 and assertions (i) and (iii) from Theorem 5.4.13, we see that we may assume that f is non-decreasing.

Fix $[\alpha,\beta] \subset [a,b]$. By 5.4.8, f is differentiable a.e., f' is integrable, and

$$g := \forall_{x\in[\alpha,\beta]}\left[f(x) - f(\alpha) - \int_\alpha^x f'(t)dt\right]$$

is non-decreasing. By 5.4.12 and 5.4.13(i), g is absolutely continuous on $[a,b]$, and by 5.4.10, $g' = 0$ (a.e.). Hence, by Lemma 5.4.14, g is constant, and since $g(\alpha) = 0$, it follows that $g(\beta) = 0$.

5.4.16. So an absolutely continuous function is an indefinite integral of its derivative. But how can we find out whether a differentiable function is absolutely continuous? Here is an answer in terms of the derivative.

Theorem. Let $f : [a,b] \to \mathbb{R}$ be continuous, and let f have a finite derivative in all but countably many points of $[a,b]$. If f' is integrable on $[a,b]$, then f is absolutely continuous on $[a,b]$.

Proof. It is sufficient to show that ${}_a\int^b f'(x)dx = f(b) - f(a)$. Let $\varepsilon > 0$. By the Vitali-Carathéodory theorem 4.2.3 there is a lower semi-continuous function g such that $f' \le g$ and ${}_a\int^b \{g(x) - f'(x)\}dx < \varepsilon$. Let E denote the exceptional set outside of which f has a finite derivative. Let $(r_n)_{n\in\mathbb{N}}$ be an enumeration of E (the most interesting case is that E is infinite), and put

$$h := \forall_{u\in[a,b]}\left[f(a) + \int_a^u g(x)dx + \varepsilon(u - a) + \varepsilon \sum_{r_n<u} 2^{-n}\right].$$

Let

$$V := \{v \in [a,b] \mid f(u) \le h(u) \text{ for all } u \in [a,v]\}.$$

Clearly, V is non-empty, since $a \in V$. Let $\alpha := \sup V$. It is not difficult to show that h is continuous from the left on $[a,b]$, which implies $\alpha \in V$. We are going to show that $\alpha = b$.

Assume $\alpha < b$. Then either

(i) $\alpha \in E$, or

(ii) $\alpha \notin E$.

(i) Suppose $\alpha \in E$, that is, $\alpha = r_m$ for some m. It is easy to see that

$$\lim_{u \downarrow \alpha} h(u) = h(\alpha) + \varepsilon \cdot 2^{-m-1} .$$

Since $\alpha \in V$ and f is continuous at α, there are points to the right of α that are still in V. Contradiction.

(ii) Since $\alpha \notin E$, f is now differentiable at α, and $O := \{x \in [a,b] \mid g(a) > f'(\alpha)-\frac{1}{2}\varepsilon\}$ is open. Choose $\delta > 0$ such that $[\alpha,\alpha+\delta] \subset O$, and such that

$$|f(u) - f(\alpha) - (u - \alpha)f'(\alpha)| \le \tfrac{1}{2}\varepsilon(u - \alpha)$$

whenever $u \in [\alpha,\alpha+\delta]$. Then $u \in [\alpha,\alpha+\delta]$ implies

$$f(u) \le f(\alpha) + (u - \alpha)f'(\alpha) + \tfrac{1}{2}\varepsilon(u - \alpha)$$

$$\le f(\alpha) + \int_\alpha^u g(x)dx + \varepsilon(u - \alpha) \le h(u) - \varepsilon \sum_{\alpha<r_n<u} 2^{-n} \le h(u) .$$

Hence $\alpha + \delta \in V$, which is again impossible.

We conclude that $\alpha = b$, and therefore $f(b) \le h(b)$. Since $\varepsilon > 0$ is arbitrary, this shows that $f(b) - f(a) \le \int_a^b f'(x)dx$. Application of this result to $-f$ gives $f(b) - f(a) \ge \int_a^b f'(x)$, and the last two inequalities combine to the desired equality. □

Exercises Section 5.4

1. Give an example of a continuous function f defined on [a,b] with $\ell(f) = \infty$. (Hint. Think of a function with many maxima and minima.)

2. Following F. Riesz we construct a continuous, strictly increasing function F defined on [0,1] such that $F(0) = 0$, $F(1) = 1$, $F' = 0$ (a.e.). We let $F_0 := \curlyvee_x x$. For $n \in \mathbb{N}$ we let F_n be the continuous, piecewise linear function with

$$F_n\left(\frac{k}{2^{n-1}}\right) = F_{n-1}\left(\frac{k}{2^{n-1}}\right) \qquad (k = 0,1,\dots,2^{n-1}) ,$$

$$F_n\left(\frac{k+\frac{1}{2}}{2^{n-1}}\right) = \tfrac{1}{4}F_n\left(\frac{k}{2^{n-1}}\right) + \tfrac{3}{4}F_n\left(\frac{k+1}{2^{n-1}}\right) \qquad (k = 0,1,\dots,2^{n-1} - 1) .$$

(i) Show that F_n is increasing and that $0 \le F_{n-1}(x) \le F_n(x)$ for $n \in \mathbb{N}$, $x \in [0,1]$.

Define $F := \lim_{n\to\infty} F_n$.

(ii) Show that F is strictly increasing .

(iii) For $n = 0,1,\ldots$, let $k_n \in \{0,1,\ldots,2^n-1\}$ be such that $a_n \le a_{n+1} < b_{n+1} \le b_n$, where

$$a_n = \frac{k_n}{2^n}, \quad b_n = \frac{k_n + 1}{2^n}.$$

Show that

$$F(b_n) - F(a_n) = F_n(b_n) - F_n(a_n) = \prod_{k=1}^{n} \frac{1 + \varepsilon_k}{2},$$

where $\varepsilon_k = \frac{1}{2}$ or $-\frac{1}{2}$ according as $b_k < b_{k-1}$ or $a_k > a_{k-1}$ $(k = 1,\ldots,n)$.

(iv) If a_n and b_n are as in (iii), then $F(b_n) - F(a_n) \to 0$ $(n \to \infty)$.

(v) Show that F is continuous.

(vi) Let $x \in (0,1)$ and assume that F is differentiable at x. Take sequences $(a_n)_{n=0,1,\ldots}$, $(b_n)_{n=0,1,\ldots}$ as in (iii) such that $a_n \le x \le b_n$. Show that

$$\frac{F(b_n) - F(a_n)}{b_n - a_n} \to F'(x) \quad \text{if } n \to \infty,$$

and that

$$F'(x) = \prod_{k=1}^{\infty} (1 + \varepsilon_k) = 0.$$

Compare Exercise 3.3-9.

3. Let $(f_k)_{k\in\mathbb{N}}$ be a sequence of non-negative, non-decreasing functions defined on $[0,1]$, and assume that $f(x) = \sum_{k=1}^{\infty} f_k(x) < \infty$ for all $x \in [0,1]$.

(i) Show that f is non-decreasing.

(ii) Show that $\sum_{k=1}^{\infty} f_k'(x) < \infty$ for almost every $x \in [0,1]$.

(iii) Let

$$g_n := f - \sum_{k=1}^{n} f_k = \sum_{k=n+1}^{\infty} f_k$$

for $n \in \mathbb{N}$, and let $(n_\ell)_{\ell\in\mathbb{N}}$ be a sequence in $\mathbb{N}$ such that $n_\ell < n_{\ell+1}$, $\sum_{\ell=1}^{\infty} g_{n_\ell}(x) < \infty$ for $x \in [0,1]$. Apply (ii) to $\sum_{\ell=1}^{\infty} g_{n_\ell}$ and conclude that $f' = \sum_{k=1}^{\infty} f_k'$ (a.e.). This is Fubini's theorem on differentiation.

4. If $\beta \in [a,b)$ and $f : [a,b] \to \mathbb{R}$ is Lebesgue integrable and continuous from the right in β, then $n \int_{\beta}^{\beta+\frac{1}{n}} f(x)dx \to f(\beta)$ if $n \to \infty$.

5. (i) Let $f : [a,b] \to \mathbb{R}$. Then f is absolutely continuous on $[a,b]$ if and only if for every $\varepsilon > 0$ there is a $\delta > 0$ such that $\sum_{n=1}^{\infty} |f(b_n) - f(a_n)| < \varepsilon$ for every sequence $((a_n, b_n))_{n\in\mathbb{N}}$ of pairwise disjoint intervals in $[a,b]$ with $\sum_{n=1}^{\infty}(b_n - a_n) < \delta$.

(ii) Assume that $f : [a,b] \to \mathbb{R}$ satisfies the following condition: for every $\varepsilon > 0$ there is a $\delta > 0$ such that for any $n = 1,2,\ldots$ and any set $\{(a_1,b_1),\ldots,(a_n,b_n)\}$ of not necessarily disjoint intervals with $\sum_{k=1}^{n}(b_k - a_k) < \delta$ we have $\sum_{k=1}^{n} |f(b_k) - f(a_k)| < \varepsilon$. Show that $\|f'\|_\infty < \infty$.

6. Let $f : [a,b] \to \mathbb{R}$ be absolutely continuous, and let E be a measurable subset of $[a,b]$. Show that $f(E)$ is measurable. (Hint. First consider the case that E is closed, then the case that E is a null set, and for the general case use Exercise 3.3-5.)

7. Give the proofs for Proposition 5.4.13.

8. We indicate an alternative proof of 5.4.15 which uses the Radon-Nikodym theorem. Let ν be the Lebesgue-Stieltjes measure generated by an absolutely continuous, non-decreasing $f : [a,b] \to \mathbb{R}$. Let Σ be the σ-algebra of Borel sets in $[a,b]$.

(i) Show that $\nu << \mu$.

(ii) Let $g : [a,b] \to \mathbb{R}$ be Σ-measurable and μ-integrable and suppose that $\nu(A) = {}_A\int g\,d\mu$ for $A \in \Sigma$. Show that

$$f(x) = f(a) + \int_a^x g(t)dt$$

for $x \in [a,b]$ and that $f' = f$ (a.e.).

9. Let $f : [a,b] \to \mathbb{R}$ be non-decreasing, and let ν be the Lebesgue-Stieltjes measure generated by f. Show that ν is absolutely continuous with respect to Lebesgue measure if and only if f is absolutely continuous.

10. Let $f : [0,1] \to \mathbb{R}$ be continuous and assume that $D_-f = 0$ with the exception of an at most countable set. Show that f is constant. (Hint. Suppose $f(0) < f(1)$, and let $f(0) < c < d < f(1)$. Let $g := \curlyvee_x [c + (d-c)x]$, and let $x_d := \sup\{x \in [0,1] \mid f(y) \le g(y)\ (y \in [0,x])\}$. Show that $(D_-f)(x_d) \ge d - c$ and that x_d increases strictly with d.)

11. Let $f : [0,1] \to \mathbb{R}$ satisfy

(i) $\limsup_{\delta \downarrow 0} f(x - \delta) \le f(x) \qquad (x \in (0,1])$,

(ii) $(D^+f)(x) \geq 0 \qquad (x \in [0,1))$.

Show that $f(0) \leq f(1)$. (Hint. Consider $\sup\{x \in [0,1] \mid f(0) \leq f(x)\}$.)

12. Let $g : [a,b] \to \mathbb{R}$ be integrable. Show that for every $\varepsilon > 0$ there is a partition $[a_0,a_1,\ldots,a_n]$ of $[a,b]$ such that

$$\sum_{k=1}^{n} \left| \int_{a_{k-1}}^{a_k} g(x)dx \right| \geq \int_a^b |g(x)|dx - \varepsilon .$$

Use this to show that

$$\mathrm{var}(f;[a,b]) = \int_a^b |f'(x)|dx$$

if $f : [a,b] \to \mathbb{R}$ is absolutely continuous.

13. Let $f : [a,b] \to \mathbb{R}$ be absolutely continuous. Show that

$$(*) \qquad \ell(f) = \int_a^b (1 + (f'(x))^2)^{\frac{1}{2}} dx$$

as follows.

(i) First assume that f' is continuous. Note that the sums occurring in the definition of $\ell(f)$ are Riemann sums for the right hand side of (*). (Hint. Mean value theorem.)

(ii) In the general case, approximate f' by continuous functions.

14. Let $f : [a,b] \to \mathbb{R}$ and $g : [a,b] \to \mathbb{R}$ be absolutely continuous on $[a,b]$. Show that $f \cdot g$ is absolutely continuous, and that $(f \cdot g)' = f \cdot g' + f' \cdot g$ (a.e.). Show that

$$\int_a^b f(x)g'(x)dx = f(b)g(b) - f(a)g(a) - \int_a^b f'(x)g(x)dx .$$

This is the "integration-by-parts" formula familiar from calculus.

15. Let $g : [a,b] \to \mathbb{R}$ be absolutely continuous, and let $f : [\alpha,\beta] \to \mathbb{R}$ be absolutely continuous, where $\alpha := \inf\{g(x) \mid x \in [a,b]\}$, $\beta := \sup\{g(x) \mid x \in [a,b]\}$. Assume that $\|f'\|_\infty < \infty$.

(i) Show that $f \circ g$ is absolutely continuous.

(ii) Let $x \in [a,b]$ be a point at which both g and $f \circ g$ are differentiable. Show that $f \circ g$ is differentiable at x and that $(f \circ g)'(x) = f'(g(x))g'(x)$ when $g'(x) \neq 0$.

(iii) Let $x \in [a,b]$ and assume $g'(x) = 0$. Show that $f \circ g$ is differentiable at x, and that $(f \circ g)'(x) = 0$.

(iv) Show that $(f \circ g)'(x) = f'(g(x))g'(x)$ for almost every x. Here we put $f'(g(x))g'(x) = 0$ when $g'(x) = 0$.

(v) Give an example of absolutely continuous functions f and g (defined on compact intervals) such that $f \circ g$ is not absolutely continuous. (Hint. Exercise 5.4-5(ii).)

16. Let E be a Lebesgue measurable subset of $\mathbb{R}$. Show that E has *density* 1 at almost every point of E, i.e., if

$$V := \left\{ x \in E \mid \lim_{h\downarrow 0, k\downarrow 0} (h+k)^{-1} \mu(E \cap (x-k, x+h)) = 1 \right\},$$

then $\mu(E \setminus V) = 0$.

17. Let $f : [a,b] \to \mathbb{R}$ be integrable. Show that for almost every $x \in [a,b]$

$$(*) \qquad \lim_{h\downarrow 0} \frac{1}{h} \int_0^h |f(x+t) - f(x)|\,dt = 0 .$$

(Hint. Let $q \in \mathbb{Q}$. There is a null set E_q in $[a,b]$ such that

$$\lim_{h\downarrow 0} \frac{1}{h} \int_0^h |f(x+t) - q|\,dt = |f(x) - q|$$

for $x \notin E_q$. Show that $(*)$ holds for $x \notin \bigcup_{q\in\mathbb{Q}} E_q$.) The points x where $(*)$ holds are called *Lebesgue points* of f; they are of importance in Fourier theory.

18. Show that

$$\lim_{t\to\infty} \int_0^t \frac{\sin x}{x}\,dx$$

exists, and determine the limit, as follows.

(i) Show that

$$\int_1^t \frac{\sin x}{x}\,dx = \cos 1 - \frac{1}{t}\cos t - \int_1^t \frac{\cos x}{x^2}\,dx$$

for $t \geq 1$, and conclude that $\lim_{t\to\infty} \int_0^t \frac{\sin x}{x}\,dx$ exists.

(ii) For $s \geq 0$ define

$$F(s) := \begin{cases} \displaystyle\int_0^\infty e^{-sx} \frac{\sin x}{x}\, dx & (s > 0)\,, \\ \displaystyle\lim_{t\to\infty} \int_0^t \frac{\sin x}{x}\, dx & (s = 0)\,. \end{cases}$$

Show that F is well-defined, that F is continuous on $[0,\infty)$ and that F is differentiable on $(0,\infty)$.

(iii) Show that $F'(s) = \dfrac{-1}{1+s^2}$ $(s > 0)$.

(iv) Show that $F(0) = \displaystyle\int_0^\infty \frac{ds}{1+s^2} = \frac{\pi}{2}$.

5.5. Change of variables in Lebesgue integrals on $\mathbb{R}$

One way to compute integrals is to find primitives. We know from calculus that often the search for primitives can be made easier by a change of variable, and we remember the formula

$$\int_a^b f(g(x))\cdot g'(x)dx = \int_{g(a)}^{g(b)} f(x)dx\,.$$

This formula subsists for Lebesgue integrals for appropriate functions f and g as shown below. Our results are by far not the most general possible, but they suffice for most purposes.

5.5.1. In this section $[a,b]$ is a fixed bounded interval in $\mathbb{R}$, g is a non-decreasing and absolutely continuous function on $[a,b]$, $\alpha := g(a)$ and $\beta := g(b)$. Measurability and integrals on $[a,b]$ or $[\alpha,\beta]$ must be taken in the Lebesgue sense, and μ stands for Lebesgue measure.

First we derive a rather technical lemma.

Lemma. Let Σ be the class of measurable subsets A of $[\alpha,\beta]$ such that

(*) $\quad h_A := \curlyvee_{x\in[a,b]} \chi_A(g(x))\cdot g'(x)$

is measurable and $\mu(A) = {}_a\int^b h_A(x)dx$. Then every measurable subset of $[\alpha,\beta]$ belongs to Σ.

Proof. By 5.4.13(ii) and 5.4.8, g is differentiable a.e. and g' is integrable. Hence, if A is a measurable subset of $[\alpha,\beta]$, then h_A as defined by $(\star)$ is defined a.e. Now we proceed step by step.

(i) Let $(A_n)_{n\in\mathbb{N}}$ be a pairwise disjoint sequence in Σ, and let $A := \bigcup_{n=1}^{\infty} A_n$. Obviously, $\sum_{n=1}^{N} h_{A_n}$ increases a.e. to h_A as $N \to \infty$. Therefore h_A is measurable, and by 2.2.11

$$\mu(A) = \lim_{N\to\infty} \sum_{n=1}^{N} \mu(A_n) = \lim_{N\to\infty} \int_a^b \sum_{n=1}^{N} h_{A_n}(x)dx = \int_a^b h_A(x)dx .$$

So h_A is integrable and $A \in \Sigma$.

(ii) Let $A,B \in \Sigma$, $A \subset B$. Since $\chi_{B\setminus A} = \chi_B - \chi_A$, $B\setminus A \in \Sigma$.

(iii) Let I be any open interval in $[\alpha,\beta]$, $I = (p,q)$, say. Let

$$\gamma := \inf\{x \in [a,b] \mid g(x) \in I\} , \quad \delta := \sup\{x \in [a,b] \mid g(x) \in I\} .$$

Since g is continuous and non-decreasing, $g(\gamma) = p$ and $g(\delta) = q$. By 5.4.15,

$$\mu(I) = q - p = g(\gamma) - g(\delta) = \int_\gamma^\delta g'(x)dx = \int_\gamma^\delta \chi_I(g(x))\cdot g'(x)dx .$$

Hence $I \in \Sigma$.

(iv) The proof that $[\alpha,\beta] \in \Sigma$ is similar to that in (iii).

(v) From (i) and (iii) it follows that Σ contains all open subsets of $[\alpha,\beta]$. Combination of this fact with (ii) and (iv) shows that all closed subsets of $[\alpha,\beta]$ are in Σ.

(vi) Finally, let A be any measurable subset of $[\alpha,\beta]$. By 3.3.10 there is a decreasing sequence $(O_n)_{n\in\mathbb{N}}$ of open sets in $[\alpha,\beta]$ and an increasing sequence $(C_n)_{n\in\mathbb{N}}$ of closed sets in $[\alpha,\beta]$ such that $C_n \subset A \subset O_n$ for all n, while $\mu(O_n\setminus C_n) \to 0$. By (v) and (iii)

$$\int_a^b h_{O_n\setminus C_n}(x)dx = \mu(O_n\setminus C_n) ,$$

and therefore $(h_{O_n\setminus C_n})_{n\in\mathbb{N}}$ decreases to 0 a.e. on $[a,b]$. This means that $(h_{C_n})_{n\in\mathbb{N}}$ increases a.e. to h_A. Hence, h_A is measurable and an application of 2.2.11 gives $\mu(A) = {}_a\int^b h_A(x)dx$. So $A \in \Sigma$. □

5.5.2. <u>Theorem</u>. Let g be an absolutely continuous non-decreasing real-valued function on [a,b]. Let $\alpha := g(a)$, $\beta := g(b)$. Let f be an integrable function on $[\alpha,\beta]$. Then the function $\forall_{x\in[a,b]} f(g(x))\cdot g'(x)$ is defined a.e. on [a,b], it is integrable on this interval, and

$$\int_\alpha^\beta f(x)dx = \int_a^b f(g(x))\cdot g'(x)dx .$$

<u>Proof</u>. First assume $f \geq 0$. By 2.1.13 there exists a sequence $(f_n)_{n\in\mathbb{N}}$ of simple functions (that is, finite linear combinations of characteristic functions of measurable subsets of $[\alpha,\beta]$) such that $0 \leq f_n \leq f_{n+1}$ $(n \in \mathbb{N})$ and $f_n \to f$. Combination of the linearity of the integral, the additivity of μ, and Lemma 5.5.1 leads to

$$\int_\alpha^\beta f_n(x)dx = \int_a^b f_n(g(x))\cdot g'(x)dx .$$

The left-hand side of this equality tends to $_\alpha\int^\beta f(x)dx$, which is finite. Since the integrands on the right-hand side increase a.e. to $\forall_{x\in[a,b]}[f(g(x))\cdot g'(x)]$, the monotone convergence theorem 2.2.11 gives all we want in this special case.

In the general case, split f as $f_1 - f_2$, where both f_1 and f_2 are non-negative and integrable, apply the special case of the preceding paragraph to both functions, and subtract. □

<u>Exercises Section 5.5</u>

1. We outline an alternative proof for Theorem 5.5.2. Let $I := [a,b]$, $J := [\alpha,\beta]$ be intervals in $\mathbb{R}$. Let μ and ν be finite measures on the Borel sets of I and J, respectively, and let g be a Borel measurable function from I onto J. We want to integrate functions f defined on J by integrating composite functions $f \circ g$ defined on I. To avoid obvious problems we assume

(*) if $E \subset I$ is a Borel set with $\mu(E) = 0$, then $\nu(g(E)) = 0$.

Define $\rho(F) := \mu(g^{-1}(F))$ for every Borel set F in J.

(i) Show that ρ is a finite measure and that $\nu \ll \rho$. If k is defined and Borel measurable on J and integrable with respect to ρ then

$$\int_J k\,d\rho = \int_I k \circ g\,d\mu .$$

(ii) Show that there is a non-negative Borel measurable function h defined on J which is integrable with respect to ρ such that

$$(**) \qquad \nu(F) = \int_F h\,d\rho = \int_I (\chi_F \cdot h) \circ g\,d\mu$$

for all Borel sets F in J.

(iii) Show that

$$\int_J f\,d\nu = \int_I (f \cdot h) \circ g\,d\mu$$

for every Borel function f integrable with respect to ν.

(iv) Specialize to the case that both μ and ν are Lebesgue measures, and assume that g is continuous and non-decreasing. Show that (*) amounts to the assumption that g be absolutely continuous, and that $h \circ g = g'$ (a.e.). What happens if g is not assumed to be non-decreasing?

2. Evaluate $C := {}_0\!\int^\infty e^{-t^2} dt$ as follows. Put

$$F(x) := \int_0^\infty \frac{e^{-x^2(1+t^2)}}{1+t^2}\,dt \qquad (x \geq 0).$$

Show that F is continuous on $[0,\infty)$, differentiable on $(0,\infty)$, and that $F'(x) = -2Ce^{-x^2}$ $(x > 0)$. Show that $F(0) = \pi/2$, $\lim_{x\to\infty} F(x) = 0$, and determine C.

3. Evaluate $C := {}_0\!\int^{\pi/2} \log \sin x\,dx$ as follows. Show that $\curlyvee_{x\in(0,\pi/2)} \log \sin x$ is integrable and that

$$C = \int_0^{\pi/2} \log \cos x\,dx = \tfrac{1}{2} \int_0^{\pi/2} \log(\tfrac{1}{2} \sin 2x)dx .$$

Now C can be determined.

4. Evaluate ${}_0\!\int^\infty t^{-\frac{1}{2}} e^{-t} dt$.

CHAPTER SIX

INTEGRATION ON THE PRODUCT OF TWO SPACES

In calculus one learns how to evaluate double integrals (integrals over subsets of $\mathbb{R}^2$) by repeated integration. Conditions permitting this procedure are rarely discussed at the same time, if only because the proper tool is not then available. This tool is the Lebesgue integral; we shall show in this final chapter how integration on products works.

We start in a rather abstract way with three sets S_1, S_2 and S_3, where $S_3 = S_1 \times S_2$. We assume that B_i is a Riesz function space of real-valued functions on S_i, that I_i is a positive linear functional on B_i, that Φ_i (the class of $\mathbb{R}^*$-valued functions on S_i) is equipped with a norm, where we now write $N_i\varphi$ for $\|\varphi\|$ if $\varphi \in \Phi_i$, and that the conditions $N(B_i,I_i,N_i)$ are satisfied for $i = 1,2,3$. The integrals I_1, I_2 and I_3 are connected by

$$(*) \qquad I_3 f = I_1 I_2 f$$

for $f \in B_3$. Here the "iterated integral" I_1I_2f has a simple interpretation: f is a function on S_3, and I_2f is a function on S_1 which has an I_1-integral. The problem is whether (*) can be extended to integrable functions on S_3. Now the integrals on the S_i are obtained from the original functionals I_i by means of processes where the norms N_i play a role. It is thus plausible that we have to assume some connection between the norms in order to get results in this direction. The appropriate condition is

$$(**) \qquad N_1N_2\varphi \le N_3\varphi \qquad (\varphi \in \Phi_3) ,$$

where $N_1N_2\varphi$ has a meaning just like I_1I_2f a moment ago. If (**) is satisfied, then (*) holds for all integrable functions f on S_3, that is, integration on the product can be carried out by repeated integration on the factors.

The integral on the product is most often constructed from integrals on the factors in such a way that (*) holds for a simple class of functions on S_3. We treat two types of construction.

In the first construction we start with two measure spaces, (S_1,Γ_1,μ_1) and (S_2,Γ_2,μ_2). It is easy to define a corresponding pair Γ_3,μ_3 on $S_3 := S_1 \times S_2$. In fact, the class Γ_3 of rectangles $A \times B$ with $A \in \Gamma_1$, $B \in \Gamma_2$ is a semiring, and if $\mu_3(A \times B) := \mu_1(A)\mu_2(B)$ for such a rectangle, then μ_3 is a measure on Γ_3. Since basic functions are finite linear combinations of characteristic functions of semiring elements, (*) now follows immediately from the definition of μ_3, and the norm condition (**) is not difficult to prove either.

The second construction works with locally compact Hausdorff spaces S_1 and S_2, with positive linear functionals I_1 and I_2 on their classes B_1 and B_2 of continuous real-valued functions with compact support. The product space $S_3 := S_1 \times S_2$ is given the product topology, and thus becomes a locally compact Hausdorff space with its own class B_3 of continuous real-valued functions of compact support. The problem how to construct I_3 neatly, is now more difficult. The solution of this problem depends on the basic result that members of B_3 can be approximated in an appropriate sense by finite linear combinations of functions $\forall_{(s,t)\in S_3}[b_1(s)\cdot b_2(t)]$ with $b_1 \in B_1$, $b_2 \in B_2$. There are now two ways to define norms on each of the spaces S_1, S_2 and S_3, and hence there are eight different combinations where (**) must be checked. Only two of these occur in practice; we treat both of them.

Important applications of the theory of product integrals occur in Fourier analysis, which has always been one of the main proving grounds for the weapons of integration. In the fourth section of this chapter we treat Fourier analysis in $L^2(\mathbb{R})$. In this case functions and their transforms belong to the same class, which is not characteristic of the theory of Fourier transforms. However, this drawback is fully compensated by the elegant symmetry which results. In the exercises of this section some of the Fourier analysis for $L^1(\mathbb{R})$ is treated.

In the final section we address a question that arises in the theory of stochastic processes. The problem is whether one can replace certain functions on a product space by measurable ones. To read this section, the reader need not know anything about stochastic processes, since we are only interested in the opportunity to apply the heavy machinery of the Radon-Nikodym theorem together with the theory of product measures.

6.1. The Fubini-Stone and Tonelli-Stone theorems

In this section we investigate how certain integrals on product spaces can be related to integrals on the factor spaces. The main result is that under appropriate conditions integration on a product can be replaced by repeated integration on the factors. Notation is necessarily pretty heavy in this part of the subject; we hope this will not obscure rather simple things.

6.1.1. In most of this chapter it is convenient to replace the notation $\|\varphi\|$ for the norm of a function φ in a situation N by $N\varphi$, or $N_1\varphi, N_2\varphi, \ldots$ as the case may be. Also, we drop brackets where possible without confusion.

Let S_1 and S_2 be non-empty sets, and $S_3 := S_1 \times S_2$. The variables in S_3 will be denoted as pairs, most of the time as (s,t), where $s \in S_1$, $t \in S_2$.

If $E \subset S_3$, $s \in S_1$, $t \in S_2$, then E_s will denote the subset $\{\tau \in S_2 \mid (s,\tau) \in E\}$ of S_2, and similarly E^t is the subset $\{\sigma \in S_1 \mid (\sigma,t) \in E\}$ of S_1. For functions we use an analogous notation: if φ is a function defined on $E \subset S_3$, then for $s \in S_1$ we denote by φ_s the function defined on E_s by $\varphi_s(\tau) := \varphi(s,\tau)$ $(\tau \in E_s)$, and if $t \in S_2$ then φ^t is defined on E^t by $\varphi^t(\sigma) := \varphi(\sigma,t)$. It is perfectly possible that E_s or E^t is empty, in which case φ_s or φ^t is defined nowhere. In the cases that we shall meet these exceptional values of s or t they will form null sets, and no serious confusion is to be expected.

Assume the following. There is a Riesz function space B_i of basic functions defined on S_i, a positive linear functional I_i on B_i, and a norm N_i on the space Φ_i of $\mathbb{R}^*$-valued functions defined on S_i such that $N(B_i, I_i, N_i)$ of Subsection 1.1.5 is satisfied, for $i = 1,2,3$. The processes of Chapters 1 and 2 lead to extensions of the integrals, which are again denoted by I_i. As in Section 2.2, Φ_i^* denotes the class of $\mathbb{R}^*$-valued functions defined a.e. on S_i, M_i^* is the class of measurable members of Φ_i^*, and L_i^* is the class of integrable members of Φ_i^*. We use the obvious extensions of N_i and I_i on Φ_i^* and L_i^*.

Let $\Phi_1^* \times \Phi_2^*$ denote the set of $\varphi \in \Phi_3^*$ such that $\varphi_s \in \Phi_2^*$ for almost all $s \in S_1$. If φ is any member of $\Phi_1^* \times \Phi_2^*$, then $N_2\varphi_s$ is well-defined for almost all $s \in S_1$; this gives a member of Φ_1^*, denoted by $N_2\varphi$, which has N_1-norm $N_1N_2\varphi$.

Let $L_1^* \times L_2^*$ denote the class of those $\varphi \in \Phi_3^*$ that satisfy

(i) $\varphi_s \in L_2^*$ for almost all $s \in S_1$,

(ii) $\Upsilon_{s \in S_1} (I_2\varphi_s) \in L_1^*$.

For a $\varphi \in L_1^* \times L_2^*$ the function in (ii) is denoted by $I_2\varphi$, and its I_1-integral by $I_1I_2\varphi$. Obviously, $L_1^* \times L_2^*$ is a subset of $\Phi_1^* \times \Phi_2^*$.

6.1.2. Here is at once one of the main theorems of the chapter.

<u>Theorem</u> (Fubini-Stone theorem). In the situation described in 6.1.1, assume the following:

(a) $B_3 \subset L_1^* \times L_2^*$,

(b) $I_3 b = I_1 I_2 b$ for all $b \in B_3$,

(c) $N_1 N_2 \varphi \le N_3 \varphi$ for all $\varphi \in \Phi_3$.

Then

(i) $\Phi_3^* = \Phi_1^* \times \Phi_2^*$,

(ii) $L_3^* \subset L_1^* \times L_2^*$ and $I_3 f = I_1 I_2 f$ for all $f \in L_3^*$.

Proof. (i) Let $\varphi \in \Phi_3$ and let $E \subset S_3$ be a null set such that φ is defined on $S_3\backslash E$. Now $\chi_E \in \Phi_3$ and, obviously, $N_1N_2\chi_E \le N_3\chi_E = 0$. Hence, by 1.1.15(ii) applied twice, E_s is a null set in S_2 for almost all $s \in S_1$. So $\varphi \in \Phi_1^* \times \Phi_2^*$, and therefore $\Phi_3^* \subset \Phi_1^* \times \Phi_2^*$.

(ii) Let $f \in L_3^*$. Take a sequence $(b_n)_{n\in\mathbb{N}}$ in B_3 such that $N_3(f - b_n) < 2^{-n}$ for all $n \in \mathbb{N}$. For every n, $f - b_n \in L_3^* \subset \Phi_3^* \subset \Phi_1^* \times \Phi_2^*$ by (i), so $N_2(f - b_n) \in \Phi_1^*$ and it is a.e. non-negative. If $g := \sum_{n=1}^{\infty} N_2(f - b_n)$, then $g \in \Phi_1^*$ and

$$N_1g \le \sum_{n=1}^{\infty} N_1N_2(f - b_n) \le \sum_{n=1}^{\infty} N_3(f - b_n) < \infty .$$

So g is finite a.e. on S_1. Let E_1 be a null set in S_1 such that g is defined and finite on $S_1\backslash E_1$.

By assumption (a), the definition of $L_1^* \times L_2^*$, and 1.1.15(v), there is a null set $F_1 \subset S_1$ such that $(b_n)_s \in L_2^*$ if $s \in S_1\backslash F_1$, $n \in \mathbb{N}$. Now if $s \in S_1\backslash(E_1 \cup F_1)$, then $(N_2(f - b_n))(s) \to 0$, which shows that f_s is approximated in the N_2-norm by the L_2^*-functions $(b_n)_s$. Hence $f_s \in L_2^*$ for such a value of s and, moreover,

$$N_1(I_2f - I_2b_n) \le N_1N_2(f - b_n) \le N_3(f - b_n) ,$$

while $N_3(f - b_n) \to 0$. Since each $I_2b_n \in L_1^*$, this shows in the first place that $I_2f \in L_1^*$, that is, $f \in L_1^* \times L_2^*$, and in the second place that

$$I_1I_2f = \lim_{n\to\infty} I_1I_2b_n = \lim_{n\to\infty} I_3b_n = I_3f . \qquad \square$$

6.1.3. There is a point in the proof of (i) that we note separately. Under the conditions of the theorem, if $E \subset S_3$ is a null set, then E_s is a null set in S_2 for almost all $s \in S_1$. We shall use this fact several times in the sequel. Another point to be noted is that for $f \in L_3^*$ there is equality in (c), so $N_1N_2f = N_3f$ in this case.

6.1.4. Our next aim is to derive a parallel for measurable functions of assertion (i) in the Fubini-Stone theorem. Here is a preliminary result.

Proposition. Assume all the conditions of Theorem 6.1.2. Let $f \in M_3^*$ and assume that there exists a sequence $(f_n)_{n\in\mathbb{N}}$ in L_3^* such that $|f| \le \sum_{n=1}^{\infty}|f_n|$ a.e. in S_3. Then $f_s \in M_2^*$ for almost all $s \in S_1$.

Proof. For $n \in \mathbb{N}$, put $h_n := \sum_{k=1}^{n}|f_k|$, and the truncated function $g_n := (f)_{h_n}$. Then $g_n \in L_3^*$ by the definition of measurability. Now 6.1.2 shows that there exists a null set $E_n \subset S_1$ such that $(g_n)_s \in L_2^* \subset M_2^*$ for $s \in S_1\backslash E_n$. Since $g_n \to f$ a.e. in S_3, there exists a null set $E \subset S_3$ such that $g_n(s,t) \to f(s,t)$ whenever $(s,t) \in S_3\backslash E$. Hence there is a null set $F \subset S_1$ such that E_s is a null set in S_2 for $s \in S_1\backslash F$. It follows that $(g_n)_s \to f_s$ a.e. in S_2 whenever $s \in S_1 \setminus (F \cup \bigcup_{n=1}^{\infty} E_n)$, which means that $f_s \in M_2^*$ for those values of s (see 2.1.3(iv)). □

6.1.5. Before announcing the Fubini-Stone result for non-negative measurable functions, we recall how the integral on S_i was extended in Section 2.2 to the (almost everywhere) non-negative members of M_i^*: each of those functions that was not in L_i^* was assigned the integral ∞. The next result shows once more how convenient this extension is.

Theorem. Assume all the conditions of Theorem 6.1.2. Let $f \in M_3^*$, $f \geq 0$ a.e. on S_3, and assume that there exists a sequence $(f_n)_{n\in\mathbb{N}}$ in L_3^* such that $f \leq \sum_{n=1}^{\infty} |f_n|$ a.e. in S_3. Then $I_2 f_s$ is well-defined for almost all $s \in S_1$ and $I_2 f := \Upsilon_{s\in S_1} I_2 f_s \in M_1^*$. Moreover, $I_2 f \geq 0$ a.e. in S_1, and $I_1 I_2 f = I_3 f$.

Proof. Combination of Proposition 6.1.4 and the remark in 6.1.3 shows that for almost every $s \in S_1$ the function f_s is an a.e. non-negative member of M_2^*. Hence, $I_2 f$ is well-defined as a member of Φ_1^*, and it is a.e. non-negative.

For each $n \in \mathbb{N}$, let $h_n := \sum_{k=1}^{n} |f_k|$ and $g_n := \min(f, h_n)$. Then $g_n \in L_3^*$, $g_n \geq 0$ a.e. in S_3, and $g_n \uparrow f$ a.e. in S_3. Hence, by the monotone convergence theorem 2.2.11, $I_3 g_n \uparrow I_3 f$. On the other hand, by the first fact noted in 6.1.3, $(g_n)_s \to f_s$ a.e. in S_2 for almost every $s \in S_1$, and since the convergence is monotone, it follows that $I_2 g_n \uparrow I_2 f$ a.e. in S_1. Now each $I_2 g_n \in L_1^*$ by 6.1.2(ii), so a final application of 2.2.11 gives $I_2 f \in M_1^*$ and $I_1 I_2 g_n \to I_1 I_2 f$.

Since $I_3 g_n = I_1 I_2 g_n$ $(n \in \mathbb{N})$ by 6.1.2, it follows that $(I_3 g_n)_{n\in\mathbb{N}}$ increases to $I_1 I_2 f$ and to $I_3 f$. Hence $I_1 I_2 f = I_3 f$. □

6.1.6. We come to the final result of this section. At first sight it just seems to be a special case of Theorem 6.1.5. In the next two sections, however, we shall see that the Tonelli-Stone theorem is the appropriate tool to justify changes in the order of integration in repeated integrals.

Theorem (Tonelli-Stone theorem). Assume all the conditions of Theorem 6.1.2. Let $f \in M_3^*$, and assume that there exists a sequence $(f_n)_{n\in\mathbb{N}}$ in L_3^* such that $|f| \leq \sum_{n=1}^{\infty} |f_n|$ a.e. in S_3. If $I_1 I_2 |f| < \infty$, then $f \in L_3^*$ and $I_3 f = I_1 I_2 f$.

Proof. Note first that $I_1 I_2 |f|$ is in fact defined, according to 6.1.5. Now apply 6.1.5 to $|f|$ to get $I_3 |f| < \infty$. It follows that $|f| \in L_3^*$, and therefore $f \in L_3^*$. Finally apply 6.1.2(ii) to f. □

6.1.7. We cannot do without the covering conditions in the last three results. This will be shown in 6.2.6, where an example will be constructed of a product space $S_3 = S_1 \times S_2$ and a non-negative measurable function f on S_3 for which it is not true that f^t is measurable for almost every $t \in S_2$. Of course, if S_3 is σ-finite, then the covering condition is automatically satisfied.

Exercises Section 6.1

In these exercises we use the notation of 6.1.1 and we assume that all conditions of Theorem 6.1.2 are satisfied.

1. For $\varphi_1 \in \Phi_1^*$, $\varphi_2 \in \Phi_2^*$ we let $(\varphi_1 \otimes \varphi_2)(s,t) := \varphi_1(s)\varphi_2(t)$ for all (s,t) for which $\varphi_1(s)$ and $\varphi_2(t)$ are defined.

(i) Assume that $b_1 \otimes b_2 \in B_3$ for $b_1 \in B_1$, $b_2 \in B_2$. Show that $f_1 \otimes f_2 \in L_3$, $I_3(f_1 \otimes f_2) = I_1f_1 \cdot I_2f_2$ for $f_1 \in L_1$, $f_2 \in L_2$.

(ii) Assume in addition that (B_i,I_i,N_i) is σ-finite (i = 1,2). Show that (B_3,I_3,N_3) is σ-finite, and that $f_1 \otimes f_2 \in L_3^*$, $I_3(f_1 \otimes f_2) = I_1f_1 \cdot I_2f_2$ for $f_1 \in L_1^*$, $f_2 \in L_2^*$.

(iii) Can the σ-finiteness condition in (ii) be deleted?

2. Does it hold that $L_1^* \times L_2^* \subset L_3^*$?

3. Can it happen that (B_3,I_3,N_3) is σ-finite, while one of the (B_i,I_i,N_i) (i = 1,2) is not?

6.2. Product measure spaces

Integrals on products are almost always constructed from integrals on the factor spaces. The first construction that we present works with measure spaces. It runs quite smoothly, because there is a natural way to define a semiring and a measure on the product of such spaces. The factors enter symmetrically into the construction; this leads to useful theorems that justify the change of the order of integration in repeated integrals.

6.2.1. Let (S_i,Γ_i,μ_i) be measure spaces, where Γ_i is a semiring of subsets of S_i, for i = 1,2. Let $S_3 := S_1 \times S_2$. Let Γ_3 be the class of *rectangles* with sides in Γ_1 and Γ_2, that is,

$$\Gamma_3 := \{A \times B \mid A \in \Gamma_1,\ B \in \Gamma_2\}\ ,$$

and define the set function μ_3 on Γ_3 by

$$\mu_3(A \times B) := \mu_1(A)\cdot\mu_2(B) \qquad (A \in \Gamma_1,\ B \in \Gamma_2)\ .$$

In connection with the definition of μ_3 we recall our convention that $0\cdot a = 0$ for any $a \in \mathbb{R}^*$; thus $\mu_3(\emptyset)$ is unambiguously defined as 0 (note that $A \times B$ is empty if at least one of A and B is empty). If we write $A \times B \in \Gamma_3$, then it is tacitly assumed that $A \in \Gamma_1$, $B \in \Gamma_2$.

6.2.2. Proposition. In the situation described in 6.2.1, (S_3,Γ_3,μ_3) is a measure space. If (S_1,Γ_1,μ_1) and (S_2,Γ_2,μ_2) are σ-finite, then so is (S_3,Γ_3,μ_3).

Proof. If $A_i \in \Gamma_1$, $B_i \in \Gamma_2$ $(i = 1,2)$, then it is easy to see that

$$(A_1 \times B_1) \cap (A_2 \times B_2) = (A_1 \cap A_2) \times (B_1 \cap B_2) \in \Omega(\Gamma_3)$$

$$(A_1 \times B_1) \setminus (A_2 \times B_2) = (A_1 \times (B_1\setminus B_2)) \cup ((A_1\setminus A_2) \times B_1) \in \Omega(\Gamma_3)\ .$$

Therefore Γ_3 is a semiring.

To show that μ_3 is a measure on Γ_3, it is sufficient to show that it is σ-additive. Let $(A_n \times B_n)_{n\in\mathbb{N}}$ be a sequence of pairwise disjoint rectangles in Γ_3, and assume that

$$\bigcup_{n=1}^{\infty} A_n \times B_n = A \times B \in \Gamma_3\ .$$

Then

$$\chi_A(s)\cdot\chi_B = \sum_{n=1}^{\infty} \chi_{A_n}(s)\cdot\chi_{B_n} \qquad (s \in S_1)\ .$$

Now fix $s \in S_1$, integrate both sides of the equality over S_2, apply the monotone convergence theorem 2.2.11, and note that $\int \chi_C\, d\mu_2 = \mu_2(C)$ for any $C \in \Gamma_2$ to get

$$\chi_A(s)\mu_2(B) = \sum_{n=1}^{\infty} \chi_{A_n}(s)\mu_2(B_n)\ .$$

In a similar way integration over S_1 gives

$$\mu_1(A)\cdot\mu_2(B) = \sum_{n=1}^{\infty} \mu_1(A_n)\cdot\mu_2(B_n)\ .$$

Hence μ_3 is a measure on Γ_3.

The proof of the σ-finiteness assertion is left to the reader. □

The measure μ_3 thus obtained is called the *product* of μ_1 and μ_2; sometimes it is denoted by $\mu_1 \times \mu_2$.

6.2.3. For each of the three measure spaces follow the route $R \to D \to A \to N \to L$ as in Section 3.2. Let B_i be the resulting class of basic functions with integral I_i for $i = 1,2,3$. Adopt the notations of 6.1.1; in particular, denote the norms by N_1, N_2 and N_3. We are going to verify the conditions of the Fubini-Stone theorem 6.1.2.

Proposition. Under the conditions stated, we have

(a) $B_3 \subset L_1^* \times L_2^*$,

(b) $I_3 b = I_1 I_2 b$ whenever $b \in B_3$,

(c) $N_1 N_2 \varphi \le N_3 \varphi$ whenever $\varphi \in \Phi_3$.

Proof. Let $C = A \times B \in \Gamma_3$, $\mu_3(C) < \infty$. It is easy to see that $\chi_C \in L_1^* \times L_2^*$, and $I_3 \chi_C = I_1 I_2 \chi_C$ (only the cases where $\mu_1(A_1) = 0$ and $\mu_2(B) = \infty$, or $\mu_1(A) = \infty$ and $\mu_2(B) = 0$, need some care). Since each member of B_3 is a finite linear combination of such characteristic functions, (a) and (b) follow.

For the proof of (c), let $\varphi \in \Phi_3$. Let $(\alpha_n)_{n \in \mathbb{N}}$ be a sequence of non-negative real numbers, $(A_n \times B_n)_{n \in \mathbb{N}}$ a sequence in Γ_3, and suppose that

$$|\varphi| \le \sum_{n=1}^{\infty} \alpha_n \chi_{A_n \times B_n} .$$

Since $N_2 \chi_{B_n} = I_2 \chi_{B_n} = \mu_2(B_n)$ for each $n \in \mathbb{N}$, it follows that

$$(N_2\varphi)(s) \le \sum_{n=1}^{\infty} N_2(\alpha_n \chi_{A_n}(s) \chi_{B_n}) = \sum_{n=1}^{\infty} \alpha_n \chi_{A_n}(s) \mu_2(B_n)$$

for each $s \in S_1$. Taking N_1-norms on both sides one obtains

$$N_1 N_2 \varphi \le \sum_{n=1}^{\infty} \alpha_n \mu_1(A_n) \mu_2(B_n) = \sum_{n=1}^{\infty} \alpha_n \mu_3(A_n \times B_n) ,$$

and now the definition of $N_3\varphi$ shows that $N_1 N_2 \varphi \le N_3 \varphi$. □

6.2.4. The proposition implies that the theorems of Section 6.1 hold in the present situation. There is an additional feature, however, due to the possibility of changing the roles of (S_1, Γ_1, μ_1) and (S_2, Γ_2, μ_2). This leads to more symmetrical results, which is nice enough. But it is more important that we get simple conditions under which the order of integration in repeated integrals may be changed (see (ii), (iv), and (v) of the theorem below).

In the statement of the theorem we use obvious extensions of the conventions of Subsection 6.1.1. For instance, the statement "$I_1 f \in L_2^*$" in (ii) means that for almost all $t \in S_2$ the function f^t belongs to L_1^*, while the function $\Upsilon_{t\in S_2} I_1 f^t$, which is therefore defined almost everywhere on S_2, belongs to L_2^*.

Theorem. Let (S_3,Γ_3,μ_3) be the product of the measure spaces (S_1,Γ_1,μ_1) and (S_2,Γ_2,μ_2). Then the following assertions hold.

(i) If $\varphi \in \Phi_3^*$, then $\varphi_s \in \Phi_2^*$ for almost all $s \in S_1$, and $\varphi^t \in \Phi_1^*$ for almost all $t \in S_2$.

(ii) If $f \in L_3^*$, then $f_s \in L_2^*$ for almost all $s \in S_1$, and $f^t \in L_1^*$ for almost all $t \in S_2$. Moreover, $I_2 f \in L_1^*$, $I_1 f \in L_2^*$, and $I_1 I_2 f = I_2 I_1 f = I_3 f$.

(iii) If $f \in M_3^*$ and if there exists a sequence $(f_n)_{n\in\mathbb{N}}$ in L_3^* such that $|f| \le \sum_{n=1}^{\infty} |f_n|$ (a.e. in S_3), then $f_s \in M_2^*$ for almost all $s \in S_1$, and $f^t \in M_1^*$ for almost all $t \in S_2$.

(iv) If f is as in (iii), and in addition $f \ge 0$ (a.e. in S_3), then $I_2 f \in M_1^*$, $I_1 f \in M_2^*$, and $I_1 I_2 f = I_2 I_1 f = I_3 f$.

(v) If f is as in (iii), and in addition $I_1 I_2 |f| < \infty$ or $I_2 I_1 |f| < \infty$, then $f \in L_3^*$.

6.2.5. Example. Take $S_1 = S_2 = \mathbb{R}$ with ordinary Lebesgue measure. In this case S_3 is just $\mathbb{R}^2$, and the product measure is usually called (two dimensional) Lebesgue measure. We have not yet specified the semirings on the factor spaces. There are several possibilities for doing this; we mention the two most natural choices:

(i) Use the semiring of nails (see 3.1.4) on both factors. Now the rectangles in the product are the cells of 3.1.5. The measure of such a rectangle is its ordinary area.

(ii) Use the σ-algebra of Lebesgue measurable sets on both factors.

Which of (i) and (ii) we use is immaterial, for we always obtain the same measure and the same measurable sets on the product. Since the factor spaces are σ-finite, the product space is also σ-finite. Hence the covering condition of 6.2.4(iii) is automatically satisfied.

It is customary to denote the integral of a function f on the product that is integrable (or measurable and non-negative) by

$$\int_{-\infty}^{\infty}\int_{-\infty}^{\infty} f(x,y)\,dx\,dy$$

or something like that. For such functions Theorem 6.2.4 asserts that

$$\int_{-\infty}^{\infty}\int_{-\infty}^{\infty} f(x,y)dxdy = \int_{-\infty}^{\infty}\left(\int_{-\infty}^{\infty} f(x,y)dy\right)dx = \int_{-\infty}^{\infty}\left(\int_{-\infty}^{\infty} f(x,y)dx\right)dy ,$$

and it is most often the last equality that we are interested in.

The restriction of two-dimensional Lebesgue measure to a rectangle $[a,b] \times [c,d]$ can also be obtained directly as the product of Lebesgue measure on its sides. Integrals with respect to this measure are denoted by

$$\int_a^b \int_c^d f(x,y)dxdy .$$

Things like ${}_0\int^\infty {}_0\int^\infty f(x,y)dxdy$ are defined in a similar way.

6.2.6. The next two examples deal with the covering condition in 6.2.4(iii) and (iv).

<u>Example</u>. Let $S_1 := [0,1]$, $\Gamma_1 := \{\emptyset, S_1\}$, $\mu_1(\emptyset) = 0$ and $\mu_1(S_1) = 1$. Let S_2 be any non-empty set, $\Gamma_2 := \{\emptyset, S_2\}$, $\mu_2(\emptyset) = 0$ and $\mu_2(S_2) = \infty$. Form the product space $S_3 := S_1 \times S_2$ with semiring Γ_3 and measure μ_3.

Let f be any real-valued function defined on S_1, and

$$f_0(s,t) := f(s) \qquad (s \in S_1,\ t \in S_2) .$$

There is only one integrable function on S_3, namely the function that is identically zero. Hence f_0 is measurable. In general, it is not true that $f_0^t = f$ is measurable for almost all $t \in S_2$ (all members of M_1^* are constant).

6.2.7. <u>Example</u>. Let $S_1 := (0,1]$, Γ_1 the class of nails in S_1, μ_1 ordinary Lebesgue measure. Let $S_2 := (0,1]$, Γ_2 the class of all subsets of S_2, and let μ_2 be defined for $A \in \Gamma_2$ by

$$\mu_2(A) := \begin{cases} \text{the number of elements of A if A is finite,} \\ \infty \ \text{otherwise.} \end{cases}$$

(This μ_2 is the counting measure of Example 3.1.11.) Form the product measure as in 6.2.1. Let D be the diagonal in S_3, that is, $D := \{(x,x) \mid 0 < x \leq 1\}$, and write $f := \chi_D$. We shall show that the non-negative function f is measurable on S_3 and that the iterated integrals I_1I_2f and I_2I_1f both exist as finite numbers, but are not equal.

Let $b \in B_3^+$ and represent b as $\sum_{n=1}^N \alpha_n\chi_{A_n\times B_n}$ where $\alpha_n \geq 0$, $A_n \in \Gamma_1$, $B_n \in \Gamma_2$, $\mu_3(A_n \times B_n) < \infty$ for $1 \leq n \leq N$, while the rectangles $A_n \times B_n$ are pairwise disjoint (see 3.2.3). Write b_n for the characteristic function of $A_n \times B_n$. It is easy to see that $(f)_b = \sum_{n=1}^{N} (f)_{\alpha_n b_n}$. Now we assert that each $(f)_{\alpha_n b_n}$ is a null function. This is obvious if $\alpha_n b_n$ is itself a null function. And if $\alpha_n b_n$ is not a null function, that is, if $\alpha_n \mu_3(A_n \times B_n) > 0$, then we must have $\mu_1(A_n) > 0$ and therefore $\mu_2(B_n) < \infty$, so B_n is a finite set. But then $(f)_{\alpha_n b_n}$ is everywhere zero except on a finite set and therefore clearly a null function. It follows that $(f)_b$ is a null function for any $b \in B_3^+$. Hence, f is measurable, and in fact it is even a locall null function.

It is not difficult to show that $I_2 f_s = 1$ for all $s \in S_1$, and that $I_1 f^t = 0$ for all $t \in S_2$. Hence $I_1 I_2 f = 1 \neq 0 = I_2 I_1 f$.

Exercises Section 6.2

1. Let (S_i, Γ_i, μ_i) be as in 6.2.1 and 6.2.2 for $i = 1,2,3$. Assume that (S_i, Γ_i, μ_i) is σ-finite for $i = 1,2$. Show that (S_3, Γ_3, μ_3) is σ-finite.

2. Let (S_i, Γ_i, μ_i) be as in 6.2.1 and 6.2.2 for $i = 1,2,3$. Show that $b_1 \otimes b_2 \in B_3$ for $b_1 \in B_1$, $b_2 \in B_2$. (For the $\otimes$-notation see Exercise 6.1-1.)

3. Let (S_i, Γ_i, μ_i) be as in 6.2.1 and 6.2.2 for $i = 1,2,3$, and assume that $E_i \subset S_i$, $N_i(\chi_{E_i}) = 0 \Rightarrow E_i \in \Gamma_i$. Show that $f_1 \otimes f_2 \in L_3^*$ for every $f_1 \in L_1^*$, $f_2 \in L_2^*$.

4. Let $f : [0,\infty) \times [0,\infty) \to \mathbb{R}$ be defined by $f := \curlyvee_{(x,y)} (x - y)e^{-|x-y|}$. Show that

$$\int_0^\infty \left(\int_0^\infty f(x,y)\,dx \right) dy = \int_0^\infty (y + 1)e^{-y}\,dy = 2 \,,$$

$$\int_0^\infty \left(\int_0^\infty f(x,y)\,dy \right) dx = -\int_0^\infty (x + 1)e^{-x}\,dx = -2 \,.$$

Does this contradict Theorem 6.2.4?

5. Evaluate

$$\int_0^\infty \frac{1 - \cos x}{x^2}\,dx = \int_0^\infty \left(\int_0^\infty (1 - \cos x)e^{-xt}\, t\,dt \right) dx \,.$$

6. Evaluate

$$\int_0^\infty \frac{\sin x}{x^p}\, dx = \frac{1}{\Gamma(p)} \int_0^\infty \left(\int_0^\infty e^{-xt}\, t^{p-1} \sin x \, dt \right) dx$$

for $1 < p < 2$.

7. For $a > 0$ evaluate

$$\int_0^\infty \frac{\sin x}{x} e^{-ax} dx = \int_0^\infty \left(\int_0^\infty \sin x \cdot e^{-(a+t)x} dt \right) dx \; .$$

8. Let f and g be integrable over $\mathbb{R}$ (ordinary Lebesgue measure), and define

$$f_1 := \curlyvee_x \int_{-\infty}^x f(t)dt \; , \quad g_1 := \curlyvee_x \int_{-\infty}^x g(t)dt \; .$$

Show that

$$\int_{-\infty}^\infty [f_1(x)g(x) + f(x)g_1(x)]dx = \int_{-\infty}^\infty f(t)dt \cdot \int_{-\infty}^\infty g(t)dt \; .$$

Compare this with Exercise 5.4-14. (Hint. The set $E := \{(x,t) \mid x \in \mathbb{R}, t \in \mathbb{R}, t \le x\}$ is measurable in $\mathbb{R}^2$, and

$$\int_{-\infty}^\infty f(x)g_1(x)dx = \int_{-\infty}^\infty \left(\int_{-\infty}^\infty f(x)g(t)\chi_E(x,t)dt \right) dx \; .$$

Interchange the integrals.)

9. (Second mean value theorem.) Let f be integrable on $[a,b]$ and let g be non-decreasing on (a,b) with $g(a + 0) = 0$. Show that there is a $\xi \in [a,b]$ such that

$$\int_a^b f(t)g(t)dt = g(b - 0) \int_\xi^b f(t)dt$$

as follows.

(i) Extend g and f by putting $g(x) := 0$ $(x \le a)$, $g(x) = g(b - 0)$ $(x \ge b)$, $f(x) = 0$ $(x \notin [a,b])$, and let $F(x) := {}_{-\infty}\int^x f(t)dt$ $(x \in \mathbb{R})$. Show that there is a $\xi \in [a,b]$ such that

$$\int_{-\infty}^\infty F d\mu_g = g(b - 0)F(\xi) \; .$$

(ii) Evaluate $\int_{-\infty}^{\infty} F\,d\mu_g$ by interchanging the integrals in

$$\int_{-\infty}^{\infty}\left(\int_{-\infty}^{\infty} f(t)\chi_{(-\infty,x]}(t)dt\right)d\mu_g(x).$$

10. Let (S,Γ,μ) be a σ-finite measure space; let $S^2 = S \times S$ and take on S^2 the product semiring $\Gamma \times \Gamma$ with product measure $\mu \times \mu$. Denote the integrals over S by $\int f(s)d\mu(s)$ (for $f \in L^1(S)$) and integrals over S^2 by $\iint F(s,t)d\mu(s)d\mu(t)$ (for $F \in L^1(S^2)$). Let $K \in L^2(S^2)$.

(i) Let $f \in L^2(S)$. Show that $\int K(s,t)f(t)d\mu(t)$ is well-defined for almost every $s \in S$.

(ii) Let $f \in L^2(S)$. Show that

$$\curlyvee_s \int K(s,t)f(t)d\mu(t) =: T_K f \in L^2(S),$$

and that

$$\|T_K f\|_2^2 \leq \iint |K(s,t)|^2\, d\mu(s)d\mu(t)\cdot\|f\|_2^2.$$

(iii) Let $K^*(s,t) := K(t,s)$ for $(s,t) \in S^2$. Show that $K^* \in L^2(S^2)$ and that $(T_K f,g) = (f,T_{K^*}g)$ for f and g in $L^2(S)$. Here $(\cdot,\cdot)$ denotes the inner product in $L^2(S)$.

(iv) Let K_1 and K_2 belong to $L^2(S^2)$, and define

$$K_1 * K_2 := \curlyvee_{(s,t)} \int K_1(s,u)K_2(u,t)d\mu(u).$$

Show that $K_1 * K_2$ is defined almost everywhere on S^2, that $K_1 * K_2 \in L^2(S^2)$ and that $T_{K_1 * K_2} = T_{K_1} \circ T_{K_2}$ (where $\circ$ denotes the usual compositions of mappings).

6.3. The topological case

A second important construction of an integral on a product space $S_3 = S_1 \times S_2$ works with locally compact Hausdorff spaces S_i and integrals I_i $(i = 1,2)$ as considered in Chapter 4. Supplied with the product topology, S_3 is also locally compact and Hausdorff. An integral I_3 on the product is readily obtained: for a continuous real-valued function b of compact support on S_3 put $I_3 b := I_1 I_2 b$. It turns out that

$I_1I_2b = I_2I_1b$ for all such b, which is a rudimentary form of the theorem on change of order of integration. To obtain more general theorems of Fubini and Tonelli type, we have to check the norm condition (c) of Theorem 6.1.2. Since there are two ways to introduce norms on each of S_1, S_2 and S_3, namely the processes $\mathcal{D}' \to \mathcal{D} \to A \to N$ and $\mathcal{D}' \to A \to N$ described in Section 4.1, several different combinations should be considered. However, only two of these are commonly used, so we treat only those two.

6.3.1. Let (S_i, T_i) be a locally compact Hausdorff space, let B_i be the class of continuous real-valued functions of compact support on S_i, and let I_i be a positive linear functional on B_i for $i = 1,2$. Let $S_3 := S_1 \times S_2$, and give S_3 the product topology T_3 of T_1 and T_2, as defined in 0.2.18. Then (S_3, T_3) is again a locally compact Hausdorff space (see 0.2.19). Let B_3 denote the class of continuous real-valued functions on S_3 that have compact support.

6.3.2. We need the following purely topological result.

Theorem. Under the conditions stated in 6.3.1, let $b \in B_3$. Then there are functions $k_i \in B_i^+$ $(i = 1,2)$ with the following property: for every $\varepsilon > 0$ there exist functions $g_{1n} \in B_1$, $g_{2n} \in B_2$ $(1 \le n \le N)$ such that

$$0 \le \sum_{n=1}^{N} g_{1n}(s)g_{2n}(t) - b(s,t) \le \varepsilon k_1(s)k_2(t)$$

for $(s,t) \in S_3$.

Proof. Since b has compact support, there exist compact sets $C_i \subset S_i$ $(i = 1,2)$ such that the support of b is contained in $C_1 \times C_2$ (Exercise 0.2-17(iii)). For $i = 1,2$, let U_i be an open set in S_i with compact closure $V_i := \bar{U}_i$ such that $C_i \subset U_i$, and let $k_i \in B_i^+$ satisfy $k_i(s_i) = 1$ for $s_i \in V_i$, $k_i(s_i) \le 1$ for $s_i \in S_i$. (The existence of k_1 and k_2 follows from Urysohn's lemma 0.2.16.)

Let $\varepsilon > 0$. We assert that we can find open sets $O_{11}, O_{12}, \ldots, O_{1p}$ in S_1, all contained in U_1, and together covering C_1, such that $|b(s,t) - b(\sigma,t)| < \frac{1}{2}\varepsilon$ for all $s \in O_{1n}$, $\sigma \in O_{1n}$, $t \in C_2$ $(1 \le n \le p)$. The proof is by two compactness arguments. First, let $s \in C_1$ be fixed. If $t \in C_2$, then by the continuity of b in (s,t) there exists an open neighborhood $O_1(t)$ of s, and an open neighborhood $O_2(t)$ of t such that b varies less than $\frac{1}{2}\varepsilon$ on the product set $O_1(t) \times O_2(t)$. Obviously, we may assume $O_1(t) \subset U_1$, $O_2(t) \subset U_2$. By compactness, C_2 may be covered by finitely many $O_2(t)$'s. Taking the intersection of the corresponding $O_1(t)$'s, we obtain an open neighborhood $O_1'(s)$ of s such that $|b(\sigma,t) - b(\sigma',t)| \le \frac{1}{2}\varepsilon$ for every $\sigma, \sigma' \in O_1'(s)$, $t \in C_2$. Next, s is allowed to vary: the neighborhoods $O_1'(s)$ form an open covering of C_1. Since C_1 is compact, there is a finite subcovering: these are the $O_{11}, O_{12}, \ldots, O_{1p}$ we are looking for.

A similar result holds for C_2: there exist open sets $O_{21},O_{22},\dots,O_{2q}$, all contained in U_2, and covering C_2, such that $|b(s,\tau) - b(s,\tau')| < \frac{1}{2}\varepsilon$ for all $\tau,\tau' \in O_{2m}$, $s \in C_1$ $(1 \le m \le q)$.

Now apply the extended version of Urysohn's lemma 0.2.17. It follows that there are $h_{11},h_{12},\dots,h_{1p}$ in B_1^+ such that $h_{1n}(s) = 0$ $(s \in O_{1n}, 1 \le n \le p)$, $\sum_{n=1}^{p} h_{1n}(s) = 1$ $(s \in C_1)$, and there are similar functions $h_{21},h_{22},\dots,h_{2q}$ in B_2^+. For each pair (n,m) with $1 \le n \le p$, $1 \le m \le q$, let

$$a_{nm} := \sup\{b(s,t) \mid s \in O_{1n},\ t \in O_{2m}\} .$$

Obviously,

$$0 \le \{a_{nm} - b(s,t)\}h_{1n}(s)h_{2m}(t) \le \varepsilon h_{1n}(s)h_{2m}(t)$$

for all $(s,t) \in S_3$. Hence

$$0 \le \sum_{n=1}^{p}\sum_{m=1}^{q} a_{nm}h_{1n}(s)h_{2m}(t) - b(s,t) \le$$

$$\le \varepsilon \sum_{n=1}^{p}\sum_{m=1}^{q} h_{1n}(s)h_{2m}(t) \le \varepsilon k_1(s)k_2(t)$$

for all $(s,t) \in S_3$. A simple change in notation gives the result. □

6.3.3. If $b \in B_3$, that is, if b is a continuous real-valued function of compact support on S_3, then for each $s \in S_1$ the function $b_s := \curlyvee_{t\in S_2}[b(s,t)]$ belongs to B_2. In fact, such a function is obviously continuous, and it has compact support because the support of b may be enclosed in a rectangle $C_1 \times C_2$ with compact sides as in the first line of the preceding proof. Hence, $\curlyvee_{s\in S_1}[I_2 b_s]$ is a function on S_1, which we denote by $I_2 b$ as in 6.1.1. The function $I_1 b$ is defined similarly on S_2.

<u>Theorem</u>. Under the conditions stated in 6.3.1, let $b \in B_3$. Then $I_2 b \in B_1$, $I_1 b \in B_2$ and $I_1 I_2 b = I_2 I_1 b$.

<u>Proof</u>. Apply Theorem 6.3.2 to b. Let k_1 and k_2 be as described there, and let $\varepsilon > 0$. Take $g_{1n} \in B_1$, $g_{2n} \in B_2$ $(1 \le n \le N)$ such that

$$0 \le \sum_{n=1}^{N} g_{1n}(s)g_{2n}(t) - b(s,t) \le \varepsilon k_1(s)k_2(t)$$

for $(s,t) \in S_3$. Since I_2 is positive and linear, it follows that

$$(*) \qquad 0 \le \sum_{n=1}^{N} I_2(g_{2n})g_{1n}(s) - I_2 b_s \le \varepsilon I_2(k_2)k_1(s)$$

for all $s \in S_1$. This shows in the first place that I_2b has compact support. And secondly it shows that I_2b can be approximated uniformly by continuous functions, whence I_2b continuous. So $I_2b \in B_1$. Similarly, $I_1b \in B_2$.

Returning to (*), we see that

$$0 \leq \sum_{n=1}^{N} I_2(g_{2n})I_1(g_{1n}) - I_1I_2b \leq \varepsilon I_2(k_2)I_1(k_1) .$$

Repeating this procedure, but now first applying I_1 and then I_2, we get a similar inequality with I_1I_2b replaced by I_2I_1b. Hence $|I_1I_2b - I_2I_1b| \leq \varepsilon I_2(k_2)\cdot I_1(k_1)$. Since $\varepsilon > 0$ is arbitrary, the repeated integrals are equal. □

6.3.4. It is now easy to obtain an appropriate positive linear functional I_3 on B_3. In fact, if I_3 is defined by

$$I_3b := I_1I_2b \qquad (b \in B_3) ,$$

then this definition is justified by the preceding theorem. Now whatever norms are used on Φ_1 and Φ_2, one always has $B_1 \subset L_1$ and $B_2 \subset L_2$. So condition (a) of the Fubini-Stone theorem 6.1.2 is satisfied in any case. This also holds for condition (b), for (b) asks exactly what the definition of I_3 gives.

6.3.5. To obtain integrals on S_1, S_2 and S_3, norms must be introduced. We consider two cases:

case (i) : on each space follow $\mathcal{T} \to \mathcal{D}' \to \mathcal{D} \to \mathcal{A} \to \mathcal{N}$,

case (ii): on each space follow $\mathcal{T} \to \mathcal{D}' \to \mathcal{A} \to \mathcal{N}$.

These processes have been described in 4.1.4 and Sections 1.3 and 1.2. In both cases the resulting norms are denoted by N_1, N_2 and N_3; this will cause no confusion. The following result shows that condition (c) of the Fubini-Stone theorem is satisfied in either case. Notation is as in 6.1.1.

<u>Proposition</u>. In cases (i) and (ii), $N_3\varphi \geq N_1N_2\varphi \quad (\varphi \in \Phi_3)$.

<u>Proof</u>. First consider case (i), which is by far the easier of the two. Let $\varphi \in \Phi_3$, and assume $N_3\varphi < \infty$. Let $(b_n)_{n\in\mathbb{N}}$ be a sequence in B_3^+ such that $|\varphi| \leq \sum_{n=1}^{\infty} b_n$. Then $0 \leq N_2\varphi_s \leq \sum_{n=1}^{\infty} N_2(b_n)_s$ for all $s \in S_1$, and therefore $N_1N_2\varphi \leq \sum_{n=1}^{\infty} N_1N_2b_n$. Since $N_1N_2b_n = N_1I_2b_n = I_1I_2b_n = I_3b_n$ for each $n \in \mathbb{N}$ by Theorem 1.1.10, Theorem 6.3.3 and the definition of I_3, the definition of $N_3\varphi$ in 1.3.3 shows that $N_1N_2\varphi \leq N_3\varphi$.

Next consider case (ii). Again, let $\varphi \in \Phi_3$ with $N_3\varphi < \infty$. By 1.3.12(ii) we have

$$N_3\varphi = \inf\{J_3(h) \mid h \in A_3, |\varphi| \leq h\} ,$$

where A_3 is defined in the obvious way. Hence we need only show $J_3(h) \geq N_1N_2h$ for all $h \in A_3$.

Let $h \in A_3$. For each $s \in S_1$, write $W(s) := \{b_s \mid b \in B_3^+, b \leq h\}$. Obviously, $W(s)$ is a directed subset of B_2^+ with $\sup_{c \in W(s)} c = h_s$. Hence $h_s \in A_2$ and by Theorem 1.3.12(iii) and 1.3.8,

$$(*) \qquad N_2(h_s) = J_2(h_s) = \sup_{c \in W(s)} I_2(c) = \sup_{b \in B_3^+, b \leq h} I_2(b_s) .$$

If $W := \{I_2b \mid b \leq h, b \in B_3^+\}$, then W is a directed subset of B_1^+ (use 6.3.3), and $(*)$ shows that $\sup_{c \in W} c = N_2h$. Hence $N_2h \in A_1$, and again by Theorem 1.3.12(iii) and 1.3.8,

$$N_1N_2h = J_1(N_2h) = \sup_{c \in W} I_1(c) = \sup_{b \in B_3^+, b \leq h} I_1I_2b = \sup_{b \in B_3^+, b \leq h} I_3b = J_3h . \qquad \square$$

6.3.6. Here is the Fubini-Tonelli result for the topological case. It is just a reformulation for the present situation of the results in Section 6.1, so no formal proof is needed.

Theorem. Let (S_i,T_i) be a locally compact Hausdorff space with a positive linear functional I_i on the class B_i of continuous real-valued functions of compact support on S_i for $i = 1,2$. Let (S_3,T_3) be the topological product of (S_1,T_1) and (S_2,T_2), and define I_3 on B_3 as in 6.3.4. Introduce norms on S_1, S_2 and S_3 as described in 6.3.5, and form spaces of integrable functions in the usual way. Then the assertions (i)-(v) of Theorem 6.2.4 hold.

Exercises Section 6.3

1. Take $S_1 = S_2 = \mathbb{R}$ with the usual topology and take for I_1 and I_2 the ordinary Riemann integral. Show that both cases discussed in 6.3.5 yield the same class of integrable functions (with the same integral) as that of 6.2.5.

2. (i) Give an example of situations $\mathcal{T}(S_i,T_i,I_i)$ $(i = 1,2)$ such that there is an f which is measurable over $S_3 = S_1 \times S_2$ and for which f^t is not measurable over S_1 for $t \in S_2$.

(ii) Give an example of situations $\mathcal{T}(S_i,T_i,I_i)$ $(i = 1,2)$ such that there is an f which is measurable and non-negative over $S_3 = S_1 \times S_2$, for which $f^t \in M_1$, $f_s \in M_2$ for $t \in S_2$, $s \in S_1$, for which $I_1f \in M_2$, $I_2f \in M_1$ and nevertheless $I_1I_2f \neq I_2I_1f$.

6.4. The Fourier transform in $L^2(\mathbb{R})$

As an application of product integrals we treat part of the theory of the Fourier transform, which is a useful tool in many parts of mathematics. This transform can be defined in many different situations; in all cases theorems of Fubini type play an important role. In the text we consider the Fourier transform on $L^2(\mathbb{R})$ only, but there are some extensions in the exercises.

6.4.1. At this point it is convenient to allow complex-valued functions in the theory, and we extend our notions accordingly.

Assume first that we have a situation $L(L,I)$ as in Section 1.1. So let S be a non-empty set, let L be a Riesz function space of real-valued functions defined on S, let I be a positive linear functional defined on L, and assume that $L1$ holds. Let f be a complex-valued function defined on S, and write $f = u + iv$, where $u := \text{Re } f$ is the real part of f, and $v := \text{Im } f$ is the imaginary part of f. If $u,v \in L$, then we define $I(f) := I(u) + iI(v)$. The space of complex-valued functions that have an I-value assigned in this way is temporarily denoted by $\tilde{L}$. (We still use I for the extended functional, for although $L \subset \tilde{L}$, the above definition yields nothing new on L, so there is no danger of confusion.) $\tilde{L}$ is a complex vector space, and the extended I is linear, that is, if $f,g \in \tilde{L}$ and $c \in \mathbb{C}$, then $f + cg \in \tilde{L}$ and $I(f + cg) = I(f) + cI(g)$. Moreover, if $f \in \tilde{L}$ and $|f| \in \tilde{L}$, then $|I(f)| \le I(|f|)$ (Exercise 6.4-1). Measurability of complex-valued functions is defined similarly by decomposition into real and imaginary parts.

It is even simpler to extend the norm to complex-valued functions in a situation $N(B,I,\|\cdot\|)$: if $\varphi : S \to \mathbb{C}$, then $\|\varphi\| := \||\varphi|\|$, where the latter is just the old norm on Φ. In such a case it is also convenient to consider complex-valued functions that are defined almost everywhere on S. All this gives no trouble at all.

Many of the theorems obtained so far for real-valued functions are valid for complex-valued functions as well, sometimes after a slight and obvious reformulation. We mention in particular Lebesgue's theorem on dominated convergence, Hölder's inequality, the approximation theorem 2.3.13 and the Fubini-Tonelli theorems. All these results will be used without further comment.

6.4.2. In the remainder of this section we use the word measure for Lebesgue measure μ on $\mathbb{R}$, or for two-dimensional Lebesgue measure $\mu \times \mu$ on $\mathbb{R}^2$. Hence the open sets in $\mathbb{R}$ or $\mathbb{R}^2$ are measurable, and so are the continuous complex-valued functions defined on $\mathbb{R}$ or $\mathbb{R}^2$. We write L^p instead of $L^p(\mathbb{R})$, and permit ourselves the usual liberty in the interpretation of the elements of these spaces (see 2.3.12) to keep the notation simple.

So L^2 is the space of (equivalence classes of) complex-valued functions on $\mathbb{R}$ that are measurable and have $\int_{-\infty}^{\infty} |f(x)|^2 dx$ finite. As in the real-valued case, an inner product can be defined in L^2, but it looks a little bit different, namely

$$(f,g) := \int_{-\infty}^{\infty} f(x)\overline{g(x)}dx \qquad (f,g \in L^2) ,$$

where the bar denotes complex conjugation. The inner product and the L^2-norm are connected in the usual way:

$$\|f\|_2 = \sqrt{(f,f)} \qquad (f \in L^2) .$$

Thus, L^2 is made into a complex Hilbert space; in particular, the space is complete with the metric $d(f,g) := \|f - g\|_2$ for $f,g \in L^2$.

The members of L^2 also have a norm as members of Φ^*; this norm for an $f \in L^2$ is denoted by $\|f\|$, as usual. Obviously, $\|f\| = {}_{-\infty}\int^{\infty}|f(x)|dx$ for all $f \in L^2$, because the functions $|f|$ are always measurable and non-negative.

6.4.3. We start by defining the Fourier transform on $L^1 \cap L^2$. This is easy. For, let $f \in L^1 \cap L^2$. If $x \in \mathbb{R}$, then the integral

$$\frac{1}{\sqrt{2\pi}} \int_{-\infty}^{\infty} e^{-iux} f(u)du$$

exists, because the integrand is measurable and has finite norm; we denote the value by $\mathcal{F}f(x)$. Thus a complex-valued function $\mathcal{F}f$ is obtained, which we call the *Fourier transform of* f.

Our first aim is to show that the Fourier transform of such an f belongs to L^2 and that it has the same L^2-norm as f. Some of the steps in the proof are isolated in separate lemmas; we start with the most subtle one.

6.4.4. Lemma. (i) Let $E \subset \mathbb{R}$ be a null set. Let $F := \{(u,v) \in \mathbb{R}^2 \mid u + v \in E\}$. Then F is a null set in $\mathbb{R}^2$.

(ii) Let $f : \mathbb{R} \to \mathbb{C}$ be measurable. Then $h := \Upsilon_{(u,v)\in\mathbb{R}^2}[f(u + v)]$ is measurable.

Proof. (i) Fix $n \in \mathbb{N}$. Let χ_n be the characteristic function of the square $\{(u,v) \in \mathbb{R}^2 \mid |u| \le n, |v| \le n\}$. We shall show that $\chi_F \cdot \chi_n$ is a null function. Let $\varepsilon > 0$, and choose an open set $O \supset E$ such that $\mu(O) < \varepsilon/2n$. It is easy to see that $g := \Upsilon_{(u,v)\in\mathbb{R}^2} \chi_O(u + v)$ is the characteristic function of an open set in $\mathbb{R}^2$, so it is measurable. Since χ_n is obviously measurable, Fubini's theorem shows that

$$\|g\cdot\chi_n\| = \int_{-\infty}^{\infty}\int_{-\infty}^{\infty} g(u,v)\chi_n(u,v)dudv = \int_{-n}^{n}\left(\int_{-n}^{n} \chi_O(u + v)du\right) dv < \varepsilon .$$

Hence $\|\chi_F \cdot \chi_n\| < \varepsilon$. Therefore $\|\chi_F \cdot \chi_n\| = 0$. Since $|\chi_F| \le \sum_{n=1}^{\infty} |\chi_F \cdot \chi_n|$, it follows that $\|\chi_F\| = 0$.

(ii) Clearly, we may assume that f is real-valued. By 2.1.13 and 1.2.4 there exists a sequence $(f_n)_{n \in \mathbb{N}}$ of continuous real-valued functions on $\mathbb{R}$ such that $f_n \to f$ (a.e.). Put $h_n := \curlyvee_{(u,v)} f_n(u + v)$ for all $n \in \mathbb{N}$. Each h_n is continuous on $\mathbb{R}^2$ and therefore measurable. Now (i) shows that $h_n \to h$ a.e. on $\mathbb{R}^2$. Hence, h is measurable. (Note also that it does not matter if f is defined only almost everywhere.) □

6.4.5. Lemma. Let $\psi : \mathbb{R} \to \mathbb{C}$ be continuous and bounded. Then

$$\lim_{\delta \downarrow 0} \frac{1}{\delta\sqrt{\pi}} \int_{-\infty}^{\infty} e^{-x^2/\delta^2} \psi(x)dx = \psi(0) .$$

Proof. Left to the reader as Exercise 6.4-2. □

6.4.6. Lemma. Let $f \in L^2$, $g \in L^2$. Then

$$\psi := \curlyvee_{t \in \mathbb{R}} \int_{-\infty}^{\infty} f(u + t)\overline{g(u)}du$$

is bounded and continuous.

Proof. Hölder's inequality 2.3.8 shows that ψ is well-defined and bounded. If f is continuous and has compact support, it is easy to show that ψ is continuous. In the general case, take a sequence $(f_n)_{n \in \mathbb{N}}$ of continuous functions of compact support (using 2.3.13 and 1.2.4) such that $\|f_n - f\|_2 \to 0$. Another application of Hölder's inequality shows that

$$\curlyvee_{t \in \mathbb{R}} \int_{-\infty}^{\infty} f_n(u + t)\overline{g(u)}du \to \psi ,$$

uniformly on $\mathbb{R}$. Hence ψ is continuous. □

6.4.7. Here is the intermediate result that we have already announced. Its proof contains several applications of the Fubini-Tonelli results, in each of which the appropriate conditions must be checked. The measurability requirements are easily seen to be verified and the rest consists of obvious estimations; we leave this to the reader as Exercise 6.4-5.

<u>Theorem</u>. Let $f \in L^1 \cap L^2$, and define

$$Ff := \mathop{\Upsilon}_{x\in\mathbb{R}} \frac{1}{\sqrt{2\pi}} \int_{-\infty}^{\infty} e^{-iux} f(u)du .$$

Then $Ff \in L^2$, and $\|Ff\|_2 = \|f\|_2$.

<u>Proof</u>. It is obvious that Ff is bounded, and it is not difficult to show that Ff is continuous (Exercise 6.4-3).

Let $\delta > 0$. Then

$$\int_{-\infty}^{\infty} e^{-\delta^2x^2} |(Ff)(x)|^2 dx =$$

$$= \frac{1}{2\pi} \int_{-\infty}^{\infty} e^{-\delta^2x^2}\left(\int_{-\infty}^{\infty} e^{-iux} f(u)du\right)\left(\int_{-\infty}^{\infty} e^{ivx} \overline{f(v)}dv\right) dx ,$$

which by an application of Fubini's theorem to the integration over u and x is equal to

$$\frac{1}{2\pi} \int_{-\infty}^{\infty} f(u)\left\{\int_{-\infty}^{\infty} e^{-\delta^2x^2-iux}\left(\int_{-\infty}^{\infty} e^{ivx} \overline{f(v)}dv\right) dx\right\}du .$$

By another application of Fubini's theorem, this time to the integration over x and v, the last expression is

$$(*) \qquad \frac{1}{2\pi} \int_{-\infty}^{\infty} f(u)\left\{\int_{-\infty}^{\infty} \overline{f(v)}\left(\int_{-\infty}^{\infty} e^{-\delta^2x^2+i(v-u)x} dx\right) dv\right\}du =$$

$$= \frac{1}{2\delta\sqrt{\pi}} \int_{-\infty}^{\infty} f(u)\left(\int_{-\infty}^{\infty} \overline{f(v)}\, e^{-(\frac{v-u}{2\delta})^2} dv\right)du =$$

$$= \frac{1}{2\delta\sqrt{\pi}} \int_{-\infty}^{\infty}\int_{-\infty}^{\infty} f(u)\overline{f(v)}\, e^{-(\frac{v-u}{2\delta})^2} du\,dv$$

where we used Exercise 6.4-4 in the first step.

Next write the right-hand side of (*) as

$$\frac{1}{2\delta\sqrt{\pi}} \int_{-\infty}^{\infty} f(u) \int_{-\infty}^{\infty} \overline{f(u+t)}\, e^{-t^2/4\delta^2} dt\,du = \frac{1}{2\delta\sqrt{\pi}} \int_{-\infty}^{\infty} e^{-t^2/4\delta^2}\psi(t)dt ,$$

where

$$\psi(t) := \int_{-\infty}^{\infty} f(u)\overline{f(u+t)}\,du \qquad (t \in \mathbb{R}),$$

(this is justified by Fubini's theorem, the result of Exercise 1.2-4, and Lemma 6.4.4). Since ψ is bounded and continuous (Lemma 6.4.6), an application of Lemma 6.4.5 and of the monotone convergence theorem gives

$$\int_{-\infty}^{\infty} |(Ff)(x)|^2\,dx = \lim_{\delta \downarrow 0} \int_{-\infty}^{\infty} e^{-\delta^2 x^2} |(Ff)(x)|^2\,dx =$$

$$= \lim_{\delta \downarrow 0} \frac{1}{2\delta\sqrt{\pi}} \int_{-\infty}^{\infty} e^{-t^2/4\delta^2} \psi(t)dt = \psi(0) = \|f\|_2^2 .$$

That is, $Ff \in L^2$ and $\|Ff\|_2 = \|f\|_2$. □

6.4.8. Obviously, $L^1 \cap L^2$ is a complex linear space, and F maps $L^1 \cap L^2$ linearly into L^2. Combining this with the fact that F preserves norms, we can extend F to all of L^2. This gives the principal result of the present section, which is always named after Plancherel. The technique we employ is similar to the one used in Section 1.1 in extending I from B to L.

Theorem (Plancherel).

(i) F can be extended in a unique way to a norm preserving linear mapping from L^2 into L^2. This extension (again denoted by F) maps L^2 onto L^2.

(ii) Let $f \in L^2$, and define

$$F_a := Y_{x\in\mathbb{R}}\left[\frac{1}{\sqrt{2\pi}} \int_{-a}^{a} e^{-iux} f(u)du\right] \qquad (a > 0) .$$

Then each $F_a \in L^2$, and $\|Ff - F_a\|_2 \to 0$ as $a \to \infty$.

(iii) Let $f \in L^2$, and define

$$f_a := Y_{u\in\mathbb{R}}\left[\frac{1}{\sqrt{2\pi}} \int_{-a}^{a} e^{iux}(Ff)(x)dx\right] \qquad (a > 0) .$$

Then each $f_a \in L^2$, and $\|f - f_a\|_2 \to 0$ as $a \to \infty$.

Proof. (i) Let $f \in L^2$. There exists a sequence of continuous complex-valued functions on $\mathbb{R}$ with compact support such that $\|f_n - f\|_2 \to 0$. Since $(f_n)_{n\in\mathbb{N}}$ is a fundamental sequence in L^2, and each $f_n \in L^1 \cap L^2$, Ff_n is defined for each $n \in \mathbb{N}$, and $(Ff_n)_{n\in\mathbb{N}}$ is a fundamental sequence in L^2. Since L^2 is complete, there exists $g \in L^2$ such that $\|Ff_n - g\|_2 \to 0$. If a norm preserving extension of F onto all of L^2 is possible at all, Ff must be equal to g. Therefore we define $Ff := g$.

As in the proof of Theorem 1.1.10 it may be shown that the function Ff thus defined does not depend on the particular sequence $(f_n)_{n\in\mathbb{N}}$ chosen in defining it. In particular, if Ff was already defined by the integral formula of 6.4.3, the process described gives nothing new. (This also justifies our using the same letter F.) It is easy to show that the extension obtained is linear and norm preserving.

The remainder of (i) is a consequence of (iii), so we treat this later.

(ii) Put $h_a := f\cdot\chi_{[-a,a]}$ for $a > 0$. Then each truncated function h_a belongs to $L^1 \cap L^2$, so $F_a = Fh_a$. Since $\|f - h_a\|_2 \to 0$ as $a \to \infty$ by Lebesgue's theorem on dominated convergence, it follows that $\|Ff - F_a\|_2 = \|F(f - h_a)\|_2 = \|f - h_a\|_2 \to 0$ as $a \to \infty$, by the definition of Ff.

(iii) Since F preserves norms, it also preserves inner products, that is,

$$(Ff,Fg) = (f,g) \qquad (f \in L^2,\ g \in L^2)$$

(Exercise 6.4-6). Apply this remark to $f \in L^2$ and $g := \chi_{[0,t]}$. The Fourier transform of g is given by a simple integral which is readily computed. It follows that

$$(*) \qquad \int_0^t f(u)du = \int_{-\infty}^{\infty} (Ff)(x)\,\frac{e^{itx} - 1}{ix}\,dx\ .$$

On the other hand, if $a > 0$, then

$$\int_0^t f_a(u)du = \int_0^t \left\{ \int_{-a}^{a} e^{iux}(Ff)(x)dx \right\} du =$$

$$= \int_{-a}^{a} (Ff)(x) \int_0^t e^{iux}\,dx = \int_{-a}^{a} (Ff)(x)\,\frac{e^{itx} - 1}{ix}\,dx$$

by Fubini's theorem. Now let $a \to \infty$. By Lebesgue's theorem the last expression tends to the right-hand side of (*). Hence

$$\lim_{a\to\infty} \int_0^t f_a(u)du = \int_0^t f(u)du\ .$$

Let h denote the Fourier transform of $\Upsilon_{x\in\mathbb{R}}(\mathcal{F}f)(-x)$. By (i) and (ii), $\|f_a - h\|_2 \to 0$ as $a \to \infty$. It is an easy consequence of Hölder's inequality that

$$\int_0^t h(u)du = \lim_{a\to\infty} \int_0^t f_a(u)du = \int_0^t f(u)du$$

for each $t \geq 0$. A similar thing holds for $t < 0$, and we conclude from Lemma 5.4.9 that $h = f$ (a.e.). Therefore $\|f - f_a\|_2 \to 0$ as $a \to \infty$, and it also follows that $\mathcal{F}$ maps L^2 onto L^2. □

Exercises Section 6.4

1. Let $\tilde{L}$ be as in 6.4.1. Show that $|I(f)| \leq I|f|$ if $f \in \tilde{L}$, $|f| \in \tilde{L}$. (Hint. Write $f = f_1 + if_2$, $I(f) = re^{i\theta}$, and note that $r = I(e^{-i\theta}(f_1 + if_2)) =$ $= I(f_1 \cos\theta + f_2 \sin\theta)$ while $|f_1 \cos\theta + f_2 \sin\theta| \leq |f|$.)

2. Let ψ be a bounded, continuous function defined on $\mathbb{R}$. Show that

$$\lim_{\delta\downarrow 0} \frac{1}{\delta\sqrt{\pi}} \int_{-\infty}^{\infty} e^{-x^2/\delta^2}\psi(x)dx = \psi(0) .$$

(Hint. Exercise 5.5-2.)

3. Let $f \in L^1$ and let

$$(\mathcal{F}f)(x) = \frac{1}{\sqrt{2\pi}} \int_{-\infty}^{\infty} e^{-iux} f(u)du$$

for $x \in \mathbb{R}$. Show that $\mathcal{F}f$ is bounded and continuous and that $\lim_{x\to\infty}(\mathcal{F}f)(x) = 0$.

4. Let

$$F(t) := \int_{-\infty}^{\infty} e^{-x^2} \cos(tx)dx .$$

Show that F is continuous and differentiable on $\mathbb{R}$. Compute F' and deduce that $F(t) = Ce^{-\frac{1}{4}t^2}$, where $C = F(0) = \sqrt{\pi}$.

5. Justify the applications of the Fubini-Tonelli theorem in the proof of Theorem 6.4.7.

6. If $f \in L^2$, $g \in L^2$, then $(Ff, Fg) = (f,g)$.

7. This exercise lines out an alternative proof of Plancherel's theorem for Fourier integrals. We use the following: If g is periodic (with period 2π) and twice continuously differentiable, then

$$g(x) = \sum_{n=-\infty}^{\infty} a_n e^{inx}$$

where

$$a_n = \frac{1}{2\pi} \int_0^{2\pi} e^{-int} g(t)dt$$

for $n = 0, \pm 1, \pm 2, \ldots$, and

$$\frac{1}{2\pi} \int_0^{2\pi} |g(x)|^2 \, dx = \sum_{n=-\infty}^{\infty} |a_n|^2 .$$

The series $\sum_{n=-\infty}^{\infty} a_n e^{inx}$ converges absolutely and uniformly.

(i) Let $f : \mathbb{R} \to \mathbb{C}$ be twice continuously differentiable, and assume f has compact support. Define $g(t) = \sum_{k=-\infty}^{\infty} f(t - 2\pi k)$. Show that g is periodic (with period 2π) and twice continuously differentiable.

(ii) Write $g(x) = \sum_{n=-\infty}^{\infty} a_n e^{inx}$ with a_n as above. Show that

$$a_n = \frac{1}{\sqrt{2\pi}} (Ff)(n)$$

so that in particular

$$\sum_{k=-\infty}^{\infty} f(2\pi k) = \frac{1}{\sqrt{2\pi}} \sum_{k=-\infty}^{\infty} (Ff)(k) .$$

(This result is known as the *Poisson summation formula*.)

(iii) Put $f_y(t) := e^{-iyt} f(t)$, $g_y(t) = \sum_{k=-\infty}^{\infty} f_y(t - 2\pi k)$ for $y \in \mathbb{R}$. Show that

$$\int_0^{2\pi} |g_y(t)|^2 \, dt = \sum_{n=-\infty}^{\infty} |(Ff)(n + y)|^2$$

and that

$$\int_0^1 \left(\int_0^{2\pi} |g_y(t)|^2 \, dt \right) dy = \int_{-\infty}^{\infty} |f(t)|^2 \, dt .$$

Then conclude that $\|f\|^2 = \|Ff\|^2$, and complete the proof of Plancherel's theorem.

8. Let $f \in L^2$ and show that a.e.

$$f = \frac{d}{dt}\left[\frac{1}{\sqrt{2\pi}} \int_{-\infty}^{\infty} (Ff)(x)\, \frac{e^{itx} - 1}{ix}\, dx\right],$$

$$Ff = \frac{d}{dt}\left[\frac{1}{\sqrt{2\pi}} \int_{-\infty}^{\infty} f(x)\, \frac{e^{-itx} - 1}{ix}\, dx\right].$$

9. Let $f \in L^1$, $g \in L^1$. Using the Fubini-Tonelli theorem, show that for almost every $x \in \mathbb{R}$ the integral $\int_{-\infty}^{\infty} f(t)g(x - t)dt$ exists. Denote $f * g = \forall_x \int_{-\infty}^{\infty} f(t)g(x-t)dt$ (*convolution* of f and g). Show that $f * g = g * f$, $f * g \in L^1$, $\|f * g\|_1 \leq \|f\|_1 \|g\|_1$, and that $(f * g) * h = f * (g * h)$ for $h \in L^1$.

10. Let $f \in L^1$, $g \in L^1$. Show that $F(f * g) = Ff \cdot Fg$ (pointwise product; also see Exercise 3).

11. Let $\lambda > 0$, $x \in \mathbb{R}$ and define

$$G := \forall_u \left(1 - \frac{|u|}{\lambda}\right) e^{-ixu}\, \chi_{[-\lambda,\lambda]}(u).$$

Show that

$$(FG)(v) = \frac{4}{\sqrt{2\pi}}\, \frac{\sin^2 \frac{1}{2}\lambda(x + v)}{\lambda(x + v)^2} \qquad (v \in \mathbb{R}),$$

and that

$$\frac{2}{\pi} \int_{-\infty}^{\infty} \frac{\sin^2 \frac{1}{2}\lambda v}{\lambda v^2}\, dv = 1.$$

12. Let f be measurable over $\mathbb{R}$ and assume that $\forall_x \frac{f(x)}{1 + x^2} \in L^1$. Assume further that

$$h^{-1} \int_0^h |f(v) - f(0)|\, dv \to 0 \qquad (h \downarrow 0).$$

Show that

$$\lim_{\lambda \to \infty} \frac{4}{\pi} \int_0^{\infty} f(v)\, \frac{\sin^2 \frac{1}{2}\lambda v}{\lambda v^2}\, dv = f(0)$$

as follows.

(i) Show that

$$\frac{\sin^2 \frac{1}{2}\lambda v}{\lambda v^2} \le \min(\tfrac{1}{4}\lambda, (\lambda v^2)^{-1})$$

for $\lambda > 0$, $v \in \mathbb{R}$, $v \neq 0$.

(ii) Let $A > 0$. Show that

$$\lim_{\lambda\to\infty} \frac{4}{\pi} \int_A^\infty |f(v) - f(0)| \frac{\sin^2 \frac{1}{2}\lambda v}{\lambda v^2} dv = 0 .$$

(iii) Show that

$$\lim_{\lambda\to\infty} \frac{4}{\pi} \int_0^{\lambda^{-1}} |f(v) - f(0)| \frac{\sin^2 \frac{1}{2}\lambda v}{\lambda v^2} dv = 0 .$$

(iv) Let $\varepsilon > 0$, and take $A > 0$ such that

$$\chi(h) := \int_0^h |f(v) - f(0)| dv \le \varepsilon h \qquad (0 \le h \le A) .$$

Show that

$$\int_{\lambda^{-1}}^A \frac{|f(v) - f(0)|}{\lambda v^2} dv = \frac{\chi(A)}{A^2\lambda} - \lambda\chi(\lambda^{-1}) + 2 \int_{\lambda^{-1}}^A \frac{\chi(v)}{\lambda v^3} dv < 3\varepsilon$$

if λ is sufficiently large.

(v) Deduce from (i)-(iv) and Exercise 11 the desired result.

13. (i) Let $f \in L^2$. Show that for $\lambda > 0$, $x \in \mathbb{R}$

$$\int_{-\lambda}^{\lambda} \left(1 - \frac{|u|}{\lambda}\right)(Ff)(u)e^{ixu} du = \frac{4}{\sqrt{2\pi}} \int_{-\infty}^{\infty} f(v) \frac{\sin^2 \frac{1}{2}\lambda(x - v)}{\lambda(x - v)^2} dv .$$

(ii) Let $f \in L^2$. Show that for almost every $x \in \mathbb{R}$

$$f(x) = \lim_{\lambda\to\infty} \frac{1}{\sqrt{2\pi}} \int_{-\lambda}^{\lambda} \left(1 - \frac{|u|}{\lambda}\right)(Ff)(u)e^{ixu} du .$$

(Hint. Exercise 5.4-17.)

(iii) Let $f \in L^1$. Show that (cf. Exercise 3) for $\lambda > 0$, $x \in \mathbb{R}$

$$\int_{-\lambda}^{\lambda} \left(1 - \frac{|u|}{\lambda}\right)(Ff)(u)e^{ixu}\,du = \frac{4}{\sqrt{2\pi}} \int_{-\infty}^{\infty} f(v)\,\frac{\sin^2 \frac{1}{2}\lambda(x-v)}{\lambda(x-v)^2}\,dv .$$

(iv) Let $f \in L^1$. Show that for almost every $x \in \mathbb{R}$

$$f(x) = \lim_{\lambda\to\infty} \frac{1}{\sqrt{2\pi}} \int_{-\lambda}^{\lambda} \left(1 - \frac{|u|}{\lambda}\right)(Ff)(u)e^{ixu}\,du .$$

14. Show that $(F(Ff))(x) = f(-x)$ for almost every $x \in \mathbb{R}$ if $f \in L^2$.

6.5. Measurable modifications of functions on product spaces

The final section of this book is devoted to a rather special question which has some importance in the theory of stochastic processes. The origin of the problem will be described in the simplest possible terms, without going into much detail. The tools we use to solve the problem are the Radon-Nikodym theorem of Section 5.2 and the theory of product measures of Section 6.2.

6.5.1. Let (S_1,Γ_1,μ_1) be a σ-finite measure space, and let L_1^* denote the collection of functions that are defined almost everywhere on S_1 and that are integrable in the sense of Section 2.2. Let $S_2 := \mathbb{R}$, let Γ_2 be the semiring of nails, let μ_2 be ordinary Lebesgue measure on Γ_2.

A mapping $\underline{x} : S_2 \to L_1^*$ is called a *stochastic process*. The functions $\curlyvee_{t\in S_2}(\underline{x}(t))(s)$ with $s \in S_1$ are called *realizations* (or *trajectories* or *sample functions*) of the process. (The variable t usually stands for time.) If $\underline{\tilde{x}}$ is another stochastic process such that for every $t \in S_2$ one has

$$\curlyvee_{s\in S_1}(\underline{x}(t))(s) = \curlyvee_{s\in S_1}(\underline{\tilde{x}}(t))(s) \qquad \text{(a.e. in } S_1) ,$$

then $\underline{\tilde{x}}$ is called a *modification* of x.

Now given a particular stochastic process, it is of interest to know whether it has a modification whose realizations are almost all measurable. This kind of questions can sometimes be answered by means of the theorem of this section.

6.5.2. Let (S_i,Γ_i,μ_i) be a σ-finite measure space, let Λ_i be the σ-algebra of measurable subsets of S_i, and let L_i^* be the collection of integrable real-valued functions defined a.e. on S_i for $i = 1,2$. Let $S_3 = S_1 \times S_2$, and form the product measure $\mu_3 = \mu_1 \times \mu_2$ as described in Section 6.2. Let Λ_3 denote the σ-algebra of measurable subsets of S_3.

If f is an integrable function on S_3, then Theorem 6.2.4 says that f^t is integrable for almost all $t \in S_2$ and that for every $A \in \Gamma_1$ the function $\curlyvee_{t\in S_2} \int_A f^t \, d\mu_1$ is integrable over S_2. Our problem is whether there is a converse: if we know that all sections f^t behave like this, must f be measurable? The answer is no (see Exercise 6.5-3), but we shall show that under appropriate conditions there is a measurable $\tilde{f}$ on the product that in many respects may replace the original function f.

6.5.3. First we formulate the theorem. Notation is as in the first paragraph of 6.5.2. For the rest we use the notation established in Sections 6.1 and 6.2.

Theorem. Let $f : S_1 \times S_2 \to \mathbb{R}^*$ satisfy the following three conditions:

(i) $N_2 N_1 f < \infty$,

(ii) $f^t \in L_1^*$ for almost all $t \in S_2$,

(iii) $\curlyvee_{t\in S_2} \int_A f^t \, d\mu_1$ is measurable over S_2 for every $A \in \Gamma_1$.

Then there exists $\tilde{f} : S_3 \to \mathbb{R}^*$ which is measurable over S_3 such that

$$\int_{A\times B} \tilde{f} \, d\mu_3 = \int_B \curlyvee_{t\in S_2}\left(\int_A f^t \, d\mu_1 \right) d\mu_2$$

for all $A \in \Lambda_1$, $B \in \Lambda_2$.

6.5.4. Most of the proof of the theorem is diverted to the following lemma, which has a rather technical proof.

Lemma. Let f satisfy the conditions of Theorem 6.5.3. Then

$$g_E := \curlyvee_{t\in S_2} \left[\int_{E^t} f^t \, d\mu_1 \right] \in L_2^*$$

for all $E \in \Lambda_3$, and the set function λ defined by

$$(*) \qquad \lambda(E) := \int g_E \, d\mu_2 \qquad (E \in \Lambda_3)$$

is a finite signed measure on Λ_3. Moreover, λ is absolutely continuous with respect to μ_3.

Proof. Let $E \in \Lambda_3$. Then $E^t \in \Lambda_1$ for almost all $t \in S_2$ by Theorem 6.2.4, and therefore g_E is defined almost everywhere in S_2. Also $N_2 g_E < \infty$, since $|g_E(t)| \le N_1 f^t$ for almost all $t \in S_2$ and $N_2 N_1 f < \infty$ by assumption (i). Hence, to show that λ is defined and finite on Λ_3, it is enough to show that g_E is measurable for each $E \in \Lambda_3$. Showing this will be the hardest task in the following.

Let Σ_0 be the collection of those $E \in \Lambda_3$ such that $g_E \in L_2^*$, or, equivalently, such that g_E is measurable, and define $\lambda(E)$ for $E \in \Lambda_3$ by the expression in (*). We want to show that Σ_0 is equal to Λ_3; the proof is step by step.

(a) Let $E = A \times B$ with $A \in \Gamma_1$, $B \in \Gamma_2$. Then $g_E(t) = {}_A\!\int f^t d\mu_1 \cdot \chi_B(t)$ for almost all $t \in S_2$. Hence, g_E is measurable in this case. Therefore Σ_0 contains all rectangles with sides in Γ_1 and Γ_2.

(b) Let $E \in \Sigma_0$, $F \in \Sigma_0$, $F \subset E$. Put $G := E \backslash F$. It is easy to see that $g_G = g_E - g_F$ almost everywhere in S_1, so g_G is measurable. Hence, Σ_0 is closed under the operation of taking these special differences.

(c) Let $(E_n)_{n \in \mathbb{N}}$ be a pairwise disjoint sequence in Σ_0, and let $E := \bigcup_{n=1}^{\infty} E_n$. To prove that $E \in \Sigma_0$, put $F_N := E \backslash \bigcup_{n=1}^{N-1} E_n$ for $N \in \mathbb{N}$, and let M be a null set in S_2 such that $E_n^t \in \Lambda_1$, $E^t \in \Lambda_1$, $f^t \in L_1^*$ for $t \in S_2 \backslash M$, $n \in \mathbb{N}$. Now let $t \in S_2 \backslash M$. For each $N \in \mathbb{N}$ we have

$$|g_E(t) - \sum_{n=1}^{N-1} g_{E_n}(t)| = |\int_{F_N^t} f^t d\mu_1| = |\int f^t \cdot \chi_{F_N^t} d\mu_1| .$$

Since $f^t \cdot \chi_{F_N^t} \to 0$ a.e. in S_1, and $f^t \in L_1^*$, it follows by dominated convergence that

$$g_E(t) = \sum_{n=1}^{\infty} g_{E_n}(t) .$$

Hence g_E is measurable over S_2, that is, $E \in \Sigma_0$. Moreover, since the E_n's are disjoint, we have

$$\sum_{n=1}^{\infty} |g_{E_n}(t)| \le \int |f^t| d\mu_1 ,$$

where the right-hand side has finite N_2-norm. Hence, by dominated convergence,

$$\lambda(E) = \int g_E d\mu_2 = \sum_{n=1}^{\infty} \int g_{E_n} d\mu_2 = \sum_{n=1}^{\infty} \lambda(E_n) .$$

(d) Since both factor spaces are σ-finite, S_3 is the union of countably many pairwise disjoint rectangles. Hence, $S_3 \in \Sigma_0$ by steps (a) and (c).

(e) Let $(E_n)_{n\in\mathbb{N}}$ be a decreasing sequence in Σ_0, that is, $E_{n+1} \subset E_n$ for all $n \in \mathbb{N}$. It follows from steps (b), (c) and (d) that $\bigcap_{n=1}^{\infty} E_n \in \Sigma_0$.

(f) Let $E \in \Lambda_3$ and assume $\mu_3(E) = 0$. Then E^t is a null set in S_1 for almost every $t \in S_2$. Hence $g_E(t) = 0$ for almost every $t \in S_2$. Therefore g_E is measurable, that is, $E \in \Sigma_0$. Obviously $\lambda(E) = 0$.

(g) Let $E \in \Lambda_3$. By Exercise 6.5.1 we can find sets E_n with $E_{n+1} \subset E_n$, $E_n = \bigcup_{k=1}^{\infty} R_{nk}$ with R_{nk} pairwise disjoint rectangles and a null set $M \subset \bigcap_{n=1}^{\infty} E_n$ such that $(\bigcap_{n=1}^{\infty} E_n) \setminus M = E$. Now $E_n \in \Sigma_0$ by (a) and (c), $\bigcap_{n=1}^{\infty} E_n \in \Sigma_0$ by (e), $M \in \Sigma_0$ by (f), and, at last, $(\bigcap_{n=1}^{\infty} E_n) \setminus M \in \Sigma_0$ by (b).

As a by-product of (c) we have already shown that λ is a signed measure on Λ_3, and (f) says that $\lambda \ll \mu$. □

6.5.5. <u>Proof of theorem.</u> Apply the Radon-Nikodym theorem (5.2.3) to the pair λ, μ_3 of the lemma. It follows that there exists $\tilde{f} \in L_3$ such that $\int_E \tilde{f}\, d\mu_3 = \lambda(E)$ for all $E \in \Lambda_3$. According to the Fubini theorem this means

$$\int \Upsilon_t\left(\int_{E^t} f^t\, d\mu_1 \right) d\mu_2 = \lambda(E) \qquad (E \in \Lambda_3),$$

and specialization to sets of the form $A \times B$ gives the result. □

<u>Exercises Section 6.5</u>

1. Let (S,Γ,μ) be a σ-finite measure space. Let F be μ-measurable. Show that there exists a sequence $(F_n)_{n\in\mathbb{N}}$ in $\Omega(\Gamma)$ and a null set M such that $F = \bigcap_{n=1}^{\infty} F_n \setminus M$. (Hint. First assume $\mu(F) < \infty$ and use Lemma 3.3.5.)

2. With notation as in 6.5.2, let (S_1,Γ_1,μ_1) be separable, i.e. there is a countable set $\Gamma_0 \subset \Gamma_2$ such that for every $E \in \Gamma_2$, $\varepsilon > 0$, there is an $E_0 \in \Gamma_0$ with $\mu(E_0 \div E) < \varepsilon$ (here $E_0 \div E = (E_0\setminus E) \cup (E\setminus E_0)$). Assume that f is as in Theorem 6.5.3. Show that we can take the $\tilde{f}$ of Theorem 6.5.3 such that $f^t = \tilde{f}^t$ (a.e. in S_1) for all $t \in S_2$.

3. Assume the notation of 6.5.2 and let $f : S_1 \times S_2 \to \mathbb{R}^*$ be non-negative a.e. and satisfy

(i) $f^t \in M_1$ for almost all $t \in S_2$,

(ii) $\Upsilon_{t\in S_2}\ {}_A\!\int f^t\,d\mu_1$ is measurable and finite-valued a.e. for all $A \in \Gamma_1$.
Show that there is an $\tilde{f} : S_1 \times S_2 \to \mathbb{R}^*$, measurable and non-negative a.e., such that

$$\int_{A\times B} f\,d\mu_3 = \int_B \Upsilon_{t\in S_2}\left(\int_A f^t\,d\mu_1\right) d\mu_2$$

for all $A \in \Gamma_1$, $B \in \Gamma_2$. (Hint. Imitate the proof of Theorem 6.5.3 and use Exercise 5.2-7.)

4. If one accepts the continuum hypothesis, one can find a bijection ξ of $[0,1]$ into the set of all elements of the long line $< \Omega$ (see Exercise 3.3.14). Introduce on $[0,1]$ an ordering $<\circ$ as follows: $x <\circ y \Leftrightarrow \xi(x) < \xi(y)$ for $x \in [0,1]$, $y \in [0,1]$. Define $E := \{(x,y) \in [0,1] \times [0,1] \mid x <\circ y\}$. Show that $\{x \in [0,1] \mid x <\circ y\}$ is countable for every $y \in [0,1]$, and that $\{y \in [0,1] \mid x <\circ y\}$ has a countable complement for every $x \in [0,1]$. Take $S_1 = S_2 = [0,1]$ with ordinary Lebesgue measure. What does $\tilde{\chi}_E$ look like ($\tilde{\chi}_E$ is the modification of χ_E according to Theorem 6.5.3), and what happens if we interchange the variables x and y ?

LIST OF CONDITIONS, PROCESSES AND SYMBOLS

A1. If $h \in A$, $0 \le \alpha < \infty$, then $\alpha h \in A$ and $J(\alpha h) = \alpha J(h)$.

A2. If $h_1, h_2 \in A$, $h_1 \le h_2$, then $J(h_1) \le J(h_2)$.

A3. If $b \in B^+$, then $I(b) = \inf \left\{ \sum_{n=1}^{\infty} J(h_n) \mid h_n \in A \ (n \in \mathbb{N}),\ b \le \sum_{n=1}^{\infty} h_n \right\}$.

A4. There exists a sequence $(b_n)_{n \in \mathbb{N}}$ in B^+ such that $\sum_{n=1}^{\infty} b_n > 0$.

A5. For every $h \in A$ with $J(h) < \infty$ and every $\varepsilon > 0$ there exists a sequence $(f_n)_{n \in \mathbb{N}}$ such that $h \le \sum_{n=1}^{\infty} f_n$ and $\sum_{n=1}^{\infty} I(f_n) < J(h) + \varepsilon$.

D1. If $(b_n)_{n \in \mathbb{N}}$ is a sequence in B^+ and $b_n \downarrow 0$, then $I(b_n) \to 0$.

D1'. If V is a non-empty subset of B^+ and $\inf_{b \in V} b = 0$, then for every $\varepsilon > 0$ there exists a finite subset $W \subset V$ such that $I(\inf_{b \in W} b) < \varepsilon$.

L1. If $(f_n)_{n \in \mathbb{N}}$ is a sequence in L^+, if $f = \sum_{n=1}^{\infty} f_n$ is finite-valued, and if $\sum_{n=1}^{\infty} I(f_n) < \infty$, then $f \in L^+$ and $I(f) = \sum_{n=1}^{\infty} I(f_n)$.

M. Constant functions are measurable.

N1. If $\varphi \in \Phi$, $\alpha \in \mathbb{R}$ then $\|\alpha\varphi\| = |\alpha| \cdot \|\varphi\|$.

N2. If $\varphi, \psi \in \Phi$, $|\varphi| \le |\psi|$, then $\|\varphi\| \le \|\psi\|$.

N3. If $(\varphi_n)_{n \in \mathbb{N}}$ is a sequence in Φ, then $\| \sum_{n=1}^{\infty} |\varphi_n| \| \le \sum_{n=1}^{\infty} \|\varphi_n\|$.

N4. If $b \in B^+$, then $\|b\| = I(b)$.

R1. $\emptyset \in \Gamma$.

R2. If $A, B \in \Gamma$, then $A \cap B \in \Omega$ and $A \setminus B \in \Omega$.

R3. $\mu(\emptyset) = 0$.

R4. If A_n $(n \in \mathbb{N})$ are pairwise disjoint elements of Γ with $\bigcup_{n=1}^{\infty} A_n \in \Gamma$, then $\mu\left(\bigcup_{n=1}^{\infty} A_n\right) = \sum_{n=1}^{\infty} \mu(A_n)$.

INDEX

Vol. 926: Geometric Techniques in Gauge Theories. Proceedings, 1981. Edited by R. Martini and E.M.de Jager. IX, 219 pages. 1982.

Vol. 927: Y. Z. Flicker, The Trace Formula and Base Change for [illegible], [illegible]II, 204 pages. 1982.

Vol. 928: Probability Measures on Groups. Proceedings 1981. Edited by H. Heyer. X, 477 pages. 1982.

Vol. 929: Ecole d'Eté de Probabilités de Saint-Flour X – 1980. Proceedings, 1980. Edited by P.L. Hennequin. X, 313 pages. 1982.

Vol. 930: P. Berthelot, L. Breen, et W. Messing, Théorie de Dieudonné Cristalline II. XI, 261 pages. 1982.

Vol. 931: D.M. Arnold, Finite Rank Torsion Free Abelian Groups and Rings. VII, 191 pages. 1982.

Vol. 932: Analytic Theory of Continued Fractions. Proceedings, 1981. Edited by W.B. Jones, W.J. Thron, and H. Waadeland. VI, 240 pages. 1982.

Vol. 933: Lie Algebras and Related Topics. Proceedings, 1981. Edited by D. Winter. VI, 236 pages. 1982.

Vol. 934: M. Sakai, Quadrature Domains. IV, 133 pages. 1982.

Vol. 935: R. Sot, Simple Morphisms in Algebraic Geometry. IV, 146 pages. 1982.

Vol. 936: S.M. Khaleelulla, Counterexamples in Topological Vector Spaces. XXI, 179 pages. 1982.

Vol. 937: E. Combet, Intégrales Exponentielles. VIII, 114 pages. 1982.

Vol. 938: Number Theory. Proceedings, 1981. Edited by K. Alladi. IX, 177 pages. 1982.

Vol. 939: Martingale Theory in Harmonic Analysis and Banach Spaces. Proceedings, 1981. Edited by J.-A. Chao and W.A. Woyczyński. VIII, 225 pages. 1982.

Vol. 940: S. Shelah, Proper Forcing. XXIX, 496 pages. 1982.

Vol. 941: A. Legrand, Homotopie des Espaces de Sections. VII, 132 pages. 1982.

Vol. 942: Theory and Applications of Singular Perturbations. Proceedings, 1981. Edited by W. Eckhaus and E.M. de Jager. V, 363 pages. 1982.

Vol. 943: V. Ancona, G. Tomassini, Modifications Analytiques. IV, 120 pages. 1982.

Vol. 944: Representations of Algebras. Workshop Proceedings, 1980. Edited by M. Auslander and E. Lluis. V, 258 pages. 1982.

Vol. 945: Measure Theory. Oberwolfach 1981, Proceedings. Edited by D. Kölzow and D. Maharam-Stone. XV, 431 pages. 1982.

Vol. 946: N. Spaltenstein, Classes Unipotentes et Sous-groupes de Borel. IX, 259 pages. 1982.

Vol. 947: Algebraic Threefolds. Proceedings, 1981. Edited by A. Conte. VII, 315 pages. 1982.

Vol. 948: Functional Analysis. Proceedings, 1981. Edited by D. Butković, H. Kraljević, and S. Kurepa. X, 239 pages. 1982.

Vol. 949: Harmonic Maps. Proceedings, 1980. Edited by R.J. Knill, M. Kalka and H.C.J. Sealey. V, 158 pages. 1982.

Vol. 950: Complex Analysis. Proceedings, 1980. Edited by J. Eells. IV, 428 pages. 1982.

Vol. 951: Advances in Non-Commutative Ring Theory. Proceedings, 1981. Edited by P.J. Fleury. V, 142 pages. 1982.

Vol. 952: Combinatorial Mathematics IX. Proceedings, 1981. Edited by E. Billington, S. Oates-Williams, and A.P. Street. XI, 443 pages. 1982.

Vol. 953: Iterative Solution of Nonlinear Systems of Equations. Proceedings, 1982. Edited by R. Ansorge, Th. Meis, and W. Törnig. VII, 202 pages. 1982.

Vol. 954: S.G. Pandit, S.G. Deo, Differential Systems Involving Impulses. VII, 102 pages. 1982.

Vol. 955: G. Gierz, Bundles of Topological Vector Spaces and Their Duality. IV, 296 pages. 1982.

Vol. 956: Group Actions and Vector Fields. Proceedings, 1981. Edited by J.B. Carrell. V, 144 pages. 1982.

Vol. 957: Differential Equations. Proceedings, 1981. Edited by D.G. de Figueiredo. VIII, 301 pages. 1982.

Vol. 958: F.R. Beyl, J. Tappe, Group Extensions, Representations, and the Schur Multiplicator. IV, 278 pages. 1982.

Vol. 959: Géométrie Algébrique Réelle et Formes Quadratiques, Proceedings, 1981. Edité par J.-L. Colliot-Thélène, M. Coste, L. Mahé, et M.-F. Roy. X, 458 pages. 1982.

Vol. 960: Multigrid Methods. Proceedings, 1981. Edited by W. Hackbusch and U. Trottenberg. VII, 652 pages. 1982.

Vol. 961: Algebraic Geometry. Proceedings, 1981. Edited by J.M. Aroca, R. Buchweitz, M. Giusti, and M. Merle. X, 500 pages. 1982.

Vol. 962: Category Theory. Proceedings, 1981. Edited by K.H. Kamps, D. Pumplün, and W. Tholen, XV, 322 pages. 1982.

Vol. 963: R. Nottrot, Optimal Processes on Manifolds. VI, 124 pages. 1982.

Vol. 964: Ordinary and Partial Differential Equations. Proceedings, 1982. Edited by W.N. Everitt and B.D. Sleeman. XVIII, 726 pages. 1982.

Vol. 965: Topics in Numerical Analysis. Proceedings, 1981. Edited by P.R. Turner. IX, 202 pages. 1982.

Vol. 966: Algebraic K-Theory. Proceedings, 1980, Part I. Edited by R.K. Dennis. VIII, 407 pages. 1982.

Vol. 967: Algebraic K-Theory. Proceedings, 1980. Part II. VIII, 409 pages. 1982.

Vol. 968: Numerical Integration of Differential Equations and Large Linear Systems. Proceedings, 1980. Edited by J. Hinze. VI, 412 pages. 1982.

Vol. 969: Combinatorial Theory. Proceedings, 1982. Edited by D. Jungnickel and K. Vedder. V, 326 pages. 1982.

Vol. 970: Twistor Geometry and Non-Linear Systems. Proceedings, 1980. Edited by H.-D. Doebner and T.D. Palev. V, 216 pages. 1982.

Vol. 971: Kleinian Groups and Related Topics. Proceedings, 1981. Edited by D.M. Gallo and R.M. Porter. V, 117 pages. 1983.

Vol. 972: Nonlinear Filtering and Stochastic Control. Proceedings, 1981. Edited by S.K. Mitter and A. Moro. VIII, 297 pages. 1983.

Vol. 973: Matrix Pencils. Proceedings, 1982. Edited by B. Kågström and A. Ruhe. XI, 293 pages. 1983.

Vol. 974: A. Draux, Polynômes Orthogonaux Formels – Applications. VI, 625 pages. 1983.

Vol. 975: Radical Banach Algebras and Automatic Continuity. Proceedings, 1981. Edited by J.M. Bachar, W.G. Bade, P.C. Curtis Jr., H.G. Dales and M.P. Thomas. VIII, 470 pages. 1983.

Vol. 976: X. Fernique, P.W. Millar, D.W. Stroock, M. Weber, Ecole d'Eté de Probabilités de Saint-Flour XI – 1981. Edited by P.L. Hennequin. XI, 465 pages. 1983.

Vol. 977: T. Parthasarathy, On Global Univalence Theorems. VIII, 106 pages. 1983.

Vol. 978: J. Ławrynowicz, J. Krzyż, Quasiconformal Mappings in the Plane. VI, 177 pages. 1983.

Vol. 979: Mathematical Theories of Optimization. Proceedings, 1981. Edited by J.P. Cecconi and T. Zolezzi. V, 268 pages. 1983.

Vol. 980: L. Breen. Fonctions thêta et théorème du cube. XIII, 115 pages. 1983.

Vol. 981: Value Distribution Theory. Proceedings, 1981. Edited by I. Laine and S. Rickman. VIII, 245 pages. 1983.

Vol. 982: Stability Problems for Stochastic Models. Proceedings, 1982. Edited by V. V. Kalashnikov and V. M. Zolotarev. XVII, 295 pages. 1983.

Vol. 983: Nonstandard Analysis-Recent Developments. Edited by A.E. Hurd. V, 213 pages. 1983.

Vol. 984: A. Bove, J. E. Lewis, C. Parenti, Propagation of Singularities for Fuchsian Operators. IV, 161 pages. 1983.

Vol. 985: Asymptotic Analysis II. Edited by F. Verhulst. VI, 497 pages. 1983.

Vol. 986: Séminaire de Probabilités XVII 1981/82. Proceedings. Edited by J. Azéma and M. Yor. V, 512 pages. 1983.

Vol. 987: C. J. Bushnell, A. Fröhlich, Gauss Sums and p-adic Division Algebras. XI, 187 pages. 1983.

Vol. 988: J. Schwermer, Kohomologie arithmetisch definierter Gruppen und Eisensteinreihen. III, 170 pages. 1983.

Vol. 989: A.B. Mingarelli, Volterra-Stieltjes Integral Equations and Generalized Ordinary Differential Expressions. XIV, 318 pages. 1983.

Vol. 990: Probability in Banach Spaces IV. Proceedings, 1982. Edited by A. Beck and K. Jacobs. V, 234 pages. 1983.

Vol. 991: Banach Space Theory and its Applications. Proceedings, 1981. Edited by A. Pietsch, N. Popa and I. Singer. X, 302 pages. 1983.

Vol. 992: Harmonic Analysis, Proceedings, 1982. Edited by G. Mauceri, F. Ricci and G. Weiss. X, 449 pages. 1983.

Vol. 993: R. D. Bourgin, Geometric Aspects of Convex Sets with the Radon-Nikodým Property. XII, 474 pages. 1983.

Vol. 994: J.-L. Journé, Calderón-Zygmund Operators, Pseudo-Differential Operators and the Cauchy Integral of Calderón. VI, 129 pages. 1983.

Vol. 995: Banach Spaces, Harmonic Analysis, and Probability Theory. Proceedings, 1980–1981. Edited by R.C. Blei and S.J. Sidney. V, 173 pages. 1983.

Vol. 996: Invariant Theory. Proceedings, 1982. Edited by F. Gherardelli. V, 159 pages. 1983.

Vol. 997: Algebraic Geometry – Open Problems. Edited by C. Ciliberto, F. Ghione and F. Orecchia. VIII, 411 pages. 1983.

Vol. 998: Recent Developments in the Algebraic, Analytical, and Topological Theory of Semigroups. Proceedings, 1981. Edited by K.H. Hofmann, H. Jürgensen and H. J. Weinert. VI, 486 pages. 1983.

Vol. 999: C. Preston, Iterates of Maps on an Interval. VII, 205 pages. 1983.

Vol. 1000: H. Hopf, Differential Geometry in the Large, VII, 184 pages. 1983.

Vol. 1001: D.A. Hejhal, The Selberg Trace Formula for PSL(2, IR). Volume 2. VIII, 806 pages. 1983.

Vol. 1002: A. Edrei, E.B. Saff, R.S. Varga, Zeros of Sections of Power Series. VIII, 115 pages. 1983.

Vol. 1003: J. Schmets, Spaces of Vector-Valued Continuous Functions. VI, 117 pages. 1983.

Vol. 1004: Universal Algebra and Lattice Theory. Proceedings, 1982. Edited by R.S. Freese and O.C. Garcia. VI, 308 pages. 1983.

Vol. 1005: Numerical Methods. Proceedings, 1982. Edited by V. Pereyra and A. Reinoza. V, 296 pages. 1983.

Vol. 1006: Abelian Group Theory. Proceedings, 1982/83. Edited by R. Göbel, L. Lady and A. Mader. XVI, 771 pages. 1983.

Vol. 1007: Geometric Dynamics. Proceedings, 1981. Edited by J. Palis Jr. IX, 827 pages. 1983.

Vol. 1008: Algebraic Geometry. Proceedings, 1981. Edited by J. Dolgachev. V, 138 pages. 1983.

Vol. 1009: T. A. Chapman, Controlled Simple Homotopy Theor Applications. III, 94 pages. 1983.

Vol. 1010: J.-E. Dies, Chaînes de Markov sur les permutatio 226 pages. 1983.

Vol. 1011: J.M. Sigal. Scattering Theory for Many-Body Qu Mechanical Systems. IV, 132 pages. 1983.

Vol. 1012: S. Kantorovitz, Spectral Theory of Banach Space ators. V, 179 pages. 1983.

Vol. 1013: Complex Analysis – Fifth Romanian-Finnish Se Part 1. Proceedings, 1981. Edited by C. Andreian Cazacu, N. B M. Jurchescu and I. Suciu. XX, 393 pages. 1983.

Vol. 1014: Complex Analysis – Fifth Romanian-Finnish Se Part 2. Proceedings, 1981. Edited by C. Andreian Cazacu, N. B M. Jurchescu and I. Suciu. XX, 334 pages. 1983.

Vol. 1015: Equations différentielles et systèmes de Pfaff d champ complexe – II. Seminar. Edited by R. Gérard et J. P. R V, 411 pages. 1983.

Vol. 1016: Algebraic Geometry. Proceedings, 1982. Edited Raynaud and T. Shioda. VIII, 528 pages. 1983.

Vol. 1017: Equadiff 82. Proceedings, 1982. Edited by H. W. Kno and K. Schmitt. XXIII, 666 pages. 1983.

Vol. 1018: Graph Theory, Łagów 1981. Proceedings, 1981. Edit M. Borowiecki, J.W. Kennedy and M.M. Sysło. X, 289 pages.

Vol. 1019: Cabal Seminar 79–81. Proceedings, 1979–81. Edit A. S. Kechris, D. A. Martin and Y. N. Moschovakis. V, 284 pages.

Vol. 1020: Non Commutative Harmonic Analysis and Lie Gr Proceedings, 1982. Edited by J. Carmona and M. Vergne. V pages. 1983.

Vol. 1021: Probability Theory and Mathematical Statistics. ceedings, 1982. Edited by K. Itô and J.V. Prokhorov. VIII, 747 p 1983.

Vol. 1022: G. Gentili, S. Salamon and J.-P. Vigué. Geometry Se "Luigi Bianchi", 1982. Edited by E. Vesentini. VI, 177 pages. 1

Vol. 1023: S. McAdam, Asymptotic Prime Divisors. IX, 118 p 1983.

Vol. 1024: Lie Group Representations I. Proceedings, 1982– Edited by R. Herb, R. Lipsman and J. Rosenberg. IX, 369 pages.

Vol. 1025: D. Tanré, Homotopie Rationnelle: Modèles de Quillen, Sullivan. X, 211 pages. 1983.

Vol. 1026: W. Plesken, Group Rings of Finite Groups Over p Integers. V, 151 pages. 1983.

Vol. 1027: M. Hasumi, Hardy Classes on Infinitely Connecte mann Surfaces. XII, 280 pages. 1983.

Vol. 1028: Séminaire d'Analyse P. Lelong – P. Dolbeault – H. S Années 1981/1983. Edité par P. Lelong, P. Dolbeault et H. S VIII, 328 pages. 1983.

Vol. 1029: Séminaire d'Algèbre Paul Dubreil et Marie-Paule Mall Proceedings, 1982. Edité par M.-P. Malliavin. V, 339 pages. 19

Vol. 1030: U. Christian, Selberg's Zeta-, L-, and Eisensteins XII, 196 pages. 1983.

Vol. 1031: Dynamics and Processes. Proceedings, 1981. Edit Ph. Blanchard and L. Streit. IX, 213 pages. 1983.

Vol. 1032: Ordinary Differential Equations and Operators. ceedings, 1982. Edited by W. N. Everitt and R. T. Lewis. XV, 521 p 1983.

Vol. 1033: Measure Theory and its Applications. Proceedings, Edited by J. M. Belley, J. Dubois and P. Morales. XV, 317 pages.

Vol. 1034: J. Musielak, Orlicz Spaces and Modular Spac 222 pages. 1983.